JEPPESEN SANDERSON

aviation yearbook 1978

(Surveying the Period January 1977 through December 1977)

8025 E. 40th Avenue, Denver, Colorado 80207, U.S.A.

AV313259 INTERNATIONAL STANDARD BOOK NUMBER 0884870200

AVIATION YEARBOOK

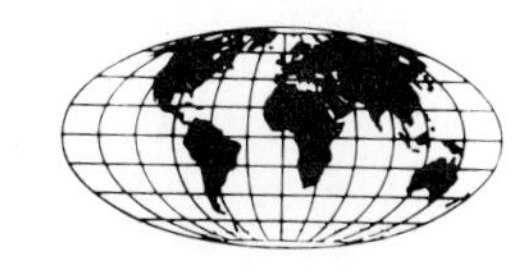

PUBLISHER'S NOTE

Your JEPPESEN SANDERSON AVIATION YEARBOOK 1978 reviews the period from January 1977 to December 1977. Articles dealing with significant topics in the world of aviation that occurred during this period have been chosen from a number of English language aviation publications. Where published material was not available for specific topics, original material has been supplied by aviation writers or the YEARBOOK staff. Certain reprinted articles have been edited for space considerations. Similarly, comparable photographs or illustrations have been substituted in the few cases where the original material was not available.

ACKNOWLEDGEMENTS

Acknowledgement of source is made with each article. The cooperation of the many publications, organizations, manufacturers, and writers represented in this volume is greatly appreciated. Organizations cooperating included the Air Line Pilots Association (AIR LINE PILOT); The Aircraft Owners and Pilots Association (THE AOPA PILOT); the Balloon Federation of America (BALLOONING); the Experimental Aircraft Association (SPORT AVIATION); Soaring Society of America (SOARING); the United States Hang Gliding Association (HANG GLIDING); the United States Parachute Association (PARACHUTIST).

TABLE OF CONTENTS

THE EDITORS

Capt. Ed Mack Miller, editor-in-chief

Author of five aviation books, 1800 articles and stories, and the recipient of 14 state and national writing awards, retired United Air Lines Captain Ed Mack Miller is uniquely qualified to serve as editor-in-chief of the JEPPESEN SANDERSON AVIATION YEARBOOK 1978. Rated in everything from DC-3s to Boeing 747s, Captain Miller has detailed operational knowledge and experience in General, Air Carrier, Military and Sport Aviation. A college teacher of journalism, Miller is also the founder and past president of the Colorado Aviation Historical Society and was selected in 1976 to receive the Colorado Wright Brothers Memorial Foundation Award.

F. T. Stent, general aviation editor

A graduate of Yale College with a Masters degree from Harvard University, Mr. Stent has served as General Aviation Editor for the first two editions of the JEPPESEN SANDERSON AVIATION YEARBOOK. An ex-Navy fighter pilot, decorated nine times for combat duty in Vietnam, Mr. Stent is now a pilot for Delta Air Lines and has also logged several thousand hours in a variety of general aviation aircraft. He has been actively writing and consulting in the aviation field for the past ten years.

Barry Schiff, air carrier editor

Barry Schiff, a Boeing 727 Captain for TWA, is the author of several books and training manuals in addition to more than 300 articles in leading aviation magazines and journals. He is also one of the few pilots in the U.S. with all seven flight instructor ratings and holds an aviation teaching credential from the California State Department of Education. As holder of numerous world class speed records, Captain Schiff was selected by the Federation Aeronautique Internationale in 1969 to receive the Louis Bleriot Air Medal. He has also been awarded a congressional commendation for his contributions to aviation, as well as a special writing citation from the Aviation/Space Writers Association.

Allan R. Scholin, military/aerospace editor

Allan R. Scholin is considered to be one of the country's leading military aviation and space experts. A professional journalist for 30 years, Mr. Scholin served as an information officer in the Air Guard and Air Force Reserve during World War II and Korea, and later worked as a writer in the Washington bureau of the National Guard. For 12 years he was also a contributing editor to *Air Progress* magazine. He is a member of the Aviation/Space Writers Association and the Society of Professional Journalists.

Dan Manningham, sport aviation editor

Dan Manningham has been an active participant, reporter, and observer in sport aviation for many years. His experience runs the gamut from sport parachuting to flying 727s and DC-8s for United Airlines. A widely read free-lance writer and contributing editor to *Business & Commercial Aviation*, Mr. Manningham first learned to fly helicopters in the military and later transitioned to heavy transports and a variety of general aviation aircrafts. A graduate of Tufts University, Manningham has thoroughly covered the broad spectrum of aviation's most unique activities and is singularly qualified to serve as the Sport Aviation Editor.

Gordon Girod, managing editor

Gordon Girod is a former faculty member of the United States Air Force Academy and several civilian universities. The holder of Bachelor and Masters degrees in English, Mr. Girod has edited several major book projects. In addition, he has written a number of teaching programs for college use. Mr. Girod was the managing editor of the first two volumes of the JEPPESEN SANDERSON AVIATION YEARBOOK.

INTRODUCTION

"Sympathy," says the wag, "is what we offer our neighbor in exchange for all the lurid details."

Well, if you are looking for sympathy, this is the wrong place; but if you're looking for the details (not lurid, but complete) on the state of aerospace in the world in 1977, you've come to the right place: the JEPPESEN SANDERSON AVIATION YEARBOOK 1978.

In addition, this year we are a little like the fellow who just got his first color television set — he wants everyone to know he's added color. Well, so do we — we've added a special section of bright and lively, light and lovely full color aviation photos. I thinks it adds just the right touch to the book.

A delicate touch, a soft brush, is needed too, this year. Some of our fine friends paid the price for all the joy that flying is. The second and third generations of the talented Loening family — a family that contributed greatly to American and world aviation — have left us (see General Aviation 78). Just at press time we learned of the death of that great Jersey gentleman, Ed Mahler, who delighted a generation of airshow buffs with his marvelous aerial performances — pleasing not only their eyes with his lomcevaks, but also their noses with his aftershave-scented smoke. We'll also miss that talented and handsome maestro of the aerobatic jet, Ormond Haydon-Baillie, another who was killed in '77.

While 1977 had its share of bad news, there was also good news. General Aviation editor Terry Stent has assembled a section that in part recounts the good news of 1977. For example, consider the absolutely thrilling little number called the Foxjet (wouldn't that be fun to have in the carport?). There's a report on the current state of development of the project between the senior statesman of the industry, Bill Lear, and Canadair: the Challenger 600. You'll enjoy "Maule on the Moose Range," and be edified by the efforts the egg-heads are making to swing the technological revolution toward the little man in aviation (T-tails and the Dowty propulsor).

Even if Heads-Up Displays aren't quite down to a Beechcraft budget yet, still and all there have been some nice advances in electronics (computer monitoring, color weather radar, and "the electronic equivalent of a complete set of Jeppesen charts") that you'll find discussed in the General Aviation section.

Our Air Carrier editor, TWA Captain Barry Schiff, crept into General Aviation this year with an article that explains the inexplicable details of VLF navigation. His report on VLF notes that "it is difficult to imagine a better method of long-range navigation."

General Aviation also gives you the nifty story of a courageous flier, Tom Kerr, who has done it the hard way.

The Air Carrier section is certainly worth a couple of nights of your time this year. Editor Schiff's selections include facts on some pretty shocking crashes, including the accident that just couldn't happen — the collision of two 747's at Tenerife. There is an article on the helicopter accident atop New York's Pan-Am building, and the cockpit recorder transcripts of the Southern DC-9 tragedy at New Hope, Georgia. (Everyone who wears wings can stand a lot taller after reading about some brave people who deserve to be in everyone's Hall of Fame after their cool, professional, final performance).

You'll find that Air Carrier contains an assessment of the Concorde's first year, and even a bit of deathless prose about the Paris Air Show (written by who else?)

There are lots of other goodies, too, on everything from a Rolls-Royce-powered 747 to Aeroflot's STOL to Narita's problems.

The Military/Aerospace section is a whopper this time. Colonel Al Scholin, our expert in this field, has picked a solid array of subjects this year — ranging from beam weapons in space to cruise missiles in nap of the earth, from evaluations of Soviet weapons to evaluations of possible space technology.

There's some no-nonsense looks at our "consistent under-estimation of the Soviet threat," some sobering comment on the B-1 cancellation, and a rational case for the manned penetrating bomber.

If you want a unique view of the air war of World War II, there's an excellent and penetrating interview between Albert Speer, Hitler's armament production minister, and Ira Eaker, the commander of the Eighth Air Force. You have the bomber comparing notes with the bombee.

For all you space aficianados, Scholin has picked coverage for you on some of Dr. Krafft Ehricke's super-imaginative ideas for solving the energy shortage.

Actually, the military/aerospace section is so rife with interesting, relevant material this year, you should just check it out for yourself!

Next we come to Sport Aviation. No way we could leave that section out. Editor Dan Manningham has been on top of the field all year long, and has culled his material to the very best. You should have a ball with this year's selections.

Hang-gliders, homebuilts, balloons, parachutes, sailplanes. They're all there. Look for your area of special interest.

One of the biggest stories in sport aviation this year was the winning of the Kremer prize. The man-powered-aircraft barrier has finally been broken, and we provide you with photos and stories about the Gossamer Condor.

Then there is Darryl Greenamyer's "hot tin project," the super-civilian F-104. You can read about his finally successful attack on the speed record, plus his record-breaking win in the Reno Air Race in '77.

Those of us who are Soaring Sams are sure to be thrilled by the audacious, record-setting, cross-country sailplane flights recorded by Karl Striedieck (who likes to sneak out early, cuddle up to a ridge, and slide a smooth thousand miles).

Finally, there's our NOTAMS section. That's the section we use for the super-neat article that we just can't fit into one of the other sections. There are good things there this year too. We have a final salute to a unique gentleman and pioneer aviator, Charles Lindbergh and even a nice, nostalgic piece on the Spruce Goose.

Well, it's all there, and — like we said at the beginning — we've tossed in a selection of color photos that will turn you on for sure if you have even a tiny set of wings in your heart.

No sense hanging around here, kids. Go ahead and leap into Aviation '77. You'll enjoy it; I'll guarantee you.

Ed Mack Miller
December 1977.

It is with regret and a deep sense of loss that we announce the death of Ed Mack Miller on January 9, 1978, from a heart attack. Many words have been written about Ed, as a pilot, as a writer, as a human being. All we can add is — "He will be missed."

—JSAY

GENERAL AVIATION

AVIATION YEARBOOK

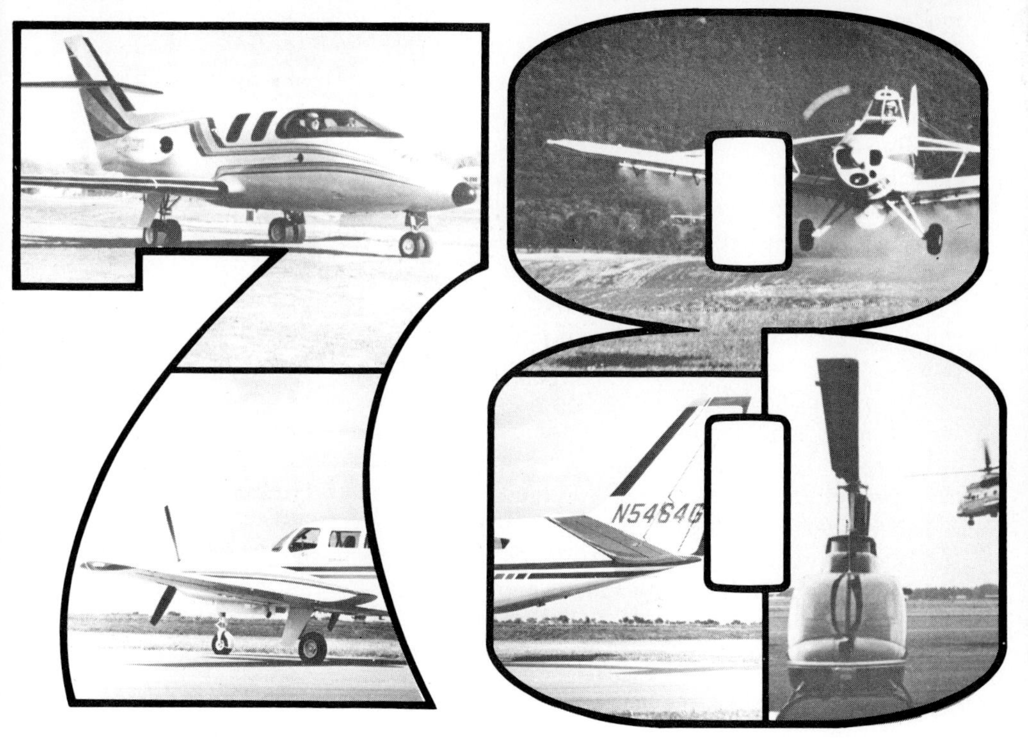

INTRODUCTION

"Won't happen to me — well, if it did, I'd get out of it!"

We all say it, every time we read a General Aviation accident report. It's always the *other* guy — unprepared, undertrained, mistake-prone, the constant klutz, never us. What if the other guy is a well-respected pilot? Probably better trained than ourselves. What if *he* becomes the subject of an accident report?

February 26, 1977. National Pilots Association president Mike Loening goes down in mountainous terrain in Utah. Some convincing evidence that all three occupants of Loening's Cessna 206 survived the crash but died of subsequent exposure. General Aviation takes note. What went wrong. . . . Could it happen that way to other G.A. pilots?. . . . To us?. . . . Could you survive a 20 degrees-below-zero temperature? What kind of rescue attempts would be made? How soon?

. . . .A spate of survival articles came out following Loening's death. Good ones that deal with wilderness and cold weather survival; rescue techniques; and the Loening crash itself. One is presented here for close reading. Perhaps from the Loening tragedy can come a new awareness of some old and important lessons.

There is an oft-repeated ritual in General Aviation heralded by pompous press releases, glossy four-color fold-out brochures and a seemingly endless stream of not so imaginative articles that announce the year's newest aircraft models. Since General Aviation aircraft design is largely evolutionary and the manufacturers are understandably reluctant to stray too far from a winner, the "new" models usually boil down to cosmetics. New paint, new name, new numbers, a few more horses, a few more dollars. Not exactly the fodder of headlines. . . .but they are good planes, and they do sell. (This is not to say that this evolutionary process is unsophisticated or undesirable. In fact there has been recent governmental interest in G.A. technological research and an excellent article covering these developments is included.)

There were however four new planes that made it either to the mock-up or prototype stage during the year and they seemed to have caused more than the usual amount of hangar gossip. First was Grumman American's new "light twin" Cougar — the four place 190-mph threat to the "heavy" singles. It's nice to look at, supposedly "priced to sell," has impressive performance, and last but not least, who can ignore the name behind it!

The second General Aviation newcomer that caused some talk in 1977 was the new Bellanca T-tail "Aries." Planned for market entry next year, the T-tail Aries still has a questionable future, (as indeed does its corporate parent) but the flight advantages of the T-tail make it one of the freshest innovations in the single engine field in a long time (a fact, incidentally, that does not appear to be lost on a couple of the biggies).

The final two General Aviation airplanes worthy of special note in 1977 are jets; the Canadair CL600 Challenger (formerly the LearStar 600) and Tony

Team Industries' Foxjet. The Canadair 600's pre-production sales have been most impressive (Federal Air Express has committed to 25!) and this will virtually assure its becoming an important addition to G.A.'s jet fleet of the late 70's. On the other hand, the tiny four-place Foxjet may never make it beyond the prototype stage but it definitely does raise many interesting questions concerning optimal sizing of corporate turbine-powered aircraft, fuel efficiency, and cost.

If development in General Aviation aircraft has come at a snail's pace, the development in avionics for small airplanes has definitely not! One needs only to remember back a few years ago to the ubiquitous Superhomers (with their fancy whistle-stop tuning!) to appreciate the advancements that have been made in this field! A happy beneficiary of transistors and space-age miniaturization, the General Aviation radio-nav industry has made spectacular advances over the past decade and 1977 was no exception. A well written article on the newest batch of light-plane black boxes has been selected and even the most staunch open-cockpit advocate will find it mouth-watering.

Oil, of course, continued to nag General Aviation in 1977. The return of 80-octane, rising prices at the pump, and a legion of soap-boxers trying to justify miles per gallon, gallons per pilot, pilots per pilots...ad nauseum per ad nauseum. Much was written about energy, (most was simply a rehash of what we all already know anyway!) so the "oil" stories this year are different...and considerably less depressing! "Maule on the Moose Range" is a look at one G.A. pilot's part in the Alaskan oil boom, and "Ketch a Chopper" covers the effects that the off-shore (Gulf coast) oil bonanza has had on the helicopter business.

Each year there are a handful of "teaching" articles that stand out above others of their genre because of subject matter, style, or ability to impart useful knowledge. This year, articles dealing with the FAA's new Enroute Flight Advisory Service and the pitfalls of power lines have been selected for their worth to the General Aviation pilot.

The General Aviation section for 1977 closes with some humor (TSOd Pencil), and a story about a pilot who was told he could never be a pilot. The kicker with this story is that when people told Tom Kerr he couldn't be a pilot he didn't listen. He couldn't. He's deaf...And a mute...And now he's a pilot.

...And that's a nice note to end General Aviation 1977 on. Happy reading!

F.T. Stent
November 1977

THE KILLING ASSUMPTION

Reprinted by Permission
THE AOPA PILOT (October 77)

by Glen Tabor

Many current philosophies and practices regarding downed aircraft and pilot survival in the Rocky Mountain states are sadly inadequate.

We pilots are guilty of exercising the placid attitude that "it can't happen to me" or, "if it did, I'd get by anyway!"

The assumption that, in the event of a crash, we can spend our downed time building an adequate shelter, scrounging water and food, and caring for our injuries with a haphazard array of tools, that only by chance we have with us, is a killing assumption.

The FAA supports this philosophy by offering their "27-cent do-it-yourself survival kit" as a bare minimum kit. It contains a metal can, a candle, a penny box of matches, three garbage bags, a dozen sugar cubes and a roll of plastic tape.

This may work for highly trained survival teams, but how about the average pilot? We need a more realistic and more conservative approach to winter survival in the mountains.

Survival, according to the popular aviation philosophy, is similar to Webster's definition — "the act of continuing to live after or in spite of. . ." — which implies that a downed pilot may have to make huge sacrifices, but he will sustain life. Hands and feet are sacrificed to frostbite (and amputation); simple injuries degenerate to serious ones; general health deteriorates until sickness takes command; suffering becomes so intense it breaks the will to live. That is a bleak way to define survival.

Suppose you are on your way from Kansas City to Salt Lake City, VFR. You are crossing the Rockies. The raw beauty of their 12,000-foot snow capped granite mountains slides quietly by your wingtips. Then your airplane is swatted down into a dark canyon and smashed against the frozen slopes.

Miraculously, everyone is alive. You have a broken leg. Your wife is unconscious with a broken arm. Your two children are dazed and in shock. Your ELT is not working.

Now you must "make-do" with the tools at hand. Will the casual clothes you brought keep out the bitter 20 below zero cold? Will the crumbled cabin work as a shelter? Can you find any firewood to burn? Is it dry enough? Who will drag it over? You can't. Your wife can't. Your children are dying because of shock and no protection against the cold. Meanwhile, your ruptured fuel tanks are pouring away the gasoline you desperately need to get frozen firewood started.

What about the first aid for your injuries? What about personal comfort and hope to spark your will to live? How are you going to signal the search plane? You will have one minute of exposure time to get the search and rescue plane's attention. It will not come until early in the third morning. What will you eat and drink for the five days it will take to get a rescue team into your location?

Slowly you begin to understand the danger of the killing assumption.

A better definition, we believe, is — "survival is keeping body and soul intact, in their entirety and in reasonably good health under adverse conditions" — to settle for anything less is foolhardy. The small cost of preparing to survive by this definition far outweighs the ultimate penalty for lesser expectations.

What will it cost in dollars, weight, bulk, time and effort to adopt this more liberal view of survival?

With today's emphasis on search and rescue, we can adopt the basic practice of staying with our airplane and waiting to be rescued. The alternative is to walk out to safety. This alternative should be followed *only* if a town is in sight and the terrain and condition of the pilot and passengers allow it. These conditions seldom apply. So plan to stay with your aircraft.

Surviving until help arrives depends on adequately filling five basic needs: first aid; shelter/warmth; signaling; food/water; and personal comfort.

First aid skills are best learned from experts. Take a good first aid course and learn what supplies you need and how to use them. Remember, you are limited by both space and weight in what you can take along, so only pack materials that you know how to use. Even if you buy a commercial kit, go through it item by item and either learn to use each of its contents, or throw out what you don't know how to use and replace it with something which will be of use to you.

Plan on a five-day supply for the expendables and make sure that you have some pain killers along. Pain contributes to shock and demoralization. Although it is difficult to control pain, it is essential that you try to do so.

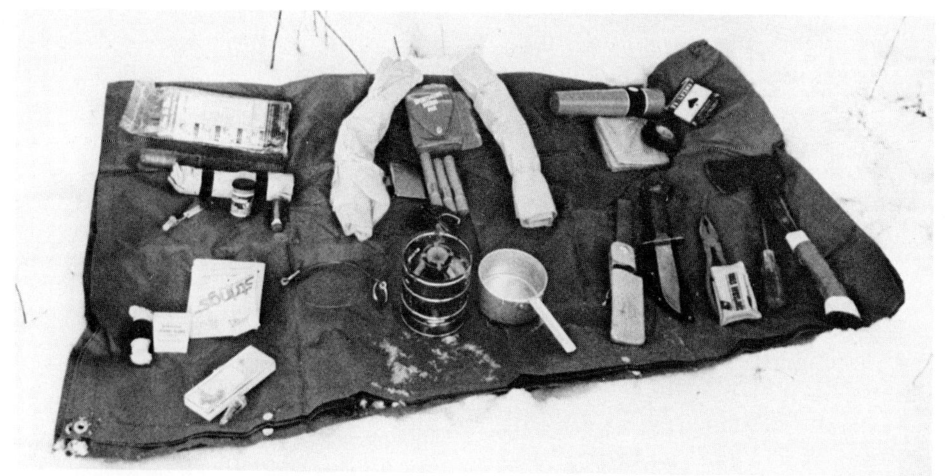

Makeshift tools may not be enough in mountain wilderness. This array, compactly packed, will give the pilot a fighting chance.

Shelter and warmth. In this area more than any other, the killing assumption takes its effect. If you had to plan on spending three days and nights in a commercial meat locker with an average temperature of -10 °F, what would you take with you? A light jacket and some frozen fire wood? A lean-to?

The question of surviving cold temperatures, even above-freezing temperatures, is laughed off far too easily. The fantasy of magically building a shelter which will keep us warm is immature. Just trying to gather the material to build a lean-to in deep powder snow can lead to exhaustion.

You may want to try to live in a snow cave. But do you know the limited temperature range in which they work? Do you know how to build one? What about the snow that melts and soaks your clothing?

Even if you know, how can you possibly build a lean-to or a snow cave with a broken arm or leg?

There is an easy and simple way to provide shelter and warmth. Take along an adequate sleeping bag and ground cloth. The airplane's cabin, if it's intact, can provide shelter and the sleeping bag, warmth. The ground cloth can be sandwiched around the sleeping bag (to look like a hot dog) if no other shelter is available.

Although the FAA insists that the cabin of an airplane is worse than being outside in a lean-to because the aluminum skin repels rather than absorbs heat, I decided to find out for myself. I would measure the rate of heat loss from a known and constant mass in each of several situations.

I enlisted the services of a "lab assistant" to help in the measurements. We began the experiment by recording time and temperature inside the cabin of my Skyhawk, outside behind a lean-to shelter, and outside with a six mph wind.

We found that heat escaped from the cabin 50% slower than from the no-shelter condition and 35% slower than behind a wind break.

We repeated the experiment for rate of heat loss in the mountains of Idaho in a survival tent (nylon with fly — three man). As it turned out, a February blizzard hit us on the first night and we were able to also try out many of the survival gadgets in my survival kit.

The heat loss for the tent was fully 100% less than remaining outside.

A snow cave was not measured because the temperature did not allow the use of one.

Our findings suggest that the intact cabin of a Cessna 172 is, in fact, superior to a wind break or the ambient conditions outside and is significant in slowing the heat loss rate over those two choices. A nylon tent with a fly was superior to the cabin.

Considering the above, the aircraft can be used to provide an

For a downed pilot, lost in the wilderness, having adequate shelter may be the difference between life and death.

adequate shelter. A tent is much better, however. If the aircraft is lost, keep in mind that mountain rescue teams recommend using an under-the-log shelter wherein pine boughs, small logs, etc. are placed against a fallen tree and more boughs used for a floor. If you're able to use this in support of your ground cloth and sleeping bag — so much the better. If not, the ground cloth and sleeping bag alone will do nicely. Above all, keep your shelter tight and snug.

Depending upon a fire to provide enough warmth to sustain life and limbs is dangerous in cold weather, especially in the high country. Just getting a fire started can be a major task, even with the right equipment. Keeping it going can be a tremendous task and keeping it fueled can exhaust you. (It can be hundreds of yards to the nearest firewood in waist-deep, wet snow).

A pocket cigarette lighter, either the new butane or the trusty flint lighter is the easiest ignitor to have with you. But if you really want to get the job done, take along a couple of fuses. These are the flares that highway patrolmen use to alert motorists to an accident or dangerous conditions. The railroad has used them for years. They burn with an extremely high temperature and last long enough to get even the stubbornest fires started. A can of lighter fluid is icing on the cake.

Locate your fire away from snowcovered limbs (snow melts — puts out the fire) and on top of bare ground or on a log "raft" on top of the snow.

If you are going to depend upon this fire to provide the warmth to keep you alive, you become a slave to it. You must feed it and care for it until rescue is accomplished. This is a great drawback.

With respect to shelter and warmth, keep in mind that protection against the wind is the primary requirement. If you lose the aircraft or if you can't get a fire started, you can still insure shelter and warmth by utilizing the ground cloth and sleeping bag — set them up to protect yourself from the wind.

My recommendation is this, take along a Polarguard or Dacron II sleeping bag and a 9 by 12 waterproof ground cloth for every other person on board, minimum. Plan your activities so that you will always keep your sleeping bag dry.

I suggest a Polarguard bag over a down bag because even though it is slightly less efficient under dry conditions, when it is wet it will offer more warmth and can be dried out more quickly over a fire.

When you are planning on traveling over extremely rugged terrain, throwing in a small tent is not a bad idea. This will give you a strong advantage over the elements.

There are other items of shelter and warmth that must also be considered. Gloves or mittens are mandatory. Waterproof boots with heavy socks are strongly recommended. Garbarge bags can be substituted for waterproof boots, but are easily nipped.

An arctic hat, balaklavia, or other cold weather hat is extremely important. Almost half of the heat loss from your body is associated with your head. This is due to the tremendous volume/flow rate of blood that goes to the head. A

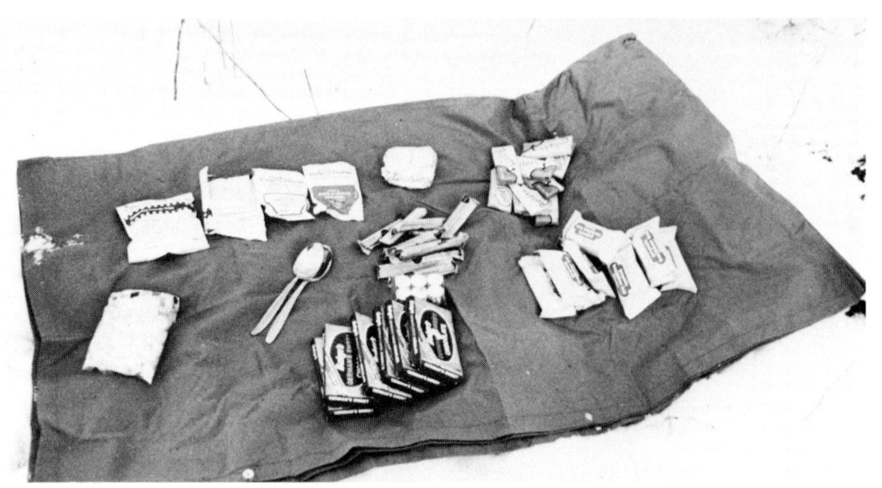

Living off the land is more than difficult in high mountain country, especially in winter, so plan to have provisions along.

proper hat can conserve that precious body heat.

To aid fellow pilots in finding us in a timely manner, we can employ some signaling devices. Obviously, our ELT comes into play at this time. It probably will be credited with getting a search plane into the general area within 24 hours. But let's help the ELT out by also having a way to show the S & R pilots exactly where we are when they fly by. Keep in mind that, should the ELT fail, these items will be the primary signaling devices.

You may have only one shot at telling someone where you are. Will you be satisfied to gamble your one shot with a fire? Can you get it started in time? Will it be bright enough in time? Will it smoke enough before the search plane passes by?

Every book that I have read on survival strongly recommends a signal mirror over all other signaling items. To discover for myself why it is recommended so highly as well as what other signal devices might offer, I enlisted the aid of two companions. We would put several signaling devices to the acid test: One man on the ground would use them while the second man and myself, all armed with walkie-talkies, would fly a search pattern over him in my Cessna 172 and observe and photograph their effectiveness.

This exercise provided a good indication of what can be expected from signal devices.

For a clear day, the signal mirror is unequaled. Its drawbacks occur on cloudy days, at dusk and dawn, and when the sun's position prevents reflecting a signal to the target. Remember, the sun and the target must be approximately on the same side of the signaler.

The aerial night flares are useless in daylight conditions. Two flares were photographed in the dusk hours and proved to be brilliant and easily spotted.

The night flare's drawbacks are obvious. They're good only from dusk to dawn. They are brief, burning only about 10 seconds. And a commercial one by SSI had two design problems which caused our ground man to injure himself. All the military flares I have seen are triggered by pulling a ring fixed to the muzzle end. The SSI flare has a trigger mechanism at the opposite end.

These two problems united to create a serious problem for our ground man. He almost "gut-shot" himself trying to signal us. His fault for not reading the instructions? Probably. But what if a novice is closest to the flares when the one search plane appears on the second or third night out?

An excellent night signaling device is offered by Honeywell, Inc. They are marketing a small, portable strobelite (Model 2700), which runs for seven (plus) hours on two size "C" batteries. The strobelite could be seen for many miles on the night I tested it and under the right conditions could probably be seen for twenty miles.

It offers an advantage over a flare in that it flashes about once per second for hours. A flare, however, can gain altitude which may be important in some applications.

Smoke bombs were good during the day. They were easy to spot and the gentle breeze created enough motion to further aid in detection. The wind also caused the smoke cloud to disperse in about one to two minutes. Obviously a strong wind would make their use ineffective.

The flame orange cross was extremely effective. Its fluttering motion due to the breeze caught our eye the moment we turned level about one mile away. It has the advantage over the mirror, the flares, and the smoke bombs in that it is not dependent on the plane-to-sun position, the wind, or the proper timing. If I had only one signal tool to choose, it would be the orange nylon cross.

My recommendation for signaling is this: Take one signal mirror, three flares or the strobelite, three smoke bombs, and the orange cross. They are small, inexpensive, and give you the benefit of hedging your bet. Remember, you may get only one shot at being seen. And our basic premise is to await rescue. Don't forget to try your communications radio on 121.5 MHz (if it is intact) when you spot a plane.

According to most experts, you can live three days without water and three weeks without food. This doesn't take into account what three weeks without eating or three days without drinking does to your desire to live. It can bring about deep depression which may contribute significantly to your death by other causes. It is best to plan on eating and drinking, perhaps sparingly, but regularly. For both eating and drinking purposes, a fire or stove is a luxury, but a most convenient one.

In the winter months, I seldom take along any liquid as it will freeze. I rely on melted snow. It is unwise to eat snow before melting it, I am told, because it wastes body heat and often contributes to a feeling of thirstiness. To melt the snow I purchased a small back-

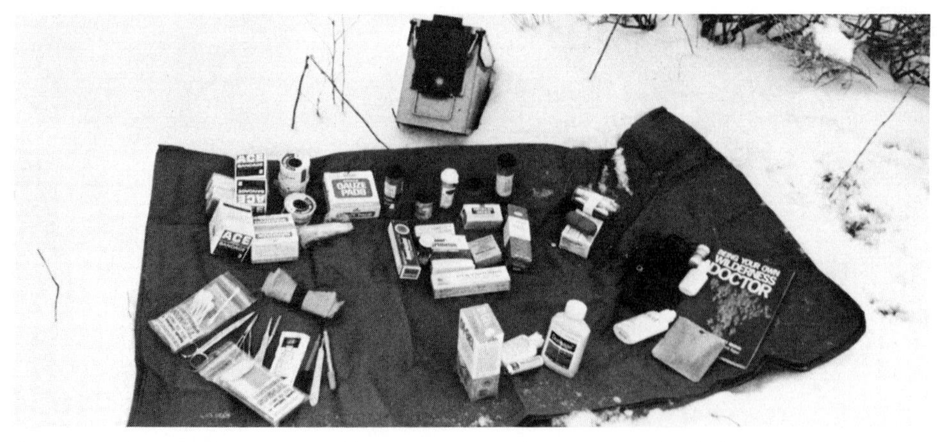

Your first priority will be first aid. These supplies can be easily stored and, if you know hwo to use them, may save your life.

packer's gasoline-fueled stove. It burns like a blow torch, lasts several hours with each fueling, costs about twenty bucks and looks rugged enough to handle a crash without damage. It also has its own small pot/cup and can use remaining gas from the aircraft as its own fuel.

A few tea bags or a pouch of coffee will provide some taste to the melted snow and will do absolute wonders for your morale. It should be noted, however, that in the case of serious injury or a possible dehydration situation, pure water (melted snow) is much better.

In the summer, a gallon of water is easy to carry. There is some support for using "Gatorade" as it is supposed to contain certain electrolytes which could be helpful, although I cannot find a specific endorsement at this time.

It is a mistake to believe that you can live off the land in the winter months here in the Rockies! If you land in the high mountains all the big game will have already vacated. The remaining squirrels and rabbits may or may not exist in the particular habitat in which you have crashed. Even if they do exist there, you may be severely limited in your ability to catch them. At best, game animals will be few and far between.

There are hundreds of food items, however, that can be taken along as part of a survival kit. The trick is to correctly trade-off perishability and food value (calories), for weight, bulk, and handling ease. Backpackers have been doing this for years. Their jobs have been made easier in the last few years by new foods and food processing techniques.

In general pack some high energy food in the form of candy, chocolate (Hershey's Tropical or Baker's Chocolate) and hard candy; bulk and warmth foods such as oatmeal and freeze-dried trail foods; hot liquids such as dried soups and hot chocolate powder; and salt. Better plan on three to five days full ration supply.

Now, with the backpackers stove, a quick cup of soup or a trail dinner can be prepared without the ordeal of first building a fire.

The next question is one of personal comfort. Why worry about personal comfort in a one-in-a-million chance that you'll need it? There is a simple answer. The answer is "morale" which translates into sustaining your "will-to-live and survive".

For me, there are a couple of things that maintain or restore my spirits when in a tight spot. A cup of hot coffee and a daily brushing of my teeth. One old salt swears that he has in his survival kit a bottle of champagne. He says it's to be used in the event of a crash to help his morale and brighten his spirits. Another says he must have a pillow. The important thing is to be sure that those few items of personal comfort, which make the difference in morale for you, are on board.

The feelings of loneliness and despair are the most severe survival stresses. Plan on having something to combat them. A good book may help pass the hours of waiting. Maintaining a log book can also help. It can be very useful to others as it may contain clues as to what to do or what not to do when it is analyzed later on.

Finally, you may wish to take along a few primary tools. The author's list includes a pair of pliers (with side cutter), a wire saw, snare wire, fish hooks, line and sinkers, dental floss, small sharpening stone, survival knife, .22 caliber survival rifle (AR-7), a steel handled hatchet and aluminum foil.

Finding a container that will package your survival paraphernalia is no small task. It should be rugged enough to take some banging around and yet be light in weight. I have never found the ideal container(s) and any suggestions would be welcomed. Currently, I use army surplus ammo cans. They are strong and waterproof — but very heavy. They can double as water pans, etc. I am exploring the idea of storing items in knapsacks or possibly a custom, multiple-pouched affair which would hang from the back of the rear seat.

Pilots are taught to plan ahead. Survival is no exception. Be prepared. Before you climb into your airplane ask yourself — "Am I prepared to survive in the terrain over which I will be passing?" Accept the fact that it *can* happen to you. Know that panic and exhaustion are real dangers of cold weather survival. Have a plan to prevent both. Know that mental obstacles can be more a factor than physical ones. Prepare a "checklist for survival" which gives you a step-by-step checklist of what to do in a survival situation. The planning will be invaluable. You will have already done most of the thinking before you are placed under the stress of a crash landing. You might take a course in survival. Several groups including the AOPA and NRA offer excellent ones.

In the event of a forced landing, will you be prepared to handle first aid, shelter and warmth, signaling, food and water, personal comfort, and the will to live?

Or, will the killing assumption kill you?

NPA PRESIDENT DIES AFTER CRASH IN UTAH

by YEARBOOK staff

Mike Loening, long-time president of the National Pilot's Association and son of aviation pioneer Grover Loening, died following the crash of his Cessna Stationair in mountainous country north of Hanna, Utah, on February 26, 1977. Also killed in the crash were Loening's son and a friend.

Loening had been in Colorado on business, had stopped in Durango to pick up his son and his friend, and was returning to Boise at the time of the accident.

Weather at the accident site was marginal VFR with gusty winds, according to the FAA. Moderate turbulence, snow showers, and moderate icing in the clouds were probable.

The aircraft came to rest on the surface of a frozen lake at 10,600 altitude. Loening was apparently attempting a forced landing at the time of the accident. Even though both wings and the right main gear had been torn from the aircraft as it struck trees in the descent, NTSB investigators determined that the flaps were full down at the time of impact, indicating landing configuration.

All occupants of the aircraft sustained only minor injuries from the accident. The medical examiner stated that all died from exposure, not injury. Temperatures in the area were estimated at -20 °F.

Investigators are suggesting a possible cause of the accident was fuel system icing. Water was found in the fuel system forward of the fuel selector location.

1978 CESSNA STATIONAIR 6

COMPUTER TO THE RESCUE

by Don Downie

Reprinted by Permission
THE AOPA PILOT (July 77)

It was late on a cold afternoon when the engine of James Kent's Cessna 150 began to sputter. Then it quit. Kent was somewhere west of Lake Tahoe VOR, in an area covered by trees. Seventeen-year-old Kent remained cool and broadcast a Mayday in the blind on 121.5. Fortunately, a high-flying executive aircraft picked up the call and relayed it to the FAA Air Traffic Control Center at Oakland.

Jim continued his silent descent and put the 150 into a river bed, hooked a wing on an aspen tree and totaled the aircraft. His shoulder harness was tight and he walked away from the wreck. Since it was near sundown, Jim turned on the airplane's still-operable rotating beacon and climbed to the top of a ridge in search of civilization. None was in sight, so he returned to the wreck and looked for shelter for the night.

Kent was dressed in lightweight summer clothing for what he had planned as a routine flight from Monterey, Calif., to Truckee-Tahoe. At the time of the mishap, student pilot Jim Kent had a total of 63.9 hours, but he was an experienced back packer and at home in the wilderness.

Captain Terry D. Jagerson (r), director of the San Bernardino County (Calif.) Sheriff's Aero Squadron, checks a DPICT computer-generated printout of a missing airplane's possible crash site, prior to taking off on a helicopter search flight.

Young Jim promptly found a hollow tree, scraped out some wood and climbed inside for a cold, uncomfortable night during which temperatures went down to 19 °F. All the young pilot's heavy clothing was hanging at his Seaside, Calif., home where his father, Lt. Col. James Kent, is Flight Surgeon at Fort Ord. The downed pilot was up-to-date on survival procedures and kept relatively warm.

During this long night the usual search and rescue (S&R) procedures were underway with takeoffs planned for first light. However, the young pilot had misjudged his position by 10 to 20 miles in his Mayday call. Yet, a rescue helicopter from Mather AFB near Sacramento was within a mile of his location when the pilot crawled out of his hollow log and figured it was time to turn on his ELT. He had "crashed" softly enough so that the ELT had not triggered, and the pilot had elected to save the batteries during the nighttime hours.

An FAA report of the crash indicates that "within five minutes after the search aircraft arrived in the area, the crashed airplane was spotted and the pilot picked up. . ." The day after the rescue, blizzard conditions existed throughout the entire Lake Tahoe area. The downed pilot's survival was directly attributed to fast action and precise thinking of personnel at the FAA Oakland Air Traffic Control Center, within whose boundaries the crash site was located.

The single, key factor that made this fast rescue possible is DPICT (computer talk for "depict"), a new and powerful search and rescue tool developed at the grass roots level by several FAA ATC center groups working independently.

Somewhere amid this rugged terrain lies the wreckage of a white Cherokee partially buried in show.

Basically, DPICT is a computer program that analyzes recorded radar data and prints out a plot of the route of a specified aircraft. This printout includes altitude (if encoder equipped), heading, time and last known position when the plane dropped off radar. Radar tapes are stored for 15 days and then reused. Thus, if you're not reported missing within two weeks, DPICT won't help you.

DPICT is a development of a number of FAA Centers and computer experts. In the successful Tahoe rescue, three data systems specialists, Wesley Montgomery, Charles Abele and Howard Seitz of the Oakland Center analyzed several radar tracks from computer tapes and determined that the only possible target was a track that ended 9-1/2 miles west of the LAT (Tahoe) VORTAC.

At the midwatch shift, military liaison specialist Jon Hancock was called in to aid in the S&R effort, using DPICT plots. After working most of the night, Hancock was able to trace the final minutes of the flight of Kent's Cessna and verify the position of the crash within a one-square-mile area.

When Hancock returned home from his unscheduled shift, his wife met him at the door with open arms and the welcome news, "They found him alive!"

Since Kent's rescue, two more men have been brought out alive through the help of DPICT techniques.

In more than 14 S&R efforts where DPICT has been the key factor in locating the crash, all other occupants had been killed on impact.

A check with Major Pete Warn at the Inland Search and Rescue Headquarters, Scott AFB, Illinois, confirmed that Kent's was the first time in his memory that the FAA's computer memory capability has brought a pilot out alive since it first was used to help find crash sites in June 1975.

Since DPICT data has become increasingly available, the Scott AFB headquarters for search and rescue has logged 34% fewer flying hours between 1975 and 1976 with 16% more missions. In 1975, the Civil Air Patrol, directed by Scott AFB, flew 72% of all Air Force directed S&R missions. 1975 totals were 555 missions and 23,400 hours, while in 1976 there were 646 missions and only 15,330 flying hours. Major Warn added that some portion of this efficiency was due to better ELTs and the use of weather satellite photographs as a planning tool for probable search areas.

The Civil Air Patrol presently includes 35,000 people, of which 15% to 20% are qualified mission search pilots. Responsibility for any S&R mission lies with the local county sheriff who may or may not request CAP assistance. And, when several counties or more than one state is involved, the logistics become complex. A close DPICT fix can eliminate much of this confusion.

A most enthusiastic supporter of the DPICT system is Capt. Terry D. Jagerson, director of the San Bernardino County Sheriff's Aero

A DPICT analysis of the airplane's final minutes of flight pinpointed its position closely enough for the wreckage to be spotted after a 20-minute helicopter search.

Squadron. He operates four helicopters, one Cessna 206 and has 12 local aircraft owners available as volunteers. Capt. Jagerson's area, the largest county in the United States, includes both desert and some high mountains. Since his area is immediately adjacent to the Los Angeles area, there are both extremely high aircraft activity and a high crash total.

"During the first 15 days of this year, we had 14 people killed in light airplane crashes in our county," Jagerson explained. "We had considerable weather during that period and almost all our bad crashes were weather-oriented."

Jagerson works closely with the Los Angeles Center and particularly Don Chaffee from whom he gets data directly on crashes in his broad area. In researching this article, we flew over a typical crash site with the captain in one of his Hughes 500 helicopters. A Cherokee Six, en route from Las Vegas, Nev., to Fullerton, Calif., arrived in Cajon Pass after dark. "The weather was very windy — 60 to 70 knots," said Jagerson.

The Los Angeles Center DPICT plot pinpointed the crash site within 100 yards, but blizzard conditions kept even ground parties from reaching the area for two days.

During a brief break in the weather on the third day, Capt. Jagerson took just 20 minutes to find the wreck.

"How do you spot a white airplane with no ELT signal when it is under three feet of snow?" we asked.

"I circled the area that was most difficult to plot on a map, looking for broken trees," explained the captain. "Finally I saw a straight line in the snow cover that later turned out to be a broken wing panel."

The victims were later removed by helicopter. Five months after the crash, the wreckage remains on a rugged hillside within a 10-minute helicopter flight from Jagerson's headquarters at the Rialto Airport.

The history of DPICT goes back to 1973 when the Kansas City Center worked up a diagnostic tool to aid in determining which radar site provided the best coverage of a given area. The result of this experimental program came out as a list of radar "hits" and, later, as a printed geographical plot of the radar track of a specified flight.

The original system was used to test the interface between the IBM 9020 computer which processes radar returns and the Raytheon CDC computer which provides the controllers' displays. This combination is now used at 15 of the 20 ATC centers throughout the nation. The five centers presently using all IBM equipment — Dallas/Fort Worth, Chicago, Cleveland, New York and Washington — do not have the capability at this time to come up with a DPICT plot. However, the FAA has an ongoing program to provide them with a DPICT capability within the next two years.

"This method is so realistic that I really had to work to locate the missing aircraft," commented Hancock.

Hancock, also a private pilot, has been working with DPICT at Oakland for more than a year and a half in addition to his regular duties. Originally, tedious manual plotting plus two manual conversions were required before the position could be transferred to a chart. Such a system was prone to human errors which have happened. Hancock worked directly with Tod Young in Denver who continued to speed up the data transfer and plotting process. DPICT is now in its 8th refinement.

L. R. Robison, Chief of the Rocky Mountain Air Traffic Division explains that "the DPICT program has been used primarily for S&R purposes with considerable success in locating transponder equipped aircraft. We have very little success with searching for primary aircraft (without transponders) due to ground clutter and weather returns interfering with the interpretation of the primary track."

Lt. Col. James D. Bigelow, Director of Operations for the California Wing of the CAP in Oakland, Calif., reports that all the searches he directs combine the use of NOAA satellite weather photographs with DPICT to narrow the search area. "We can see the weather status within 30 minutes of the time the pilot was in distress," said Bigelow. "We can better determine the height of the top of the clouds to see if a pilot could fly over a storm system, or was faced with staying at lower altitudes. . ."

During 1976, California Wing flew 58 search missions for lost aircraft, and all but two were found in 48 hours. The year before there were 46 missions conducted by CAP pilots with an average of more than one week to find the crash sites. Some were never found.

A potential breakthrough with weather and DPICT is now under development by Tod Young at the Denver Center in which weather-oriented radar "hits" associated with hail, heavy rain or cumulus build-ups may be reported on mosaic radar (radar from several receivers). If this weather data can be extracted from the same DPICT tapes used for tracing the aircraft, DPICT will make another important step forward toward a complete, accurate, fast S&R tool.

With DPICT, the FAA is now an increasingly vital part of the search and rescue business, but implications of the new technique are not without possible problems. For example, who establishes that an aircraft is really missing and authorizes the expensive computer search? Presently this is being done on an informal basis with Scott AFB and directly with various local law enforcement search organizations. A national standard of priorities is needed since each of the 15 FAA centers involved is working with a minimum of direction.

From a legal standpoint, the problem is awesome. There will be some who will consider DPICT an invasion of privacy. This could be one reason that DPICT has not received the public credit that it richly deserves. Perhaps you never knew before that "Big Daddy" is really watching you, but in the area of search and rescue, this isn't so bad, is it?

COUGAR: A PIREP

by Bill Cox

Reprinted by Permission
PLANE & PILOT (September 77)

When Grumman American set out in 1973 to build a true economy twin to replace the Twin Comanche, there were those in the aviation press who questioned the wisdom of such an undertaking. Memories of the Twin Comanche's slow death by a combination of slander and misunderstanding was all too fresh. It seemed the market for a "little" twin had evaporated, and no one wanted to see Grumman American recreate the mistakes that contributed to the PA-30/39's demise.

As some readers probably already knew, aviation writers often are called born cynics anyway (among other things). Those who doubted Grumman American would ever get the airplane into production have had to revise their thinking. The Grumman Cougar is here and very likely will be rolling off the line at a rate of five or six a month by the time you read this. Grumman American Vice-President Roy Garrison has plans to build 100 airplanes in the first year with future production rates contingent upon the success of the initial lot.

At this writing, the Cougar has not yet been certificated, though GA has received permission from the FAA for press demo flights. I was lucky enough to be the first of that group entrusted with the left seat for a day of exploring the Cougar's flight characteristics. In company with project pilot Lloyd Bingham, better known as Bing, I flew the airplane from Grumman American's new plant at Savannah, Ga.

Prior to the flight, I discussed the Cougar concept with VP Roy Garrison. "What we hope to do with the Cougar," Roy told me, "is bring back the owner-operated twin. We hope to catch some of the step-up market of pilots who used to transition from Arrows and 182s to Bonanzas and 210s. Now those pilots can buy a Cougar for about the same price, realize roughly the same performance and economy, yet fly with the extra margin of safety offered by an additional engine."

Besides tapping the private owner market, Garrison feels the company's new twin should make an excellent trainer that will find happiness in the hands of flight school operators. Initially, Grumman American plans to sell the Cougar through 25 of its 126 dealers. As the airplane matures, GA will expand the twin dealer organization to 70 or 80.

There's always a certain atmosphere of excitement associated with flying a totally new airplane, though with the first two machines having something like 400 hours of test time, there are few surprises left. For reasons best known only to my subconscious, the original prototype Cougar looked vaguely familiar the first time I saw it during a photo session in August '76. That's hard to figure, because the GA-7 has little resemblance to existing twins. There's some Baron look to the fuselage, and the very conventional wing (a 2063A-415 airfoil for aerodynamics buffs) could pass for many of the current crop of multi lifting surfaces. But the Cougar is its own airplane, owing very little to what has gone before.

At first glance, the production prototype I flew looks physically similar to the original "boiler plate" prototype. There are a number of notable differences, though. The sliding canopy inherited from the Tiger has given way to a more conventional door on the right side. Another door on the left was considered but abandoned in favor of economic construction. So as not to be too conventional, the prototype's inward-retracting main gear has been replaced by an outboard-folding system. Although it's not visually apparent, the production version also is four inches wider (44 inches across inside the cabin) and nearly two inches taller at the top of the fuselage. Walking around the airplane during the preflight, I couldn't help but note how big the Cougar really is for a supposedly "little" twin. The airplane's taller than the Baron, longer than either the Baron or 310 and has a wingspan equal to a 310's. It's perhaps significant that both the twin Cessna and the B-55 weigh well over 5,000 pounds, so the Cougar realizes some weight economies that aren't obvious from its size.

The production prototype was moderately well-equipped, including a full rack of Narco Centerline avionics, a Century III autopilot and the optional nose baggage compartment. Though the latter is not intended as a stash for golf clubs or fishing poles, the front luggage area looks adequate for two good-sized suitcases as long as they don't weigh over 75 pounds. Battery access is through the compartment floor, and avionics may be serviced through any of the three plates mounted to the back of the compartment.

Aft baggage goes aboard through a Tiger-type door on the right side. The rear cabin floor is stressed for 200 pounds and can accommodate a child's seat filled with up to 120 pounds of child.

Because of the test airplane's mission to prove a concept, rather than please a buyer, there were still a few loose ends in evidence. Aileron and flap hinges looked better suited to a DC-3, they're that beefy. Bingham suggested this might be a likely place for aerodynamic cleanup on final production airplanes. The interior was in surprisingly good shape considering that it had been removed and reinstalled several times. Seats and side panels are done in durable vinyls and configured similar to those in the Tiger. The front seats are buckets, while the rear pair are combined in a single bench-type seat stylized to resemble individual buckets.

Standard fuel tanks on the Cougar hold 59 gallons per side, and the airplane I flew had tabs showing 45 gallons, an early experiment. Finished airplanes will have a tab to indicate 39 and 49 gallons. Finally, airspeed was indicated on a special flight test instrument calibrated for accuracy down to one knot increments.

When Bingham and I climbed aboard for the flight, the airplane's fully-loaded weight was 3,525 pounds. That included 380 pounds of pilots, 480 pounds of fuel (80 gallons) and 20 pounds of camera gear. Working backwards, I came up with an empty weight of 2,645 pounds. For normal four passenger flying, Grumman American intends for the Cougar to carry 78 useable gallons, and that works out just about right. N530GA was a little heavier than a typical production airplane because of flight test instrumentation, but it's reasonable to assume a fully-equipped useful load of about 1,180-1200 pounds. Subtract 500 pounds for fuel, tow bar, tool kit and other miscellaneous extras that most owners carry and you're left with 700 pounds of payload; two normal couples plus a weekend's worth of baggage or four businessmen with toothbrushes.

Engine start with the diminutive Lycomings is uncomplicated by considerations of fuel injection. The engines are simple, reliable, 2,000 hour TBO, carbureted 0-320-D1Ds, basically the same mill that powered the last of the Tri-Pacers, the current Warrior 161 and Cessna's new Skyhawk 100. Fuel pump switches mounted on the selector console between the seats go on to fill the lines, and annuniciator lights shine through the switches to remind you they're on. The Lycomings have a spark boosting impulse coupling hooked to the left mag, so start consists of master and left mag on, prime as necessary, clear the prop and punch the start button. Once the engine is running smoothly, you click in the other mag and the alternator.

There's no interconnect between rudder and aileron on the Cougar, meaning you can taxi a crooked line without working the yoke back and forth. Pedal pressures are only moderate to minimize the need for asymetric thrust maneuvering on the ground.

Runup doesn't differ much from other twins. First power setting is 1800 rpm for the mag and carb heat check, then up to 2200 for prop cycle and back to 1500 for the feather and alternator test.

Acceleration on takeoff is fairly good, and there's little question that the Cougar is ready to fly at 70 knots. I learned later to roll in just a shade of aft elevator trim to make the airplane rotate and liftoff automatically at 75 knots. Gear is electrohydraulic, i.e. an electric motor driving a hydraulic pump, so you'll have no worries about getting the wheels up if one engine suddenly winds down and you elect to fly it rather than park it. Some older economy twins relied on an engine-driven pump to retract the rollers, a system that worked just fine till that engine quit. Then, you were usually left with a choice of pumping the wheels up by hand (if you had time) or hoping the trees were soft. Fortunately, on the Cougar, nothing short of a full electical failure will keep the wheels from folding outboard into the wells. If the electrons keep flowing normally, retraction takes about 10 seconds.

My climb checks proved the Cougar probably could make a solid 1100 fpm on a sea level standard day at gross. Temperature at ground level was a sultry 95 degrees on the day of my flights, driving Savannah's density altitude to 2300 feet. Under those conditions, sea level performance wasn't in the cards. Nevertheless, I saw 1,000 fpm on the VSI and timed a climb between 1,000 and 5,000 MSL in 4:20. The manual for the Cougar hasn't been written yet, but Bing mentioned it will probably suggest a cruise climb at 25 inches and 2500 revs, using 110 knots. Accordingly, I pushed the nose over from a best rate 97 knots and established the recommended climb power. After some hesitation, the Cougar settled on 700-800 fpm, not bad on a hot day with an inversion working.

If one engine goes to sleep, though, the airplane isn't quite so eager. Optimum rate at VYSE of 88 knots is listed as 280 fpm. Like the Twin Comanche before it and the old 160 Apache before that, the Cougar has a limited single-engine ceiling. Though the final number isn't firm yet, it will likely be about 5,000 feet, and that's just not going to cut it out West, especially in the summer. Grumman American isn't too concerned about the problem, because they don't intend to bill the Cougar as the ultimate twin. Besides converting some single-engine fans to twins, they hope to steal a bunch of buyers away from Piper's Seneca and Cessna's 337, and lure a few Baron, 210 and Aztec drivers. Grumman frankly admits that the current Cougar isn't the plane for you if you like to practice your engine-out procedures over the Colorado Rockies. The company's intent was to please most of the twin pilots most of the time at an affordable price. Those who need more capability should be willing to pay more bucks.

One thing you can't buy at any price is better asymmetric thrust handling. My first reaction when Roy Garrison told me the Cougar's VMC was below stall at all altitudes right down to sea level was that Roy was a little overenthusiastic. While it's true VMC reduces with altitude and most normally-aspirated twins will reach a height where there's insufficient power remaining to cause an uncontrollable yaw, I couldn't believe the Cougar could be that docile at low altitude. Or could it?

It could. To prove it, check pilot Lloyd Bingham had me shut down the left engine at 2,000 feet, drop gear and flaps, lever in full power on the opposite side and bring the nose up to a stall. Sure enough, the Cougar simply refused to break hard left. There was a lazy yaw toward the port side when the wing stopped flying, but it was nothing to write home about and could be easily controlled with opposite rudder and aileron.

To drive the point home, Lloyd gave me both engines back; then suggested I try a steeply-banked departure stall to the left. I knew what he was planning, but I was still unprepared for the airplane's reaction. When he chopped the bottom engine, the Cougar sagged sluggishly to the left in a totally matter-of-fact manner without any tendency to roll viciously out of control. My instinctive reaction (chop the top engine, push the nose over and add top rudder) was hardly necessary. The Cougar simply wasn't about to roll onto its back.

Just so I wouldn't forget, Lloyd also showed me a single-engine takoff. After I made a full stop, Bing took over from the right seat, set the left engine at zero thrust, and pushed all the right levers full forward. It took several hard jabs of right brake to hold the centerline under the nosewheel, but the Cougar accelerated deliberately to 70 knots and Bing very gently lifted it off. While we were definitely flying, the summer heat at Savannah made positive climb a little questionable and Bing elected to leave the gear down and plunk it back on the runway rather than retract the feet and take a chance on letting Grumman's primary demonstrator dig up the asphalt. I doubt if the airplane would have had any trouble limping around the patch, but Bing wasn't about to ding his favorite airplane.

Back at altitude, I was a little surprised to discover the Couger is faster than the old Twin Comanche, despite the fact the latter is a smaller, lighter design. Using 2700 rpm with 22 inches at 7500 feet, the Cougar showed me an indicated 145 knots for a true 167 knots (192 mph), slightly better than the 165 knots factory claim. Lloyd explained that performance figures really haven't been established yet because he and his staff of test pilots haven't had time to pin down firm numbers.

They have determined, however, that fuel consumption at 75 per cent cruise will run 8.6 gph per engine or

Grumman American's new twin is four years in the making, but flight test proved it was worth the wait. Cougar carries four at 190 mph+.

17.2 gph total. If 167 knots is a representative cruise (and Bingham seems to think it's pretty close) the Cougar delivers 9.7 nmpg, making it the most fuel efficient production twin ever built. (The Wing Derringer never made it to production.) There was no fuel flow meter on the production prototype, so I couldn't verify the estimated consumption, but past experience in Twin Comanches suggests that may be about right. A fuel flow gauge won't be standard equipment, incidentally, but probably will become available as an option.

Letdowns are facilitated by a gear speed of 145 knots and an initial 10 degrees of flap deflection at the same limit. The remaining 18 degrees of electric flaps can go out below 110 knots. Because the first 20 degrees are almost pure lift, there's virtually no pitch change until you add the final eight degrees. Then, the nose gradually drops lower.

Bing recommends normal approaches at 75 knots, though I tried finals down to 65 without any problem. Full flap flares at that speed demand a shot of power to avoid running out of elevator, but the rollouts can be ridiculously short. The Cougar's landing characteristics are about as conventional as possible. Power-off landings don't present the high sink rate prevalent on some heavier twins, and full flap slips are permissible if it's necessary to increase the descent. The best technique is still to hold a little power right into the flare; then, squeeze the throttles back and ease the airplane onto the runway.

Of course, the bottom line for most readers is how much you must pay for the world's most efficient production twin. That amount is not as much as you'd think. The basic GA-7 sans all extras sells for $69,900. Because people don't normally purchase twins in a bare bones condition, the airplane nearly everyone will start with will be the Cougar at $75,500. That extra $5600 makes the airplane a flyaway machine, adding a Narco Com 120/Nav 121, gyros and vacuum system, dual controls, a clock, OAT, VSI and turn coordinator, landing and strobe lights, a heated pitot, sun visors and a tinted windshield plus a few other lesser extras. The first real live "For Sale" Cougars will have a fairly limited option list to get the airplane to the dealers as quickly as possible. Present production prototypes are equipped with Narco Centerline radios and Edo-Aire Century autopilots, and those are the only avionics that will be available for the first few months after certification. Eventually, the Cougar will be certified with King, Collins and Bendix radios, probably in that order. Roy Garrison told me he also plans to offer radar (the fiberglass nosecone is already designed for it), prop autosync and a full list of de-ice options as soon as he can get the FAA's blessing on the basic airplane.

The production prototype I flew was worth $93,850 as equipped, and that's probably a price representative of what many buyers will pay. If that still sounds a little high, consider first that those are 1978 prices, not 1977 dollars. Consider also what it cost to buy a fully-equipped V-35 Bonanza in 1977 bucks. According to the ADSA Aircraft Bluebook, average list for a new V-tail was $90,350 in early '77. The same source indicates prices have escalated around 10 percent a year since 1970. Using that guideline, a '78 model V-35 (fully equipped) will sell for right at $100,000.

Comparing the Cougar's all up pricetag to those of competitive light twins in probable 1978 dollars, the Seneca II will bring $115,000, and a Cessna Skymaster will demand about $120,000. Beech's model 76 twin hasn't been priced yet, but it's unlikely that airplane will sell for too much less than a Seneca II. The model 76 will feature a pair of more expensive 0-360 Lycomings (compared to the Cougar's 0-320's) and will retain the Sierra's two-door design, so it's hard to imagine a fully-equipped price much below $105,000.

It may be a little premature to speculate on future versions of the Cougar before the initial design is even certified. There are, however, a number of very logical variations on the basic Cougar theme. The Twin Comanche found happiness, at least temporarily, with turbocharged 160's and a blown Cougar very well might wind up with those engines. Six seats in the existing cabin certainly isn't far off, though Grumman American undoubtedly would up the power to a pair of 200s or 210s to haul the extra load. Pressurization also is a remote possibility with the Cougar's super-strong bonded honeycomb structure; remote because the Cougar's fuselage is basically a box rather than a cone, and boxes are notoriously difficult to pressurize.

Roy Garrison told me there's even a possibility Grumman American might remove the little 160s from the wings and hang a single, high horsepower, blown engine on the nose.

In the here and now, however, Grumman's new cat has its work cut out for it. Not only is Grumman taking on the most successful twin in the world, the Seneca II, they'll also be slugging it out toe-to-toe with one of the industry's living legends, the Bonanza (not to mention the formidable Centurion).

While many pilots no doubt will stick with their single simply because it is a single, and they have no need to complicate their lives with an extra engine, GA is hoping there's another large contingent of aviators who'd rather switch than fight.

Add to this the inevitable rush of flight school FBOs looking for the cheapest possible airplane to use as a twin trainer, and there's every reason to believe Grumman's gamble is really more of a sure thing.

TOMAHAWK TAKES T-TAIL

by Dennis Shattuck

Edited for Space Reprinted by Permission PRIVATE PILOT (December 77)

When Piper sales managers introduced the company's new T-tailed trainer, the PA-38 Tomahawk, to their assembled dealers and distributors, the classy little two-seater received a standing ovation. "Just what we wanted," the dealers enthused to everyone who asked.

As they learned the specifications and operating costs involved, they cheered even more. "Now I can get rid of my 150 trainer and teach people to fly in a Piper," several dealers added.

Piper introduced the Tomahawk and a batch of other new items at a festive dealer gathering at Florida's Disneyworld recently, and the news of the new trainer was like laying out a filet mignon banquet for a starving man. After the visual, remote presentation of photos of the plane had stimulated their hunger pangs, the sight and feel of the actual plane in the reception area caused the dealers' taste buds to stand up and salivate. This is one plane that looks better in actual view than it does in the photos, so the dealers nearly gobbled up the two display models. If they'd been ready to fly home, no doubt they'd have disappeared immediately.

That Piper dealers have been starved for a true two-seat trainer is no secret...they haven't had one since the 108-hp Colt was dropped from production in 1963. The subsequent Cherokee was really a four-seater with only two seats installed, but its 150-hp engine was not thrifty enough for everyday competition against the ubiquitous Cessna 150. Even the Colt wasn't a good match; it was really a Tri-Pacer with only two seats and 108 hp.

The new trainer should prove more than competitive, however, not only with Cessna's 150, but with Beech's

Tomahawk trainer is Piper's latest entry in the pilot training sweepstakes. Tall T-tail and high aspect ratio wing identify it.

forthcoming Model 77 (a remarkable look-alike), and Grumman American's Lynx. All use the durable and long-lived Lycoming 0-235 four-cylinder engine, though in slightly varying strengths. The Tomahawk has 112 hp at 2600 rpm, the new Cessna 152 has 110 hp at 2550; the Grumman has 115 hp, but the Beech trainer's power hasn't been announced. No doubt fuel consumption and engine maintenance will be virtually identical, so airframe costs will spell the difference.

The Piper Tomahawk is already certified and ready to go. Production will begin in January, with first deliveries scheduled for February, 1978. The plane will be built in Lock Haven, Pennnsylvania, Piper's headquarters (all Cherokee series including Seneca II are built in Vero Beach, Florida), and the production goal is 1000 units during its first year.

"The Tomahawk is the first new trainer in 30 years," said Piper President Lynn Helms, at its introduction. "It's designed to make flight instruction both enjoyable and profitable...it puts the thrill, practicality and pleasure back into learning to fly."

Piper surveyed 10,000 flight instructors twice, and from those 20,000 forms developed a set of parameters that a new trainer must meet. Helms himself had a big hand in its development, flying and sampling each prototype to see if it met his standards, too.

The Tomahawk has been 10 years in development, the project originally begun at the behest of "Pug" Piper, son of the company's founder. As late as two years ago, the plane was still flying with conventional, fuselage-mounted stabilator. But, it probably sprouted its T-tail about 1-1/2 years ago when Piper engineers also made the change on the big single-engine Lance.

(As neither the press nor the dealers were permitted to fly the new trainer at the sales meeting, a report on its handling qualities and flight characteristics will be scheduled as soon as the airplane is available.)

Two doors and good visibility were two of the most mentioned features wanted by flight instructors, so the Tomahawk was so constructed. It has roomy, wide-opening doors on each side that open up into the curved roof to make entrance and exit an easy exercise. A large windshield and an equally large rear window contribute to the bubble-canopy effect and extremely good visibility. The metal roof gives just enough shade to prevent broiling under a hot sun.

The cockpit also has plenty of hip, shoulder, foot and elbow room. In fact, its the roomiest cockpit of all today's two-seaters. The panel is conventional in its layout, and there appears to be sufficient space for full IFR avionics and instrumentation. There's a 20-cubic-foot baggage area behind the individually adjustable seats. A console between the seats carries the manual flap handle and trim wheel. The fuel selector and gauges are unique: they're just above the throttle and mixture controls, and the tip of the selector knob points toward the gauge of the tank (left or right wing tanks, 30 gallons usable fuel) selected. It would be very hard to run out with this system.

Another interesting feature Piper is including on the Tomahawk is a two-year or 2000-hour, whichever comes first, warranty on factory-installed avionics. This will include removal and reinstallation. The list of avionics manufacturers approved for this program includes King, Narco and Collins. This program is a result of Piper's insistence that avionics makers thoroughly test all units before they are delivered for factory installation. Though it currently applies only to the Tomahawk trainer, it can be safely forecast for inclusion on larger Piper products in the future.

The Tomahawk comes out a nice-looking ship at first appearance, its T-tail lending a bit of modernity to the overall view. Of conventional design and configuration, the Tomahawk has flat spring-legged main gear with an oleo strut nosewheel, a low wing with good aspect ratio for efficiency and dihedral for stability, and a large cabin for passenger comfort.

The construction is all-metal monocoque, very similar to that of the current Cherokee series, though it should be mentioned that the Tomahawk shares almost nothing with that Piper line.

The tail is a separate elevator/-stabilizer arrangement with supporting structure inside the vertical fin. It is both mass and dynamically balanced for easy action at all airspeeds. The wing has a fat laminar flow airfoil section, much like that on the Cherokee series, but is an adaptation of the NASA-developed GA(W) wing. Its ratio of chord to span is much finer, with only 125 square feet of area for 34 feet of length (a Cessna 152 has 159 square feet and 33.17 feet span), which gives it a high aspect ratio and less drag per pound of lift developed. Some of the specifics of the PA-38 trainer are: Gross weight, 1670 pounds; empty weight (standard), 1064 pounds; useful load with full fuel, 426 pounds; fuel capacity, 32 gallons (30 usable); baggage capacity, 100 pounds; wing loading, 13.36 pounds per square foot; power loading, 14.9 pounds per horsepower.

Performance figures given by the factory include: Maximum speed, 113 knots; cruise speed at 75% power, 8800 feet, 109 knots; cruise at 65% power, 11,500 feet, 102 knots; stall speed with flaps up, 48 knots; with 30° flaps down, 46 knots; cruise range at 75% power, 402 nautical miles; at 65% power, 436 n.m.; rate of climb at sea level, 700 feet per minute; service ceiling, 12,850 feet; takeoff ground roll, 945 feet; takeoff over 50-foot obstacle, 1400 feet; landing over 50-foot obstacle, 1374 feet; landing ground roll, 642 feet.

Price is expected to be within a few hundred dollars of that for a comparably-equipped Cessna 152, the prime competitor for this new Piper Tomahawk trainer. A 1978 Cessna 152-II with Nav-Pac lists for $20,635; a similarly equipped PA-38 would be around $22,000.

NEW BELLANCA WITH A T-TAIL

by Ned Papineau

Reprinted by Permission
AIR PROGRESS (March 77)

As we slipped into the pattern and trimmed the nose up to slow down, a voice on the Unicom said:

"51 Alpha Golf, what kind of airplane is that?"

"It's a Bellanca," I said.

"Boy, I've never seen a Bellanca that looked like *that*!" said the stranger.

And neither has anyone else. In fact, until a few weeks ago, nobody was sure the airplane would be a Bellanca. It could just as well have made its debut as a Greenwood, or Anderson-Greenwood, or Aries T-250, which is the name stencilled on its fuselage at the moment.

The airplane, a true T-tail, was designed by Marvin Greenwood, Chairman of Anderson, Greenwood & Co., a valve manufacturing company of Houston, Texas. Now, how did a valve manufacturer get into the aircraft business? Well, it seems that the company was originally formed to do just that — design and build airplanes — back in 1945. And they did build a pretty little twin-boom pusher-type plane, called the AG-14, which was also designed by Marvin Greenwood. But the timing was wrong. Between the Korean war, which created a shortage of critical materials, and the recession of that period, enough sales could not be generated to sustain production. In 1952, Anderson, Greenwood & Co. gave up on airplanes and turned instead to aeronautical, missile and military work. Later, the company set its engineers to designing and producing a variety of pressure and safety relief valves, including those used on nuclear reactors.

But the desire to create small aircraft still burned in Greenwood's heart, so a few years ago he began designing an airplane that would be roomy, fast, and supremely easy to handle. The prototype of Greenwood's design went through the FAA certification process in a mere 37 days, which indicates that the documentation on the airplane's static and flight tests was both thorough and comprehensive.

With the bird in hand, the next question was, how could it best be produced and marketed? Bellanca had the production facilities and the distribution capability, and more importantly Bellanca needed some financing if it was to achieve future growth. And so a bargin was struck. Marvin Greenwood's T-tail design will be produced in Bellanca's Minnesota plant and marketed as a Bellanca; Anderson, Greenwood & Co. will raise the necessary financing by private means that will fund production of the airplane and give Bellanca a much-needed shot-in-the-arm as well. While all of the terms of the agreement have not yet been settled, it seems likely that some time in the future Anderson, Greenwood & Co., along with other investors, may own some healthy shares of Bellanca stock.

All this, of course, is predicated on the salability of the new design. Does the airplane have what it takes to be competitive in today's market? To get the answer to that question, I journeyed to Houston to see how the airplane performed.

The T-tail was waiting in the company hangar at Hook's Field, painted in shiny white with a distinctive trim. I stood back and eyed it. The fuselage is formed of nearly straight lines, which gives the plane a slightly angular look in profile. The top of the fuselage seems at first to be very flat; it appears that way because of the total absence of even the smallest dorsal fin. The tail, of course, hits you at once. Unlike Rockwell's 112 and 114 models, which have a mid-line T-tail, the new Bellanca's "T" crosses at the top of the vertical fin just like a jet tail.

Greenwood says he decided on a T-tail design because it requires minimum trim changes and produces greater longitudinal stability, more positive control and lower interference drag.

The plane has a 64A313.5 NACA airfoil with modified leading edge, which produces low drag and good stall characteristics. Drag is further reduced on the T-tail by the use of butt skin joints, flush rivets, flush and sealed nose gear doors, and flush fuel vents, drains and engine breather.

Greenwood chose the reliable Lycoming 0-540, 250 hp engine to power the T-tail's 3150 lbs. gross wt., and cools it through a central air inlet in the front cowling. The cowl itself is fiberglass, made in two sections for easy removal.

All these things suggested speed; now it was time to go upstairs and see what the bird could do. With Marv Greenwood in back and Test Pilot Jack Burden in the right seat, we cranked up and taxied to the active. Run-up is standard, with the usual cycling of the constant-speed prop, mag check, trim and fuel selector check. The wing tanks carry 80 gallons (76 usable) with fuel flow from both tanks for easy fuel management. The selector does have a detent to select either tank singly, but this would only be used if a vent became plugged, or a fuel line ceased to function.

"What airspeed should I use for rotation?" I asked Jack Burden.

Burden has spent 360 hours in the T-tail and knows it like his own skin. He has flown every test the airplane required for certification, and a few he devised himself. Now his eyes twinkled and he said:

"Oh . . . don't worry about the airspeed. Just feel the yoke, and when it's ready, it'll fly itself off."

He was right. I tentatively applied the tiniest bit of back pressure and suddenly we were airborne. I got one quick look at the A.S. as we lifted off and it was rapidly winding up from 60 IAS/54K to 80 IAS/70K as we climbed off the runway. Jack suggested 125 IAS, 25" mp and 2500 rpm for climb; with these figures, three aboard, and half tanks the VSI showed 1300 fpm, which is

60 fpm more than Greenwood claims for the plane at max gross.

There is one thing different about the T-tail bird on takeoff — TORQUE! No I don't mean spiral slip-stream or P-factor; I mean plain old torque that's reminiscent of the old radials. It takes some right boot to keep things straight down the runway. Jack Burden explained that it's due to an unusually large wash-in in the airplane's rigging. It causes no problem, of course, but is surprising in an engine with only 250 horses.

The trim wheel is located just below the panel near the pilot's right knee; on the prototype plane, it was a little stiff so I pushed on it fairly hard to trim for level flight. That was a mistake! Trim on the airplane is extremely responsive — a large movement of the trim wheel produces an instant and outrageous change in pitch. Undoubtedly the T-tail configuration has a lot to do with pitch sensitivity, as well as the stabilator/anti-servo tab combination. Once you get the feel of the minute forces required to trim the airplane, it's an easy task to set up level flight, or constant speed climbs/descents. Just to make things easier, the trim wheel has numbers marked on it for reference points.

With servo tabs on the ailerons, only light control pressure is needed for any degree of bank. I tried 55°-60° banks and the airplane handled beautifully through all of them. Stalls are so docile a child could probably fly through one just fine. I did several power-off stalls where the nose simply bobbed gently, and recovered with power in about 50 ft. of altitude. Even without adding power, the T-tail recovers from a power-off stall in a clean configuration in less than 200 ft.; it simply starts flying again the minute you release back-pressure. Unlike most airplanes, you can actually *feel* that tail surface flying deep into the stall and the rudders are completely responsive throughout the maneuver. With partial power, the T-tail simply doesn't want to stall at all; by the time I got the yoke full back the plane was approaching what looked like close to a 60° pitch angle, which Greenwood and Burden confirmed later; even then, the stall was perfectly gentle and absolutely straightforward.

Now it was time to check cruising speed, which Greenwood quotes as 208 mph/176K at 75% power. I had assumed he meant true airspeed, since that seems to be the way aircraft designers figure things. But at 24" mp and 2400 rpm (about 78%) we showed 210 mph *indicated* airspeed, (184K/IAS) and at 23" and 2300 rpm, (about 63%) a solid 205 IAS/180 K/IAS. Later, I asked Jack Burden to fly the plane on a short hop to another airport for lunch, and in his capable hands the 250 hp T-tail showed a top speed of 218 IAS/192 K/IAS which is three mph more than the specifications claim at max gross. All of this airwork was done at a relatively low 3,000 ft. altitude, which means that up higher you can expect some outstanding speeds from this bird.

As fast as it is, I expected the airplane to take awhile to slow down, but Greenwood thought of that too. V_{le} is a high 168 IAS/145 KIAS and when the gear comes down it acts just like a speed brake!

The airplane slows down dramatically, with perfect stability. Quarter flaps can be added at gear-down speed, and full flaps at 128 IAS/112 KIAS. However, it's best to hold full flaps until *short* final; the 35° Fowler flaps are highly effective as I found out on the first landing, which by the way was totally uneventful. Despite its short legs, the airplane has no quirks at all in ground effect; it simply lands — rather firmly, but very easily. Again, you can literally feel the T-tail flying right to the ground, and if there *is* any mistake you could make on landing, it might be too much movement of the elevator in the flare. The transition into the flare requires only a minute change in back-pressure, compared to most airplanes; a little too much pressure keeps the plane flying just above the runway.

All told, I flew the T-tail design a little over three hours and even with three of us aboard, I never felt cramped or hemmed in for space. At 45" across, elbow to elbow, the airplane is wider than the Bonanza and only a couple of inches shorter in over-all length. The windows are immense, by small aircraft standards, and the wrap-around windshield provides the best visibility, including the space *above* the airplane, of anything outside of a canopy job. The seats are comfortable, and the two in front are even adjustable. The prototype was finished in a handsome blue nubby fabric that made the interior quite cheerful.

As for noise, prototype aircraft are traditionally noisy, since few of them are insulated or sealed as they will be when production begins. I could tell there were 250 horses under the cowl, but the noise level was not bad at all, and will be even less when the plane is ready for sale.

Currently, Greenwood thinks the T-tail will sell for about $50,000 without avionics, but this price may change if the cost of materials goes up in the coming year.

With a useful load of 1300 lbs., a range of close to 1000 miles at 75% power, and speeds that top 200 IAS/176KIAS, the airplane beats any single in production today except the Viking. The T-tail plane was designed and will be built to comply with FAR Part 23-13, with

all the structural integrity required by recent amendments. It also has an unusually large CG envelope which makes it difficult to load the airplane out of limits so long as it's within max gross.

If everything works out as planned with Bellanca's production timetable, it seems very likely that a year from now the new T-tail Aries 250 will be giving competitors something to think seriously about. It's fast, easy to fly and the only really *new* design to come out in several years. By anybody's standards, the T-tail is a fine airplane.

Which leads one to wonder what will now happen to the Bellanca Viking? With the same speed, range and less than $1,000 difference between the price of the wooden-winged and partially fiberglass Viking and the all-metal Aries T-250, which airplane will the buyers opt for? Or is it possible that the Aries T-250 will eventually replace the Viking, bringing Bellanca's production methods in line with the other aircraft manufacturers who build birds with all-metal skins?

Like all attempts to see into the future, the picture still remains a network of conjecture; only time will tell what Bellanca has in mind for its new T-tail airplane.

And, you, Mr. Greenwood — what else do you have up your designing sleeve?

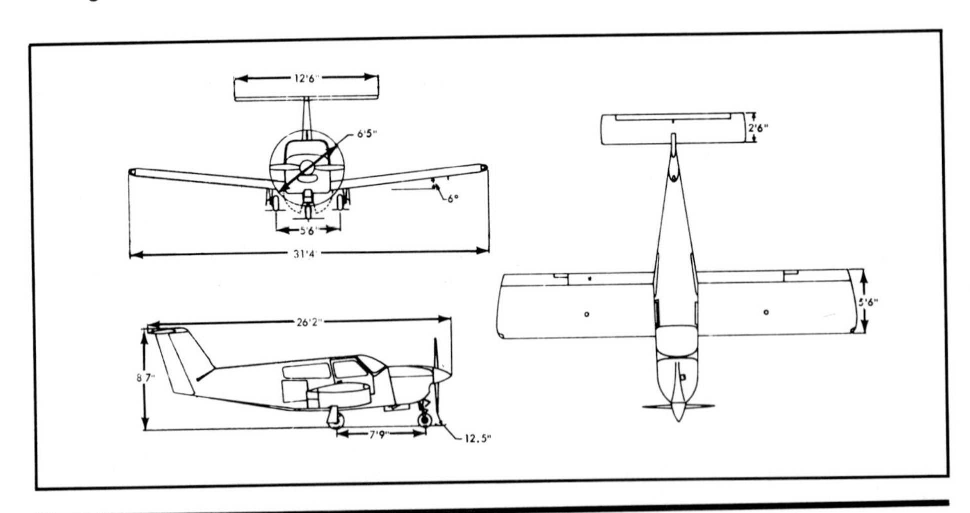

FOXJET GOAL IS JETTING ON 9¢ A MILE

Reprinted by Permission
CANADIAN AVIATION (June 77)

Development of a four-place, twin-engine business jet to cost about $350,000 equipped has been announced by a Bloomington, Minn. company.

The airplane is called the Foxjet ST-600 and, according to the developers, Tony Team Industries, will operate for an astoudingly low fuel cost of 9 cents per statute mile. . .less than most cars and less than any other jet.

"This aircraft was designed specifically in response to the crucial fuel problem world wide," said Tony Fox, president of Tony Team Industries.

He also said the aircraft was designed to meet FAR 23 regulations for single-pilot jet aircraft.

Fox, who claims to hold several patents on propulsion and rocketry designs, said the Foxjet would be powered by two Williams Research Corp. WR19-3 turbofans of 570 lbs. thrust each. The engine weighs 141 lbs. and is said to be the world's smallest turbofan.

Design empty weight of the Foxjet is 2,131 lbs., maximum take-off weight is 4,181 lbs.

Fuel capacity is 1,200 lbs., Fox said, producing a range with four occupants at 36,000 ft. of 1,215 nm at 277 kts. with 30-mins. reserve at 10,000 ft. Cabin pressurization is 7.4, providing a 10,000-ft cabin altitude at 40,000 ft., although twin engine service ceiling is 39,000 ft.

Landing gear is electric/hydraulic, with a pneumatic back-up system. The aircraft will carry its own electrically powered one-man portable towing system.

Fox said the airplane was designed to operate from almost any runway that other aircraft of its size and weight can handle, including sod. Takeoff run is 1,400 ft. and the aircraft has dual low pressure main wheels.

Other features of the airplane, Fox said, are: anti-skid brakes, wraparound bird-proof windshield, weather radar, electric or pneumatic de-icers on all flight surface leading edges, dual navcoms, autopilot, all-weather capability.

Chief engineer on the project is Emerson W. Stevens, an aircraft design engineer at Bell Aerospace for 27 years.

FINAL CL600 DESIGN SET

by Hugh Whittington CANADIAN AVIATION (May 77) **Reprinted by Permission**

Final configuration of the Canadair CL 600 Challenger — formerly the LearStar 600 — twin turbofan, widebody business jet, has the airplane with a T-tail and an AC electrical system.

The Challenger is also the first wide-body business jet, and the first transport category aircraft to be built around the supercritical airfoil. Canadair said the supercritical wing was the most significant aerodynamic advance since area rule, and the first major subsonic wing advance since the 1930s. Basically, the design of the supercritical wing moves the shock wave further aft and weakens it, increasing the critical Mach number from 0.75 to 0.85. The advantages are increased lift, reduced drag, and the design allows a thicker wing for more fuel and lighter stucture weight.

As of April 1, Canadair has customer names beside the first 80 serial numbers, with 73 firm orders — including three Canadian airplanes and 25 windowless cargo versions for Federal Express Corp., the Memphis-based small cargo express airline. The 25 Federal Express aircraft were financed by a $115 million loan — 90% of the total value — from the Canadian Export Development Corp. None of the other 46 firm orders has received EDC assistance.

Serial numbers 1 to 63 cost $4,375,000; 64 to 80, $4,975,000; and 81 to 100, an estimated $5,675,000 (all figures are in U.S. dollars). Canadian corporate operators can count on an extra $1,248,500 in Federal Sales and Excise taxes.

Development costs through to the first customer airplane are estimated at $130 million, of which $50 million is guaranteed by the Federal government and $20 million by the Quebec government. Customer deposits of approximately $250,000 gain interest at the prime Canadian rate until 18 months before delivery, at which time the interest earned can be applied against the total cost, or taken back in cash.

Customer deliveries are scheduled to begin about mid-1979, after a flight test program conducted in the western U.S. Testing will be done there because of more favorable weather.

Canadair executive vice-president Harry Halton, who is in charge of the design, said the T-tail configuration was chosen as a result of wind tunnel testing and other analytical data. The wind tunnel tests had shown longitudinal stability problems when the horizontal stabilizer was in the conventional position at the bottom of the tail fin.

"At this point, we do not believe the airplane will require a stick pusher," Halton said. "There will be a shaker; decision has not been made on whether there will be a pusher."

Halton said the original DC electrical system was changed to a constant speed AC system because of pilot suggestions. The system is powered by two engine driven generators of three-phase, 400 cycle, 115/200 volt rated at 30 KVA. Primary power can also be supplied by the standard-equipment APU or a deployable air driven generator. The air driven generator is automatically deployed from the bottom of the fuselage if both engine driven generators fail and the APU generator is inoperative.

The flight control system is powered hydraulically by three fully independent systems with no manual reversion. Two of these, providing general services as well as flight control, are powered by engine driven variable-displacement pumps. The third system, providing flight control only, has AC electrical motor driven pumps, with electrical power also supplied by the engines. All three systems operate at all times. Halton said there will be no control degradation if one system is lost, but there will be some degradation if two systems are lost, as some control systems — flaps, landing gear, brakes and control wheel steering — are powered by only two of the systems. However, in the event of a total engine failure, hydraulic power would be supplied by the APU — it must be on for all take-offs and landings, making it a no-go item — or a deployable ram-air turbine which would drop from below the fuselage.

Halton said Canadair was still trying to determine whether the ram-air turbine would spool up to speed in quick time at speeds as low as 110 kts. If it would, he said, then the APU could be off for take-offs and landings.

Halton added that the necessity for all these "contrivances" — APU and ram-air turbine — was dependent to a large extent on how much windmilling could be expected from a dead engine. If the engines windmilled at sufficient speed to maintain the hydraulic capacity required, he said, then the ram-air turbine would be unnecessary.

"However, there will be no way of determining this until we get into the actual flight testing," he said. "The ram-air turbine will certainly be on the first airplanes, but we may be able to drop it later."

Spoilers are comprised of three panels on each wing. The two inboard panels are descent spoilers; the outboard is for lateral control. All six boards operate as lift dumpers when the weight is on the main gear and the thrust levers are closed.

Halton said Canadair would ask the FAA to permit manual all-boards-up operation at altitude in case of emergency.

"Deployment of the spoilers is of major concern to the Canadian Department of Transport," Halton said. (The Challenger will be certificated jointly by the FAA and DoT.) "We don't yet know how it will all be covered, but it will be the safest spoiler system in the world," he added. (Spoilers have been a concern of the DoT since the July, 1970 crash of an Air Canada McDonnell Douglas DC-8-63 near Toronto after a hard landing caused by inadvertent manual operation of the spoilers at about 60 ft on the approach.)

The Lycoming ALF 502L turbofan engines were certificated last year at 6,700 lbs. thrust. It will be increased to 7,500 lbs. with the addition of a second stage to the low pressure compressor, which will provide 20% more thrust at the same turbine temperatures as the 6,700-lb. engine. The 5:1 bypass ratio provides 75% of the thrust from fan air and 25% from the core. Reversers are fan air only. Engine start is by air.

A Lycoming spokesman said like all jet engines, the 502L would get its lowest thrust specific fuel consumption at 36,000 ft. But, he said, the engine had been run at 75,000 ft. in a test cell "and performed well." The engine provides 700 lbs. thrust M 0.08 at 49,000 ft.

The core of the 502L is the T-55 engine which powers the Chinook helicopter, and has three million hours. The engine is fully modular — fan, gas producer, combustor turbine, and accessory gearbox — and any one module can be replaced while the original is being repaired. TBOs are 10,000 hrs. on the fan, 6,000 hrs. on the compressor, and 4,000 hrs. on the core, with the first hot section inspection at 1,500 hrs.

Lycoming is offering a two-year or 1,000-hour parts and labor warranty and guarantees a maximum $5,000 hot section inspection. The airplane and engines will be certified to meet FAR 36, with 78 pndb at take-off, 90 on approach and 87 at sideline.

Canadair hopes to have the Challenger certificated for flight at 49,000 ft., but admitted that it might have problems. The 9 psi pressure differential would maintain an 8,000-ft. cabin at FL 490.

Pilots will be comfortable at all altitudes, however. Halton, remarking that the biggest pilot complaint was cold shoulders on long flights, said the Challenger's wide cockpit would put the pilots further from the side windows, while heating air would be directed between pilot and window.

The full 14,610 lbs. of fuel is carried in the wing: 711 U.S. gallons in the left and right wings and 756 U.S. gallons in a transfer tank in the wing centre section, with single-point refuelling and a capacitance-type direct reading fuel quantity system. The original jet pumps have been replaced by two AC motor-driven pumps in the feeder tanks. The pumps are canister-mounted and can be removed on the ground without draining the tanks. A fuel/oil heat exchanger heats the fuel before it hits the pump, preventing icing.

Factory-installed standard avionics consist of dual Sperry SPZ 600 Flight Director, with 5-in. ADI and HSI; dual SPZ 600 autopilot; dual Navcoms, transponder, DME, C-14 compass and intercom system; and single radio altimeter, ADF, air data computer and weather radar. High Frequency radio is not standard; however, the HF antenna is.

One pilot complained about the C-14 compass, saying that it was not certificated in Northern Canada, and recommended that anyone planning to operate a Challenger in the Arctic should have the C-12 compass installed instead.

A typical 11-passenger executive interior would include two facing single seats separated by a table at each side of the fuselage; two facing double seats separated by a table on the starboard side, and a three-place divan on the port side. The cabin would also contain buffet and bar, cabinet, lavatory and wardrobe. Cabin headroom is 6 ft. 1 in.

The 30-passenger commuter interior has four-abreast seating, 16 on one side and 14 on the other.

Cargo version contains extra seats for cockpit observer and courier, front luggage compartment for crew belongings, a 96-in. cargo door, 9g barrier net, and 750 cu. ft. of container volume.

The price of the aircraft also covers type rating for two captains, airframe and engine ground school for two crew, training for two mechanics, flight training in customer aircraft, one year of computerized maintenance, first hot section labor by Lycoming, and manuals. Designated flight schools will have simulators.

Product support will be through two factory service centres in the U.S. and one in Europe with other selected facilities appointed as service stations.

LEAR-CANADAIR RELATIONSHIP OFTEN CONTROVERSIAL

The relationship between William P. Lear, who conceived the idea which has become the Challenger, and Canadair Ltd., which is finalizing the design and building the airplane, has been stormy and controversial.

At the time of the National Business Aircraft Association convention at Denver last September, it appeared as though the controversy could severely damage the airplane's credibility.

Outlining the history of the Canadair/Lear relationship, Canadair president F. R. Kearns told *Canadian Aviation* in an interview that Canadair first approached Lear

when it heard that he had conceived of mating the super-critical wing with new technology high bypass ratio engines, and took an option on exclusive rights to the manufacture and marketing of the proposed airplane. Lear, he said, would work with Canadair on an "as required" basis.

However, Kearns admitted, there were two problem areas that Canadair had failed to realize or recognize.

"The number one cause of the problem was that it was a big transition for Lear to work with Canadair," Kearns said. "For sixty years, he's always been in charge and we failed to recognize the immensity of the task to get him to work with someone else who had final authority.

"The second thing we failed to recognize was that Lear is a workaholic. He has to work twelve hours a day, seven days a week and the work on this program was not enough to keep him busy."

Last year,Kearns continued, Lear asked for something to do at his facility at Reno, Nevada. So Canadair funded some development work, and that evolved into a design which Lear called the Allegro, featuring a new wing, T-tail and engines buried in the aft fuselage. (It was this design that Lear was showing at NBAA and which caused the controversy there.)

"We told him that when he finished with the work (the Allegro) and put it through the wind tunnel, we'd compare data to see if we should adapt some of the features of the Allegro or — if our design bombed out — we'd switch to the Allegro.

"At the end of January this year, our aircraft was looking good and meeting its specifications and guarantees, and his data was looking bad — the fuselage-buried engines were giving trouble and the wing was showing less than predicted. We said at that time there was no way we would consider either his wing or the buried engine concept.

"But the main thing was, we were not looking at the Allegro design as competition, but as a complement to our design. Bill saw it as competition."

Today, he said, Bill Lear has accepted the fact that the Allegro will not be pursued by him or Canadair, so they are going to try to reach an agreement on something for him to do which will not be in competition with the Challenger.

Lear is already working on a new airframe design which, if feasible, would be a Canadair project because "anything he does in the aircraft business is for our account" under a contractual relationship.

As for the name change from LearStar to Challenger, it was not based on any disagreement with Lear, Kearns said. It was purely a marketing decision because customers kept confusing the LearStar with the JetStar and Learjet.

MAULE ON THE MOOSE RANGE

by Elsa Pedersen

Reprinted by Permission
PRIVATE PILOT (July 77)

Walt Pedersen's Maule M-4 Rocket is the most important piece of equipment in his seismic trail clean-up and erosion control business. Without the plane, it would be impossible to take a crew to some of the trails in the remote Alaska wilderness. His major jobs have been on the Kenai National Moose Range, where seismic work is permitted only in winter, and the clean-up and repair must be done in the summer. Much of the northern Moose Range is swamp and lake country, making a floatplane the best transportation.

The first major oil strike in Alaska was in 1957, 18 miles from Walt's homestead at the confluence of the Kenai and Moose rivers. Before then, the chief source of his flying business was hunters — moose hunters he took to the lakes in the foothills of the Kenai Mountains, and sheep and goat hunters he landed on the lakes formed by melting glaciers among the mountain peaks.

The oil strike at Swanson River changed that. A frenzy of seismic exploration erupted, so that within two years, long, straight trails were carved through forest and swamp-land of the western Kenai Peninsula. Officials of the federally administered Moose Range, concerned about land damage and erosion, restricted seismic crews to winter work when snow and frozen ground prevented most of the damage. Crews also were required to repair any surface damage and clean up garbage and litter left behind.

Pedersen, from his observations while delivering freight and carrying crewmen for the seismic camps, was aware of the problems and entered bids to various geophysical exploration companies. So for 16 years seismic trail clean-up and repair has been his major summer activity.

When a company contacts him about a seismic trail job, he is given a shot-hole map to guide him on his bid. Walt flies over the area, checking distance between shot-holes marked by a wide stain of plainly visible grey drilling mud. He studies the terrain to ascertain how much trail is in the woods, and how much crosses swampland. He locates lakes where he can land with a minimum of hiking to the trails.

Because his Maule is a four-place plane, Walt usually limits his crew to 3 and works along with them. They start out from Moose River early each morning and return at night.

Every bit of seismic trail must be covered on foot. Survey tape flags on wire stems, laths giving survey information, dropped tools, cigarette packages, pop, beer and engine oil cans, all are picked up.

Litter is deposited into the shot-hole and spilled mud is broken into small pieces, then shoveled into the hole. Mud dust is covered with tufts of moss from the nearby earth.

When the shot-hole is in a swamp, concealment is more difficult. Then moss must be plucked from a wide area, with care so the surface is not conspicuously denuded.

The hardest part of swamp work is traveling from hole to hole. Since the seismic work was done when the surface was frozen, there is usually little or no damage except the shot-hole and the surrounding drilling mud. Walt sometimes has to fly over the area and take bearings on the shot-holes to guide the surface crew.

Near the lakes the swamp is sometimes a vast floating pad of muskeg, broken by holes waist-deep with icy water. It is frightening and even perilous to cross these places on foot, carrying the cumbersome tools. Walt has a plastic folding boat he lashes to the plane pontoon. Its smooth bottom slides easily over the mossy swamp surface with the tools inside, as the men pull it with a rope.

Once the shot-hole area has been cleaned, Walt plugs the hole with a wire and canvas plug inserted with a special tool that holds the wire arms together as the plug is pushed into the hole. Loose earth is tamped on top of the plug to the surface. By the time the canvas has rotted, roots from the growth around the hole penetrate the fill and will bind the soil to make a permanent plug.

When the seismic trail has been cut through woods, clean-up and repair become more difficult. In addition to the shot-hole, much more damage is usually done. The bulldozer operator sometimes pushes aside trees which entwine in the branches of standing trees and do not fall to earth. These "leaners" must be felled so they will rot, and if possible the circular pad of roots and sod at their base must be dropped back on the ground.

In the winter, when bulldozers are used as snowplows to clear the trail for the seismic vehicles, the forest floor is sometimes gouged. These places are usually on a slope, and are an invitation to erosion by melting snow and spring rains. To repair them, Walt and his crew place dead trees across the trail and, if necessary, dig ditches to lead the run-off water harmlessly into the woods.

With increased oil exploratory activity and production, in recent years the State of Alaska and authorities in charge of Federal land have shown more concern about protecting the wilderness. Only rarely are seismic crews permitted to cut new trails — usually they are limited to trails cut years ago.

Seismic exploration methods also have improved. Instead of blasting deep shotholes, the crews gather subsurface information by use of vibration equipment that can be operated on old seismic trails and on roads and highways with no damage.

These improvements mean that Walt is no longer called upon to supply cleanup crews, and he now uses his Maule mostly for photographic work in Alaska's wilderness, and to prospect for gold in lakes at the base of Kenai Mountain glaciers.

TECHNOLOGY COMES TO GENERAL AVIATION

Reprinted by Permission
FLIGHT INTERNATIONAL (August 27, 1977) Edited for Space

Technical advances are supporting a revolution in fighter design and promise even quieter and more economical airliners. General aviation is now caught up in the momentum: the US Government, for example, is subsidising its industry to the tune of $10 million a year in research leading to better and safer aeroplanes. The TECHNICAL EDITOR reports from Nasa's Langley Research Centre.

For many years light aviation has remained on a plateau, with technical advances taking a very low place. A standing joke among some private pilots is that the word "improvement" refers to new cheatlines on next year's models. A cynical judgement, perhaps, but one recognizing that small aeroplanes have never been candidates for the kind of technology developed for the services and the airlines. But the general-aviation spectrum is wide, and as speed, size and costs go up there is more incentive for continuous improvement so that the cheatline gibe hardly applies.

First, advances in aerodynamics are becoming attractive to companies and corporations which want longer range to free their executives from inconvenient or even nonexistent airline schedules. As organisations become increasingly transnational, large numbers of quite junior staff need to travel frequently between divisions or subsidiaries. In particular these companies are seeking to extend the range of their current bizjets to give transatlantic capability without the now crippling costs of size and power growth. Secondly, better performance at relatively low cost is also in prospect for the huge volume of light aircraft. Partly it will arise from new production techniques permitting lighter structures or a better surface finish, and partly from a better understanding of airframe shapes and how they can influence performance and handling. Thirdly, opportunities are opening up for better propulsion efficiencies in the form of specialised engines, advanced propellers and new devices such as ducted fans and propulsors. Rising fuel cost is likely to accelerate the disappearance of older and less economical aeroplanes. Fourthly, aviation authorities continue to be concerned about the accident rate, which remains high.

Most of the impetus behind this new-found concern for light aircraft comes from Nasa, which some years ago took a renewed interest in aeronautics in response to criticism that it had abandoned aviation for the glamour of space. The pressure in fact came from congressional committees, which about 1970 called for the US Government to increase its involvement in general aviation.

Two factors which immediately pointed to the possibility of worthwhile advances at reasonable cost were the advent of supercritical aerofoils and the availability of big computers in which iterative calculations (i.e. successive computations based on assumptions that become more accurate with each result) could be run through rapidly. Though indispensible for commercial and military aircraft, they were largely inaccessible to most general-aviation companies because of the difficulty of absorbing expensive research on relatively low-cost aeroplanes. Nasa's job was not to act as a consultant to the companies on specific projects but rather to develop technology applicable to any design.

Supercritical aerofoils, developed by Dr Richard Whitcomb at Langley, appeared in 1965. They are superior to "conventional" sections for high-subsonic aircraft because, for the same thickness, they can go faster for the same drag and compressibility effects.

But the supercritical wing was also seen to have advantages for light aircraft. While the drag-rise benefit was of course no longer appropriate at cruising speeds of Mach 0.2 or thereabouts, the new sections, with their blunt leading edges, generate more lift with lower drag than their predecessors. They have typical maximum lift coefficients of 1.7 clean and 3.0 with flaps down. A line of investigation independent from the high-Mach work produced in 1972 the GA(W)-1 (General Aviation-Whitcomb) section, which met all four requirements set for it: cruise drag about the same as that of the equivalent conventional section; climb gradient increased by 50 percent to improve single-engined safety and make for greater tolerance of inaccurate flying under these conditions; maximum unflapped lift up by 30 per cent; and a gentle and consistent stall.

A development of this aerofoil is the GA(W)-2, a 13 percent section with a cruise lift coefficient of 0.4 which has been flown on a Beech Sundowner. The programme is being conducted by the Ohio State University. Potential benefits on the GA(W)-2 are a better lift/drag ratio and a higher maximum lift coefficient. The better L/D permits a faster cruise on the same power and longer range at a higher weight, while the greater lift lowers stalling speed and improves rate of climb.

Reduction of the airframe drag is a natural extension of Nasa's aerofoil research and goes hand in hand with new methods of predicting accurately the lift and drag of new designs while they are still on the drawing board. In July 1975 a combined Nasa/industry/-university meeting at Kansas University showed that drag could be reduced drastically, particularly in propeller-driven aeroplanes. It also concluded that the existing drag-prediction methods were inadequate to anticipate accurately the characteristics of new designs.

How much is really known about potential drag reduction for typical high-performance business aircraft? The earlier twins sported an abundance of mushroom rivet heads, large external aerials, suspect wing/fuselage filleting, lapped skin joints, air inlets at poor aerodynamic locations, and perhaps large-diameter exhaust pipes protruding at right angles to the local air flow.

The source of most drag on any light aeroplane is the wing, and improvements can be made through the new aerofoils mentioned above, together with reduction of drag from lift and skin friction and interference with fuselage and nacelles. Quite large reductions in lift-associated drag are possible with devices which interact with and reduce the strength of the vortices shed by the wingtips. Winglets (also developed by Whitcomb) were the most distinguishing feature of the external modifications which would have turned the basically transcontinental GII into the transatlantic GIII. The Vari-Eze also has tip surfaces which reduce the vortex intensity to permit more range.

Strict bookeeping is needed to account for the contributions from each part of the airframe, and as a result of this important meeting Nasa has begun to investigate the drag of complete airframes in a programme called Atlit (Advanced Technology Light Twin). The idea was to build a new twin-engined aeroplane embodying the most up-to-date drag-reduction techniques, including a new aerofoil. During the design stage Nasa would apply new computer-based analytical methods to predict lift and drag. The aeroplane would then be tunnel-tested and flown to see how well it matched the calculated performance.

Designing a new aeroplane from scratch was impracticable, so Langley bought the prototype Seneca I and modified it in association with Piper, which was responsible for the airframe, Robertson (design of a new wing) and Kansas University, which integrated all the changes to ensure their compatibility with one another. The original parallel-chord wing was replaced by a new tapered wing of greater span with GA(W)-1 section, spoilers and Fowler flaps. In the 20ft X 40ft Langley wind-tunnel, lift and drag were both found to be greater than predicted. The greater lift was attributable to the immersion of part of the wing in the propeller slipstream (the effect of which had been underestimated in the calculations) and the excess drag was put down to an ill-fitting door, which affected the flow around the wing-fuselage junction. Other rather large drag sources were the flap extension mechanism, which was not well streamlined, and the presence of chordwise reinforcing strips over the wing.

A large part of Langley's work is concerned with safety, and falls into two main categories: the development of safer structures, seats and body restraints, and the investigation of stall/spin-resistant aircraft configurations.

In order to study the behaviour of aircraft on impact with the ground, Langley modified its huge gantry which a decade ago was used by astronauts during simulated Moon landings in the Lunar Landing Research Vehicle. This has become the Impact Dynamics Research Facility, in which airframes restrained by cables are allowed to fall and hit the ground at accurately known attitudes and speeds. Comprehensive instrumentation includes ciné cameras to record the event from inside and out. Dummy figures simulating the human frame in weight, inertia and stiffness are strapped in so that their behaviour on impact can be measured and filmed. Airframes, even old ones, are not easy to come by and for Langley at least it was a stroke of good fortune when in 1972 Piper's plant at Lock Haven was flooded, drowning some 40 completed airframes, principally Navajos. Although the immersion was of short duration, and no sea-water was involved, the FAA was concerned that traces of acid from leaking industrial sources might weaken the airframes, and called for their scrapping. Thus Nasa was able to buy up about $5-1/4 million worth of aeroplanes for less than $50,000, and this has been the mainstay of the controlled crash programme.

Nasa analysis indicates that the most common type of stall/spin accident terminates in a crash with the nose pitched down at 40°, and with one wing about 20° down. Spinal injuries predominate in this type of accident, and the outcome of the crashworthiness programme may in the near term be a special energy-absorbent seat.

The GA work by Nasa in stall/spin accidents is a fallout from investigations over the past 35 years at Langley for the military. The emphasis is now on air-superiority fighters which must remain tractable at very high angles of attack to provide a reliable gun platform. It is this effort in particular that is bearing fruit for the light-aircraft industry.

Langley is seeking to develop a set of criteria that manufacturers can use to ensure a design that is reluctant to spin even with gross mishandling. While no one solution is applicable to all aeroplanes, certain characteristics appear to confer better qualities than others. For example one of the most important factors appears to be the position of the tail; T-tails are more spin resistant than fuselage-mounted tailplanes, but the influence of wing position — low wing, mid or high — is still not fully understood. Investigation into five typical aeroplanes appears to show that the conventional fuselage-mounted tailplane cannot be made to work predictably. Fuselage shape in particular influences the behaviour of the aeroplane at the stall, as does the shape of the fuselage/wing fillet. Rounded fuselage shapes appear to give less satisfactory characteristics than angular ones.

Eventually theories have to be tried out and, since spinning real aeroplanes can be dangerous, Nasa is now using large scale (quarter size or even bigger) radio-controlled models suitably weighted and ballasted so that the mass and inertia bear the same relation to the aerodynamic forces and moments as they do on the full-size aeroplane.

But finally comes the acid test, spin trials on the full size aeroplane. Langley has a Yankee which is capable of being modified for spin trials. The most important part of its work will be to show correlation between results of tests in spin-tunnels and with radio-controlled models, and behaviour of the actual aeroplane.

For existing aeroplanes or those likely to be less resistant to spinning, Nasa is investigating the ingredients of a really good stall-warning device, which may later become the basis of a stick-shaker or stick-pusher. The technique, first developed in the 1950s by Dr August Raspet (whose name is associated with low-drag aerodynamics), is to measure the fluctuations in the boundary layer at the trailing edge of a wing as the airflow is beginning to break down. This research is being done in association with Mississippi State University. The system uses an acoustic sensor, effectively a microphone, to measure the noise increase (due to turbulence) over the trailing edge as the wing approaches the stall. This position was chosen because it is less sensitive to engine power setting, yaw angle, flap position or wing roughness than any other location on the aeroplane.

In the early 1960s the FAA began to be concerned that numbers of GA aircraft were being flown quite near their structural limit, and asked Langley to find out how people actually used aeroplanes. Typical operations are being examined for excutive aircraft, trainers, crop-sprayers, fish-spotters and forest firefighters. Aerobatic aircraft are soon to come under scrutiny also. A number of aeroplanes in each group were fitted with Vg recorders to measure speed and normal acceleration. The results are disturbing, and show that the FAA's suspicions were justified. In firefighting activities for example the design strength (which the pilot may use) is exceeded about 10 per cent of the time. Nasa and the FAA are also concerned that some aeroplanes are being used in operations for which they were not designed, e.g. executive aeroplanes employed for the low-level survey of pipelines. The results of this work may show up partly as new design criteria and partly as revised operational procedures or limitations.

The turbine powerplant continues its inexorable drive to displace the piston engine, and has succeeded to the extent that relatively expensive turboprops are now attractive alternatives to supercharged reciprocating engines in the large business twins. But as size decreases the traditional advantages of turbines are less apparent, and can be secured only with increasing attention to the specialised needs of general aviation.

Recently Nasa's Lewis Research Centre has turned its attention to the propulsion needs of the light aviation industry in the closing years of the next decade. On July 20 it awarded 10-month contracts to Teledyne CAE, AirResearch, Allison and Williams Research to participate in its Gate (General Aviation Turbine Engine) programme. The companies will examine prospects for turbine engines up to 1,000 h.p. in shaft form for helicopters and propeller-driven aircraft and up to 1,500lb for fan propulsion.

Initially Nasa wants to find out what sizes of engine are likely to be most acceptable for the fixed-wing and helicopter applications, so Phase 1 of the effort will be a market survey. In Phase 2 the companies will analyze the technical and economic trade-offs (e.g. pressure ratios and temperatures against fuel consumption) and will identify any special technology which will need to be developed. This might be, for example, special combustor design (perhaps not duplicated in bigger engines) to ensure low smoke emission. Paper engines will be compared in Phase 3 with current turbine and piston powerplants to see if they are likely to be technically and economically better. If the prospects are bright enough, Phase 4 will recommend a technology development program to examine key components such as combustors, and perhaps to build demonstrator engines.

Few people, however, have sufficient imagination to envisage aviation without reciprocating engines. But they are increasingly coming under fire because of noise, particularly in the larger, 250-425 h.p., sizes. Probably little further can be done to quieten the powerplant itself without heavy performance penalties, but there is mounting interest in quiet propellers. Two propulsive innovations from Britain are the ducted propulsor from Dowty and a new family of propeller aerofoils from ARA, the Aircraft Research Association in Bedford.

Dowty's ducted propulsor aims at replacing the propeller with a multi-blade, variable-pitch ducted fan, which at a single stroke will reduce by some 15dBA the noise generated by a conventional propeller producing the same thrust. Initial flight tests under way in an Islander have shown big reductions in noise.

The aerofoil research work at ARA has resulted in a family of propeller sections generating more thrust for the same size. Alternatively, and this seems to be the main application for light aircraft, propellers with the new sections can generate the same thrust with a much smaller chord and consequent weight saving. This also results in lower hub stresses, which in turn permit a simpler and lighter hub.

The Lewis Research Centre, which is studying new propellers for the Mach 0.8 quiet and economic turboprop transports being examined by the big US airframe companies, is hoping shortly to launch its GAP (General Aviation Propeller) programme. Again, the idea is to see what advances are possible, and whether newly developed aerodynamics and technology can economically be applied to reduce noise. With the noise legislators looking threateningly at general aviation, designers are faced with the dilemma of having to decide whether to proceed with the known compromise of lower tip-speeds or to wait for the less certain promises of longer-term technology.

Cabin noise also is exercising acoustics engineers. The noise level in small aeroplanes is higher, or at least as high as, that in any other type of aircraft or public-service vehicle. The present emphasis is on noise transmission through the airframe and its control. It includes the study of noise sources themselves — not just engines and propellers, but also airframe noise. A lot of noise is transmitted to the cabin through the structure, and analysis of noise paths might indicate how the structure can be modified to provide better attenuation. To assess airframe noise in isolation from powerplant effects, Nasa hired Bob Hoover to carry out a series of flights in a Shrike Commander with the engines stopped, a routine familiar to thousands of air-show visitors.

Engine noise is the dominant factor and major reductions in cabin noise from this cause can come only from special engine mountings. Vibration damping and isolation technology has been an important aspect of helicopter development, and Nasa expects that much of it will be applicable to light aeroplanes.

After a slow start, solid-state electronics have spilt over. A number of manufacturers have thought it profitable to launch developments specially for general aviation. Although transistors came to the airlines in 1958, it was not until the early 1970s that VHF transceivers — which comprise most of the electronics in the simplest GA aircraft established themselves. But rapid developments (particularly in microprocessors and displays) for commercial and military users seem also likely to find application. It may be that the microprocessor can economically integrate computing and sensor information associated with communication, navigation, aircraft control and engine management, and Nasa has a program under way to see how this may be done and whether it is practicable.

It is likely that the larger business aeroplanes of the 1980s will be called upon to fly their owners to any airfield on the globe. A simplified form of the low-frequency Omega Earth-based beacon navigation system may be attractive here. Somewhat further away is satellite navigation which, despite its apparently exotic implications, may offer navigation fixes and height information superior to those of any other equipment.

Nowadays even single-engined aeroplanes need full airways capability. An important development here, since the majority of business aircraft in the United States are still singles, is the appearance of suitable weather radars for these types. Bendix has developed a radar with pedestal-mounted display and a scanning head installed in a streamlined pod on the leading edge or under the wing. Such a development has been made possible by impressive advances in radar during the past decade.

Autopilots represent a sizeable part of the cost of a light aeroplane, and in a joint exercise with Kansas University Nasa is studying ways in which they can be improved and some of their disadvantages removed. As part of this effort to overcome system disadvantages, Nasa and Kansas University have been experimenting with small supplementary control surfaces operated by local servos, entirely separate from the manual flying-control surfaces and circuits. This technique has a number of advantages: there is no stability augmentation or autopilot feedback to the pilot; full-time stability augmentation is feasible; the two functions (autopilot and stability augmentation) can be separate or combined as desired; the response time is improved; the system is potentially lighter because it does not need to move the control runs and because it acts on small surfaces sized to the relatively small moments required during landing or take-off; and the pilot can fly the aeroplane independently of the system.

Nasa is aware that one of its chief problems in general aviation is credibility. While the business and corporate sectors approve technology because it promises more convenience, many light aircraft pilots — whose flying comes out of their pockets — are suspicious of Nasa's motives.

INNOVATIVE AVIONICS

by David Underwood

Reprinted by Permission
CANADIAN AVIATION (July 77)

A favored crutch of the avionics sales brochure writer is the good old "major breakthrough", sometimes elaborated into "major technological breakthrough" and now and again — perhaps after a few lunchtime martinis — escalated to the level of "space age break-through". In fact, claims of breakthroughs appear so often these days in sales brochures that you almost start to wonder when they'll run out of things through which to break.

But disappointingly, too many of the breakthroughs turn out to be more cosmetic than electronic, and only rarely do many of them result in the pilot seeing anything really significantly better or different than he saw on the previous model or, for that matter, the one before that. As a case in point, I've heard it said that over the past 30 years of flight directors nothing has *fundamentally* changed since Sperry introduced the Zero Reader in the late 'forties. A sweeping statement, perhaps, but there's more than a grain of truth there when you think about it.

Now don't misunderstand me. There's no question that new technology like LSI chips has brought tremendous improvements in reliability, size, power consumption, and so forth. But over the past several years the avionics industry seems to have been slow to use this technology towards innovative *operational* approaches to the civil problem. This is less so in the military and research fields, but in civil aviation, progress has been evolutionary, rather than revolutionary.

As one example, Head Up Displays (HUD) have long been recognized as the answer to the demanding transition from instruments to the outside world at low limits — but where would you buy one for civil use today? As another example, NASA has been flying a TV-type flight director display as part of its Microwave Landing System evaluation. This display carries an electronic map of the terminal area, including the ILS/MLS localizer, and a symbol for the aircraft. But it also shows a three segment curving arc ahead of the aircraft. This is a *predictive vector*, and it tells the pilot where he will be if he holds his present bank angle for the next 90 seconds, with each segment representing 30 seconds of flight. If he levelled his wings, the vector would straighten out ahead, and if he then slowly added left bank, he would slowly bring the vector around onto the localizer. Think about this gadget the next time you are doing an ILS intercept in choppy weather — and also think about the fact that it was developed for the U.S. SST program, if you can remember how long ago that was. Who sells Predictive Vector Flight Directors? No one, and that's the pity.

But if Head Up Displays and PVFDs aren't moving, there are encouraging signs that industry is emerging from the avionic doldrums, and several innovative items are starting to set trends for the future.

The one with the biggest visual impact has to be RCA's color weather radar, coupled with the company's X-Y display. To some, color may seem somewhat gimmicky, but an actual demonstration ride (or even a "flight" in RCA's demonstration van, appropriately named Primus I) will convince the skeptics. Manufacturers' claims notwithstanding, it isn't easy to detect heavy turbulence "contoured" areas on present radar displays, as the introduction of cyclic and similar modes tends to bear out. With color it's simple — you just keep well clear of the red patches on the screen. The total weather picture is immediately recognizable, with no need for the experienced interpretation that was sometimes called for in the past. Almost unanimously, I think, pilots will ask, "Why didn't anyone do this before?" — surely the acid test of a true breakthrough.

Over the next several months, we can expect the other radar manufacturers to announce their color radars (Bendix has been the first) and no doubt others will follow RCA's X-Y scan approach. To understand the X-Y scan, remember that conventional weather radar displays show the radar beam scanning ahead in windshield wiper fashion, with the apex at the bottom centre of the scope. And, like VOR radials, this gives higher accuracy or definition at the origin than at the maximum range. Also, any numbers used on the display to indicate range, etc., follow the same curvature as the scan itself.

RCA has done three things here. First, the radar picture is not generated from the bottom centre but horizontally and vertically — hence X-Y like a domestic television. Then RCA has quadrupled the resolution, which makes for a much sharper picture (and a marked reduction in the "square cloud" effect common to digital radars). This crisper definition pays off in the third step, which is the future use of the radar screen to present check lists and other flight data — and one day, hopefully, electronic airways and terminal charts.

Does all this cost more? Yes, it does, but after all VORs cost more than NDBs, yet no one questions which is better.

RCA's announcement triggered a similar statement from Bendix that it also has an X-Y color radar, which will be introduced this fall. Aimed at the medium range general aviation market, the Bendix system is similar in general to RCA's but for those who already have certain components of their BX-2000 avionics system, the Bendix display will also provide electronic track waypoint and heading data superimposed on the radar picture, so that pilots can assess weather avoidance alternatives in relation to their planned route. The Bendix cathode ray tube (CRT) can also be used for displaying checklists and other data, and will offer the handy feature of allowing the pilot to insert his own personal operational data

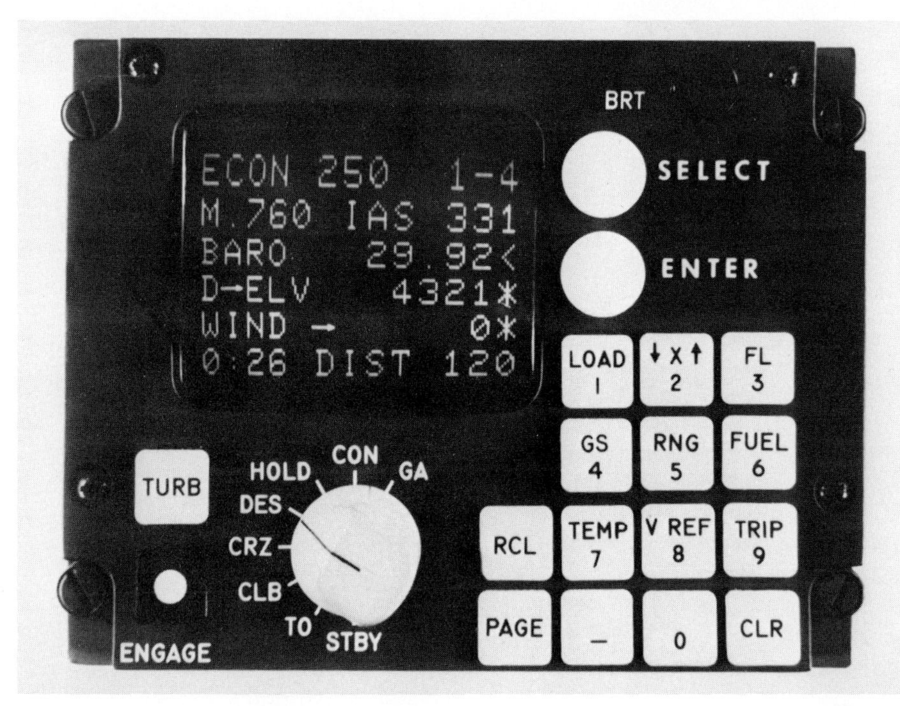

into the computer memory, such as special procedures, telephone numbers, etc., in addition to the standard formats for the specific aircraft type. Doubtless RCA will now reciprocate and offer the same thing.

But all the foregoing is of little consolation to the single engine pilot, for whom radar has been an oft-promised, but unfulfilled, dream.

Two systems now on the market could change this.

First, Bendix has introduced the RDR-160 — a 14.9-lb. total weight radar whose combined transceiver/antenna unit weighs 9.4 lbs. and measures 9 ins. long and 10 ins. in diameter. The RDR-160 has already been pod-mounted in a Cessna Skymaster and Centurion, and wing pods are now being developed for Beech Bonanzas, and similar types. No short cuts have been taken — it's a full 160-mile digital display radar — which, apart from its impressively small size, is conventional in every other way.

By contrast, the Ryan Stormscope is very unconventional.

Conventional weather radar operates on the principle that severity of turbulence is directly proportional to the intensity of the rain, and it's the rain that gives the radar return. The Stormscope operates on the principle that turbulence is proportional to electrical discharges, from severe static to lightning. The system uses the aircraft's ADF antenna to pick up electrical discharges (making the ADF good for *something* when storms are around) which it then processess on a 360° display at ranges up to 200 miles. Actually, the ranges are called "pseudo ranges", because a more intense storm will appear closer on the display than it actually is. That in itself is a useful safety factor. The Stormscope is intended as a weather avoidance tool, not something to be used to penetrate a squall line, and it represents precisely the sort of innovative thinking that general aviation needs.

The LSI chip technology has brought with it great computing power, but a non-flying computer engineer expressed polite surprise when I described to him the present state of general aviation Area Navigation (RNAV) computers with their 10 or 20 waypoints, distance/time to go, groundspeed, heading, and such-like readouts. Surprise, because to him they appeared so *primitive* in the light of what the computer industry is doing in other fields. An it's true enough. The ballyhoo surrounding, say, a new DME readout which also gives you groundspeed and time to station makes the achievement sound like a minor miracle. But when you think about it, the arithmetic involved is pretty elementary and you can get the same answers almost as quickly with a stopwatch and the back of a cigarette package.

But more powerful computers are coming, and the pacemaker could be Sperry's Tern 100 system, which literally tells you everything about going anywhere. This tell-all capability is located in the Disc Memory Unit, capable of storing every known navigation aid, airway, waypoint, RNAV procedure, Standard Departure and Arrival procedure and suchlike that exists *throughout the world.* Simply put, the Tern is the electronic equivalent of a complete global set of Jeppesen charts — and those who have spent a large chunk of their lives plodding through Jeppesen amendments will be intrigued to learn that Sperry has a tape cassette device which you carry aboard every 28 days to "talk" to the Tern, and which updates its worldwide data bank in five minutes flat.

In addition to the standard published data, you can also store 50 of your own commonly flown flight plans on the disc, as well as 204 waypoints that are special to your own operation.

In operation, you insert your flight plan data in ATC language and the system will then take over, either supplying commands to the flight director for manual flying or coupling directly to the autopilot, and all the time automatically selecting and tuning navaids as required and taking in data from other sources, e.g. Omega/VLF, INS, etc. (the Tern CDU can, in fact, replace the standard INS CDU). Finally it does the whole thing in vertical as well as lateral navigation terms.

It's a business pilot's dream, and even my computer friend allows that it does appear to bring aviation up to date again.

Sperry has been flying the system for a year-and-a-half in a Boeing 707, and is currently installing Tern in its King Air, and expects to have systems for Gulfstream IIs and Lockheed JetStars later this year.

What Sperry's Tern does for navigation, Lear Siegler's Performance Data Computer System does for aircraft and engine performance — described these days as "total energy management". Aimed initally at the airline jet market, the system continuously monitors ambient temperature/altitude/airspeed, etc., and aircraft data, e.g. present fuel weight, EPR, etc., against the aircraft handbook/flight manual data stored in its memory, and then automatically positions target "bugs" on the airspeed and EPR indicators to which the pilots fly to achieve optimum performance. The system operates continuously throughout take-off, climb, cruise, and descent, and also covers holding, diversion, engine out, and go-around situations. In other words, whereever you're airborne, it will tell you the optimum way to fly to achieve fuel economy.

Now of course, pilots can, and already do, the same comparisons themselves from the available data but it would be impossible to do the whole thing continuously like the computer or with its speed and accuracy — hence the development of the computer approach.

As said, the prime interest right now is with the airlines (not surprisingly, when you consider Air Canada's fuel bill of over $520,000 a *day*). But as the cost of fuel goes up — and it surely will — business jet pilots would be well advised to keep a close watch on developments such as this in the months ahead.

Over the years, Canada's avionics industry has produced more than its fair share of innovations — the original R-Theta nav display and the Crash Position Indicator being two of many.

Canadian Marconi's pre-eminence in the airborne Omega field won it a major retrofit contract for Pan Am's 707 fleet earlier this year and Litton — whose name is usually associated with commercial INS — has developed a very promising airborne search radar system which will go into DND's Tracker aircraft, and is being considered by a number of foreign nations for their maritime surveillance aircraft. Computing Devices continues to be primarily engaged in military systems, although its Thrust Measuring System — which allows both maintenance personnel and pilots to achieve optimum engine performance — is being viewed with interest by commercial customers. Leigh Instruments — which started life as the Crash Position Indicator manufacturer but has since diversified into a variety of areas including computer-based marine traffic control systems — is still coming up with original concepts like an intriguing ice detector for hovering or slow moving helicopters (a tricky problem, that) and an inflight stress recorder which weighs 2 ounces, is less than a cubic inch in size, records better than 100 hours of flight data, and requires absolutely no electrical power to operate it, being driven by the slight flexing of the structure being measured.

Overall, then, there's a lot of potential innovative power available in the avionics industry. Its up to us as users to ensure not only that the industry knows what we want but, equally important; that we're ready to pay for it when they can deliver.

VLF NAVIGATION

by Barry Schiff

Reprinted by Permission
THE AOPA PILOT (July 77)

In the beginning, there were heaven and earth. And fortunately for man (who came later), there was a useful predictability to the apparent motion of the celestial bodies. Thus was born one of the earliest methods of global navigation.

The first popular, electronic method was LORAN, an acronym derived from its formal name. LOng RAnge Navigation. An on-board receiver/processor is used to determine the difference in time between the reception of signals from a pair of station transmitters. The result is a hyperbolically curved line of position (LOP) that must be plotted on a special chart.

The procedure is then repeated to obtain another curved LOP which results in a reasonably accurate fix (usually within 10 miles)

Obtaining such a fix, however, is a time-consuming affair, subject to misinterpretation by the pilot/navigator and frequently impossible when needed the most because LORAN signals often do not extend far enough.

The next significant improvement was Doppler, a self-contained radar system that constantly measures groundspeed and drift. Given these variables and some input from the compass, the on-board computer can "reckon" continuously the progress of an aircraft with respect to any chosen course. But Doppler is not more precise than its weakest link, the compass. Magnetic and precessional errors accumulate during flight, resulting in a loss of system accuracy.

Then came the magic of the inertial navigation system (INS), a marvelous collection of accelerometers, gyros and other electronic wizardry that brilliantly measures every change in aircraft speed and direction. An on-board computer digests these changes to keep track very accurately of en route progress.

INS is the ultimate navigation system — if you can afford it. Prices range from $80,000 to $112,000 *plus* an expensive installation *plus* maintenance which averages about $3.25 per hour per unit.

But a recent newcomer, VLF Navigation, is arousing considerable interest in general aviation circles. It is much less expensive than INS (prices begin at $20,000), does not accumulate errors en route (and is, therefore, often more accurate than INS), requires about as much maintenance as a transistor radio, and at least one model (the GNS-500A) recently was approved for domestic IFR, en route navigation (on or off airways).

VLF navigation utilizes very low frequency radio signals in the 3 to 30 kHz band. The advantage of such low frequencies is that they closely follow the earth's surface for vast distances (depending on the amount of radiated power). A receiver situated in a Pennsylvania valley can easily receive signals originating in Norway, for example. VLF reception is not subject to line-of-sight limitations as is VHF (30-300 MHz).

VLF navigation is based on "phase measurement," a principle that is not difficult to understand so long as some simplification is allowed.

Figure 1 shows a 10-kHz (10,000 cycles per second) VLF radio wave traveling from left to right. Each complete wave (the solid line) is one cycle. Frequency, by definition, is the number of these cyles produced in one second. Since radio waves are known to travel at a constant 166,925 nm/second, it is easy to determine the precise length of a complete cycle. In other words, if the speed of radio wave transmission (166,925 nm/second) is divided by the frequency (10,000 cycles/-second), the result is the precise length of each cycle (or wave) which, in this case, is 16.7 nm.

Now assume that at the beginning of flight, the aircraft is at point A in the diagram. The pilot turns on the VLF navigation system and, by punching the appropriate buttons, informs the computer of his known location (using longitude and latitude). In the meantime, the system very carefully measures the *phase* of the signal being received. In other words, the system determines precisely where along the wave the aircraft is situated. A minute later, the system determines that the aircraft has moved to a different phase of the cycle (point B) and knows, therefore, precisely how far the aircraft has moved either toward or away from the transmitter.

In the example, the aircraft has flown along one-fourth of a wave length and has traveled, therefore 1/4th of 16.7 nm, or 4.2 miles away from the transmitter. This provides one circular line of position. Simultaneously, the system "phase measures" the waves from one or more additional transmitters to obtain a fix.

In reality, this process takes place constantly, thus providing a con-

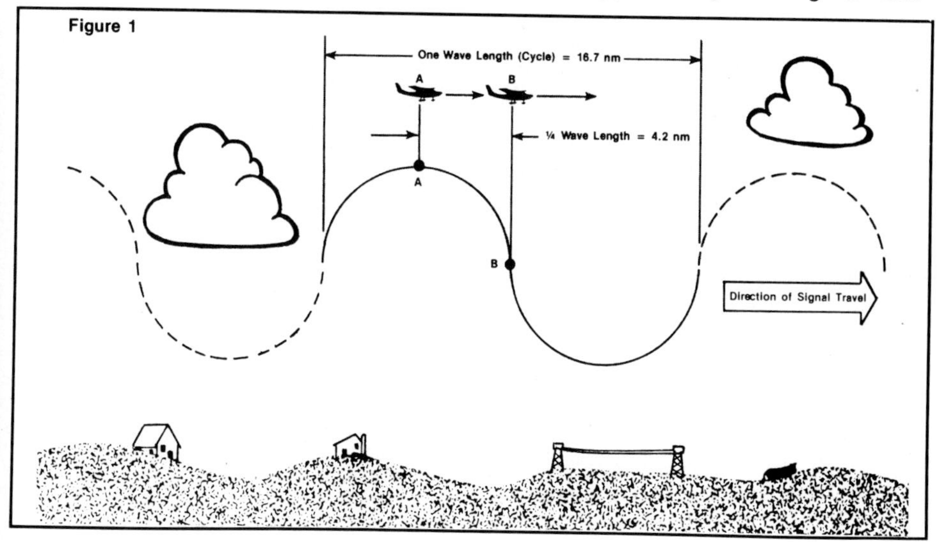

The GNS-500 VLF/Omega control display unit, besides allowing the pilot to input his initial geographical fix and select waypoints, can display such information as latitude/longitude, waypoint displays, time and distance to waypoints, drift angle, groundspeed, actual wind, etc. If that's not graphic enough, a course deviation indicator is also available.

tinuous, digital display of aircraft position. And by measuring the rate at which position changes, the system computer also provides instantaneous groundspeed as well as a plethora of other navigational data (described later).

One weakness of the system is its inability to determine position from scratch. To provide meaningful information, the system must be told (via push-buttons) of its starting position and be allowed to monitor aircraft progress from that point. In this respect, VLF and INS are similar.

Anyone who knows much about radio wave propagation realizes that VLF signals travel between the earth's surface and the ionosphere, one of the atmosphere's numerous upper layers. And since the height of the electrically charged ionosphere is constantly changing from day to night and with the seasons, this has an effect on the *apparent speed* at which radio waves travel. This phenomenon is called the "diurnal shift" and requires compensation by the VLF system. Otherwise, unacceptable errors would accumulate.

For this reason, the pilot must tell the VLF computer (at the beginning of flight) the time of day (GMT) and date. The computer, which has been programmed with an accurate "picture" of the very predictable diurnal shift, can then correct all incoming signals to eliminate errors caused by the varying height of the ionosphere.

Figure 2, for example, shows an aircraft flying over Southern Nevada in the late afternoon. The airborne VLF system is receiving and processing VLF signals from Hawaii, the state of Washington, Norway and Panama. Since the computer knows the time of day, the date and the precise location of each transmitter, it corrects each incoming signal for the varying effect of the ionosphere. The Norwegian signal, for example, travels over a portion of the earth that is predominantly dark (a lowered ionosphere), whereas the Hawaiian signal travels (at this time) along a path of daylight (a raised ionosphere). The computer corrects these and all other incoming signals to a common denominator so that the apparent speed of them all is identical.

A few years ago, such a "diurnal computation" would have required a computer the size of a small house; today, that same function is completed by microcircuitry that can be held effortlessly with your fingertips.

The signals used in VHF navigation originate from two independent, global transmission networks. The first consists of nine communications stations (most of which are operated by the U.S. Navy) for the purpose of maintaining worldwide fleet communications. Each transmitter has an assigned frequency between 14 and 24 kHz and is extremely powerful (as much as one million watts of radiated power) which results in reception ranges of up to 10,000 nm.

And since VLF navigation systems are not equipped with audio circuitry to allow eavesdropping on military communications, the Navy

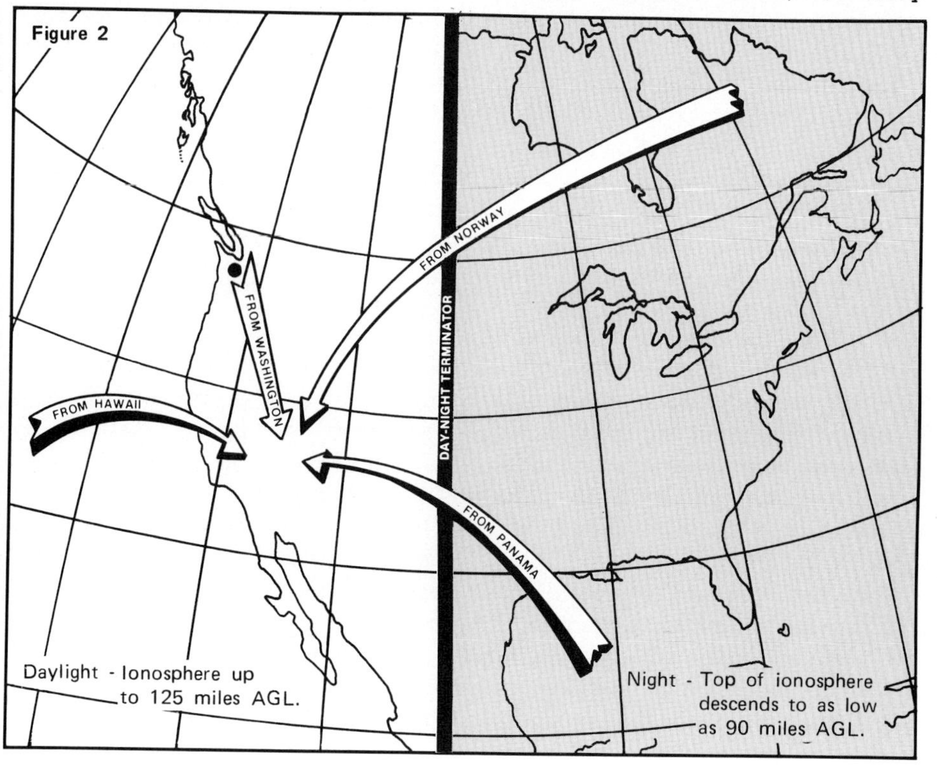

does not object to the use of its transmissions by civilian users. (There was some controversy about this a few years ago, but that problem has been resolved.)

The second VLF network consists of eight Omega Navigation transmitters operated and maintained by the U.S. Coast Guard. Originally, this system of long-range navigation was developed for surface vessels and submarines, but the development of lightweight, digital computers allows the use of Omega in aerial navigation.

To the pilot using VLF navigation, there is essentially no difference between the nine communications stations and the eight Omega stations. As far as he is concerned, there is a total of 17 VLF transmitters spread around the world, any combination of which can be used for navigation.

The beauty of VLF is that the pilot is not required to tune and/or identify any of the stations. It's all done automatically when the system is activated. The GNS-500A (manufactured by Global Navigation), for example, continuously monitors all 17 stations and selects the eight most suitable signals for navigation. Every 10 seconds, the system reviews the quality of all incoming signals and makes the appropriate substitutions whenever necessary, even when one of the stations being used suddenly goes off the air.

The pilot, however, is unaware of this internal search for perfection. If he desires, he can query the computer as to which stations are being received and the reception quality of each. Obediently, the coded answers will be displayed digitally on the control display unit (CDU).

Although it might be disconcerting to know that the closest station being used for navigation may be several thousand miles away, this need not be so. The proximity of an aircraft to a VLF transmitter has no effect on system accuracy.

VLF is not without its Achilles heel: solar flares, for example. These eruptions on the surface of the sun raise havoc with VFL signals and can be responsible for errors of up to five miles. Fortunately, however, these infrequent disturbances last for only a few minutes at a time.

Otherwise, system accuracy is nothing less than fantastic. Typically, VLF errors are measured in fractions of a mile. When the FAA flew a Global GNS-500A for purposes of bestowing an STC on the installation, the error at the end of a two-hour flight was .1 nm. An Air Force VC-135, equipped with every modern form of electronic, long-range navigation equipment ran extensive tests of VLF and found system tests of VLF and found system errors to average .7 nm, as well as or slightly better than INS. A virtue of VLF worth repeating is that errors do not accumulate during a long flight as they do with Doppler or INS.

The control display unit (CDU) is the business end of a VLF system and is similar in appearance and operation to that of INS.

After the pilot turns on the system, he prepares for flight by inserting present position, time and date into the computer. If the proper keyboard buttons have been pushed, the CDU will acknowledge its readiness for flight.

The system can then be loaded with up to 10 en route waypoints to define the planned flight, or his procedure can be delayed until after departure. Like INS, VLF requires little navigational knowledge or talent to be a system "programmer."

Once underway, present position is constantly shown in the left (latitude) and right (longitude) data displays. Or if he desires, the pilot can rotate the display selector switch to obtain waypoint displays, time and distance to any selected waypoint (based on existing wind), drift angle, groundspeed, actual track, desired course, distance off-course, parallel track, and actual wind. Virtually every navigational chore is possible.

And when the pilot gets lonely, he can even have a limited en route conversation with the computer (really!) to obtain a phenomenal amount of data (including the direct distance between any two points on earth).

At this point, it is important to differentiate between the two types of VLF systems presently available. One type utilizes both the Navy's communication stations and the Coast Guard's Omega network. Thus far, only three such units are in production. Canadian Marconi's CMA-734, Communications Components' Ontrac III and Global Navigation's GNS-500A. In addition to these, a dozen manufacturers have chosen to produce pure Omega systems, devices that receive only Omega signals.

The combination systems (referred to as VLF/Omega) offer the greatest geographic coverage. Since 17 stations blanket the earth with VLF signals, navigational capability is virtually assured anywhere in the world.

This is not necessarily true of the pure Omega systems that rely on fewer, relatively low-powered, transmitters (only 10,000 watts), each of which is scheduled for a maintenance shutdown during a specific month of the year. During July of every year, for example, when the Omega station in Norway is taken off the air, total North Atlantic coverage is questionable. But even when all eight Omega stations are on-line, there remain several large, remote areas of the world that lack adequate coverage.

The Navy's communications stations have a staggered maintenance schedule with each station going off the air several hours a week, but the loss of one such station rarely has an adverse effect on VLF/Omega navigation. (The remaining stations still offer a sufficient overlap of VLF coverage.)

Temporary loss of adequate signal coverage is not a serious problem, however; VLF systems will automatically dead reckon (based on last known wind) until signal acquisition is resumed.

Why, then, are the airlines replacing their obsolescent LORAN units with only Omega instead of VLF/Omega? Simple. It's a matter of money. Since VLF has yet to be approved as a means of *primary* oceanic navigation for the carriers, this chore must be accomplished by something that has been approved, namely Doppler or INS. The Omega systems are being purchased only to replace the aging LORAN as a means of providing an occasional fix to cross-check Doppler accuracy.

Once VLF is approved as a primary method of air carrier navigation (and it eventually will be), there is little doubt that the airlines will convert totally to VLF/Omega and/or INS.

In the meantime, if a general aviation operator has to choose between VLF/Omega and pure Omega, it would apear that VLF/Omega is the best choice even though — on the average — it is a bit more costly. The expanded, redundant coverage makes it a better buy.

After being exposed to VLF (or INS), it is difficult to imagine a better method of long-range navigation. (Where have we heard that before?)

But you don't have to exercise your imagination; the Department of Defense is already developing its futuristic GPS (Global Positioning System). Nicknamed "Navstar," this system will incorporate 24 earth-orbiting satellites to provide horizontal *and* vertical positioning information to within 10 meters (33 feet) anywhere on earth. In other words, Navstar will not only tell you whose house you're in, but also what floor you're on with more accuracy than most barometric altimeters.

Although the initial purpose of Navstar is to guide weapons delivery systems, there is little doubt that civilian applications will evolve. With the degree of accuracy expected from Navstar, the system could be used to establish a Category I ILS approach to every runway in the world without the need for expensive ground facilities.

Eventually, according to some experts, a soldier in the field (or a general aviation pilot) may be equipped with a GPS receiver similar in appearance to a pocket calculator. All he'd have to do is flip the switch and watch the LEDs advertise his precise location and elevation — to within 33 feet.

Now that, my friends, will be utlimate navigation system (until something better comes along).

THE FAA CHALLENGE

by Langhorne M. Bond

This article by the recently appointed administrator of the Federal Aviation Administration was written in response to several questions posed by AOPA.

Reprinted by Permission
THE AOPA PILOT (October 77)

The first months in any new job are essentially a learning experience. And the FAA presents a particularly formidable challenge. It's a very big, diverse, complex and far-flung organization with many and varied responsibilities.

Not that I have had to start from scratch, fortunately. Over the past dozen or so years, I have viewed the agency from a number of different positions. My first experience was as a student and then private pilot. Later, I was on the task force that drafted the basic legislation that created the Department of Transportation of which the FAA is a part.

Subsequently, I served as Special Assistant to the first Secretary of Transportation, Alan Boyd. Most recently, my contacts with the agency have been as transportation secretary for the State of Illinois.

So, quite naturally, all of these experiences have influenced my thinking about the FAA and the job of the FAA administrator. The result, essentially, is an outsider's view and I think this can be a very useful management tool. It provides a certain objectivity which I hope can be preserved in the months and years ahead.

Similarly, I believe I have acquired a good working knowledge of general aviation over the years, as both a pilot and a user of its many services. No one has to convince me of the value of this segment of aviation or the need to keep it healthy and viable.

I began flying for a couple of reasons while attending the University of Virginia law school in 1962. Of immediate concern at the time were my plans for a career in aviation law, and I thought pilot training would be a tremendous help in that area. But my pilot ambitions had even deeper roots than that. My father was a pilot and flying was something that was just accepted around our house.

I have long felt that flying has a great deal to offer the average person, even those who are not going to use it in business or become full-time professionals. You don't have to be an Olympic athlete or an All-American halfback to fly an airplane. Most people can learn to do it and do it safely. It's a great builder of confidence.

Having mentioned this, I don't want to give the impression that flying is no more difficult than riding a bicycle or driving an automobile — although as an amateur automobile racer, I believe our general perception of the skill and judgment required to be a safe driver is dangerously distorted. Still, flying is even more demanding; there is less margin for error, and fewer opportunities to learn from our mistakes.

Looking back on my own flying experience, however, I can recall several incidents that helped to make me a better pilot. For example, I groundlooped an airplane as a student pilot when the wind got under the upwind wing of the Super Cub I was flying and spun me around in a tight little circle. It proved more of an embarrassment than enything else. I ended up facing in the same direction I had come from. Fortunately, there was no damage to the aircraft and I never made the same mistake again.

Also impressed on my memory is my first FAA flight check, which I failed. I flew right over the check point without seeing it on a cross-country out of Richmond, Va., and the FAA inspector sent me back to Charlottesville for a little more work on dead reckoning. Next time I took the test, I passed.

Most recently, my flight activity has been limited to the right seat. I probably logged over 1,000 hours in the copilot's position during my four years with the Illinois DOT. We had something like 18 general aviation aircraft in our fleet and used them extensively on state business.

My Illinois experience greatly enhanced my appreciation of the general aviation aircraft as a business tool. We found that government, as well as industry, works better when it capitalizes on general aviation's unique capabilities. Our aircraft returned their investment many times over.

For example, we used our aircraft to show prospective industrial clients around the state. Using Jet Rangers or other aircraft, we could cover a dozen or so really good industrial sites and have the prospect on the way back to his out-of-state home, all in the same day. We felt this was a very enterprising way to display the merits of the State of Illinois.

Other important uses of our aircraft were aerial survey and aerial photography. We did a great deal of work for the state in these areas, including surveys of highway locations, water resources and agricultural products. We found the airplane could do these jobs faster, better and — most importantly — cheaper than any other means.

General aviation aircraft even affected the way our agency was organized and did business. In the case of our state airport inspectors, to cite one example, we discovered that it was far more economical to have them operate out of Springfield using light, single-engine airplanes than to have them based at field offices at various locations around the state.

My years in Illinois also helped shape my views on the federal government's role, vis-a-vis the states, in promoting aviation and transportation generally. I don't think the federal government can do it all or should do it all. State governments represent an important source of additional revenue and should be viewed in that light rather than as a bureaucratic competitor. The federal/state partnership has worked extremely well in the highway program and there is no reason why this success cannot be repeated in aviation.

In Illinois, we had a very active program of promoting aviation development and the result is what I consider the best airport system of any state. We had a $100 million bond issue for capital improvements and we used this money both for matching federal grants and for carrying out our own projects with local agencies, generally on a 50-50 basis. Frankly, we found we could do the job quicker and cheaper without federal involvement because of the simplified administrative process and the fact that we could design to less complex standards.

Also, in going it alone, we were able to put the money into projects which were not eligible for federal assistance such as access roads, parking lots, terminal improvements, and the like. I think there probably are some good reasons for the federal limitation on funding these kinds of projects but, insofar as an airport user is concerned, a muddy parking lot can be almost as much of a problem as a too-short runway.

I think the whole facility has to be developed, and many of the smaller airports simply do not have the revenue-generating potential to undertake all the needed improvements. This is where the state government can play an important role as we did in Illinois.

On the matter of federal/state partnerships, the FAA has a demonstration program underway with four states (Arizona, Pennsylvania, South Dakota and Michigan) that makes them responsible for administering federal grants for the development of their general aviation airports.

We are trying to establish the capability of the states to manage these funds in a more cost-effective manner than the federal government. If we find they can, and I have every confidence that we will, there is no reason why the program shouldn't be expanded to include most, if not all, the states.

Getting the most for our money is something that vitally concerns all of us today — in government as well as in the private sector. And the matter seems to be particularly relevant to general aviation. Capital costs, training costs, insurance costs and — perhaps the most critical of all — fuel costs show a steady upward trend.

And like it or not, the cost of using the system also is going to rise. It's unrealistic to expect that it won't in a time when the concept of a balanced budget is a primary objective of the Carter administration. But the FAA is committed to the concept that any new or additional user charges must be fair and equitable and based on an accurate assessment of the valuable contribution that general aviation makes in meeting the nation's total transportation needs.

Similarly, equipment and other requirements for operating in the system must be tailored to aircraft use and must recognize the difference between a rancher operating from his own sod strip in, say, Montana and a businessman who regularly flies into busy hub areas like Chicago, New York and Washington. The rancher in our example should be able to get along with a minimum of avionics, but the businessman must have the same basic capabilities as the airlines.

Cost effectiveness also must be stressed in evaluating the various services provided by the FAA to the aviation community. We must strive as never before to maintain the service level while holding the line on cost. In fact, I suspect there are many areas where service might be improved at the same time that costs are reduced.

This is not as paradoxical as it sounds. I believe the previously mentioned demonstration program for the administration of airport grants will help us stretch available funding in that area. Another example is the flight service station modernization program, which would permit the agency to make substantial cuts in overhead and labor costs without diminishing vital services to general aviation pilots.

The present FSS network badly needs an overhaul. Hardly anyone disputes that fact. It has grown up piecemeal to meet the demands of a developing air transportation system, which in many instances has passed it by. New, automated techniques — similar to those employed at the Atlanta FSS — must be implemented or we can anticipate a tripling of the FSS workforce and operating budget over the next decade as the system tries to keep pace with traffic growth.

But change never comes easily. Plans for consolidating FSS locations into a nucleus of hub locations have aroused considerable opposition, and the Congress has restricted us from proceeding with this aspect of our program through Fiscal Year 1979.

However, we are proceeding with the evaluation of the hub FSS concept at Leesburg, Va., and we plan to automate approximately 40 of our busiest FSSs over the next several years at the same time we are reducing the hours of operation at 80 to 85 less busy facilities. These actions will keep the modernization program on track and move us closer to our ultimate objective.

As a corollary to providing better service at less cost, we also must give greater attention to the concept of providing no service where none is required. This is a favorite theme of AOPA's I know, and one that strikes me as having considerable merit. One example of what the agency already has done in this area is our revised control tower criteria. It has resulted in a 75% reduction in the number of towers the agency was planning to build over the next decade. I expect we can find other areas where similar savings might be achieved.

One area where I think service can be improved with very favorable cost-benefit — as well as humanitarian — results is the collection and dissemination of weather data for general aviation pilots. As was recently pointed out in a letter to the secretaries of Commerce and Transportation by AOPA President John Baker and other general aviation leaders, approximately four out of every 10 fatal general aviation accidents involve weather as a primary or contributory cause. It's also a cause or factor in two out of 10 non-fatal accidents.

The FAA has a number of programs underway which are designed to facilitate the flow of weather information to pilots. Of particular significance since it relates to the point raised by Mr. Baker and his colleagues is the test of the Aviation Automated Weather Observation System (AV-AWOS) at Patrick Henry Airport in Newport News, Va. Although this particular installation was designed in connection with our FSS modernization program, the agency also is working on a lower-cost version for use at airports where weather data is not available at the present time.

Also worth mentioning is the Enroute Flight Advisory Service — Flight Watch — which uses a specially-trained FSS specialist to funnel pertinent weather information to en route pilots over a discrete frequency. We expect to complete implementation of this highly-successful program at 43 to 44 target FSS locations by year's end, providing virtually nationwide coverage.

But along with our efforts to improve the availability of weather information, I think we also must intensify our educational efforts, because I'm not at all convinced that poor judgment isn't as much a factor in many of these accidents as the lack of accurate and timely weather data.

AOPA, I know, has been active in this field for a great many years and FAA's own Accident Prevention Program continues to yield encouraging results. Still, when one reviews the daily accident log, one can't help but conclude that much more needs to be done. My own personal feeling is that a great many general aviation accidents are avoidable through the exercise of simple prudence.

Looking to the future, I see general aviation continuing to expand despite rising costs and increasing environmental and energy pressures. Enactment of regulatory reform — an idea whose time has arrived — will contribute to this growth with commuter airlines and air taxi operators being the primary beneficiaries. Overall, the FAA's latest "Aviation Forecasts" projects a 64% increase in both the size of the general aviation fleet and the number of flight hours by the end of the 1980s. Total aircraft will rise from 177,000 to 291,000 and flight hours from 36.7 million to 60.1 million.

These numbers are quite imposing; some may even find them frightening. But I see no need to impose artificial constraints on general aviation growth and I, for one, have no intention of initiating any such actions. But I do believe there are certain natural constraints that may influence our future projections more than any of us now realize. These include cost factors, the environment and the energy situation.

Another serious problem facing general aviation is the growing shortage of reliever airports in our large metropolitan areas. Stated in its simplest terms, the problem is that we are losing general aviation airports where we need them most — in our large urban centers.

I know from my own experience in the Chicago area that there are very difficult problems involved in keeping reliever airports open. Trying to expand or improve them is equally difficult, and finding locations for new fields is all but impossible. It's not a question of money but of public attitudes and, unless we can change those attitudes, general aviation will have an increasingly difficult time gaining access to the nation's major market places.

But I don't feel that any of the problems confronting general aviation now and in the years ahead are insoluble. As already indicated, I think the future is very promising. It's up to us to see that this promise is realized and I believe that, working together, we can.

NEW EN ROUTE FLIGHT SERVICE

by Joel Hamm
Edited for Space
Reprinted by Permission
PLANE & PILOT (March 77)

On a sunny summer morning, when there's not a cloud in the sky, you can call your local FAA flight service station and learn that the particular portion of airspace through which you intend to fly is about to be occupied by numerous lines and clusters of severe thunderstorms, accompanied by extreme rain showers, gale force winds, baseball-sized hail, and of course, low ceilings and visibilities.

That, at least, is the opinion of many pilots who are less than pleased with the FAA's pilot weather briefing program. Pilots have long criticized the accuracy and timeliness (or lack thereof) of the weather information available at flight service stations, being especially resentful of the tendency of many briefers to present the worst possible outlook of flying conditions.

Pilots would be less critical of the briefer if they better understood the purpose and limitations of a preflight weather briefing. Obviously, it is not possible, to state with absolute certainty what conditions a pilot will encounter in flight. Utilizing forecast and observed data, the person at the FSS counter can only present a picture of expected weather, emphasizing possible hazards to flight.

Weather forecasting is not, to say the least, an exact science. Forecasters can prophesy the immediate meteorological future in general terms.

Forecast materials, thus, tend to be highly vague as to locations, time frames, and precise ceiling and visibility values. They are based on large scale air mass movements and are intended to cover wide geographic areas. Since local terrain features can drastically affect weather patterns, actual flight conditions can be vastly and unexpectedly different from those forecast for the area.

Observed data, in the form of teletype reports and facsimile charts, are similarly limited in their relevence to en route conditions. By the time they are collected, charted, disseminated, and posted, surface observations can be several hours old.

If taking off into anything but perfectly clear skies seems akin to playing Russian roulette, rest assured that a new FSS program, offering continuous inflight weather advisories, promises to significantly improve the accuracy and timeliness of weather information available to the pilot.

The FAA, guardian of General Aviation pilots that it is, is aware of the deficiencies in the weather briefing system, and has been seeking ways to improve the quality of briefing services.

It was decided to designate certain flight service stations to act as central collecting and distributing points for pilot weather reports (PIREPS), and to equip these stations with live weather radar displays and all the tools necessary to provide pilots with accurate, up to date weather information.

Four such stations have been operating on the West Coast for several years and have been credited with a significant decrease in weather related accidents in that area. The system is titled Enroute Flight Advisory Service or EFAS for short.

This network is being expanded to include approximately 40 stations across the country. Each will have one or more remote transceivers to provide unbroken radio coverage over all but the Rocky Mountain area. A complete listing of EFAS facilities and remote sites is published in the Airman's Information Manual. Stations providing the service answer to the radio call sign "Flightwatch" on a universal frequency of 122.0 MHz.

Pilots are encouraged to monitor this frequency and participate in what is hoped will be a continuous exchange of weather information between pilots and EFAS specialists. (This frequency should not, however, be used for non-weather communications such as flight plans, airport advisories, etc.)

The idea behind EFAS is to give the specialist the sole responsibility of maintaining a constant watch on weather conditions, over a large area and providing current weather information to en route pilots.

Personnel selected as Flightwatch specialists receive a thorough course in weather analysis at the FAA's Academy. The positions are equipped with live weather radar displays, many of which are capable of selecting presentations from multiple radar sites. The EFAS specialist is provided with a "hotline" telephone link to the aviation forecaster at the National Weather Service Office.

With pilot cooperation, the EFAS program can do much to enhance the safety and efficiency of flying. With his knowledge of weather conditions over a wide area, the EFAS specialist can provide invaluable assistance to aircraft in distress — the pilot caught in weather who needs a diversionary route; the VFR caught on top looking for a hole, or the pilot caught in ice or turbulence in need of a better route or altitude.

The success of the program depends on pilot cooperation. In spite of weather radar and fancy gadgetry, the Flightwatch specialist relies primarily on PIREPs as his source of information. Pilots should make frequent reports of flying conditions, be they good, bad, or indifferent. PIREPs, to be useful, must be clear, concise, and include details on cloud cover, type of clouds, height of bases and/or tops, precipitation, ice, turbulence, restrictions to visibility, storm activity, or any pertinent factors.

EFAS is a weather dialogue. It can succeed only if pilots are willing to give as much information as they expect to receive.

"KETCH A CHOPPER:" A WAY OF LIFE ON THE GULF COAST

by Frank McGuire
Edited for Space

Reprinted by Permission
ROTOR & WING (April 77)

There's as much money in supporting oil and gas activity in the Gulf of Mexico as there is mud in Louisiana. At least 4,000 people are transported by helicopter to and from oil rigs every day. And the ever-increasing demand is opening the door to other experienced and well-financed operators.

There is no possibility whatever of describing the helicopter activity around the Gulf of Mexico without describing the environment. The environment greatly dictates the type of equipment used and abused, determines the kind of men who operate and ride in the aircraft, and decides the conditions under which the equipment will be used and the men will ride.

Driving from Metairie to Baton Rouge, Lafayette, St. Martinville, New Iberia, Morgan City, Houma and back to New Orleans, one passes places like Grosse Tete, Bayou Black and Bayou Blue, passing trash compactor trucks painted red, white and blue and carrying a big sign proclaiming: "America, We Love You."

This is part of "The Oil Patch" where helicopters often seem as numerous as pickup trucks, and are certainly as necessary. Nobody flies by helicopter, instead they "ketch a chopper."

Helicopter pilots in the offshore support business do not look like the spiffy necktie-draped specimens in the helicopter advertisements. They are casual but very professional, inspiring confidence in their passengers even though not looking like airline pilots. They are, after all, working pilots in the oil patch.

And the oilfield hands — the "roughnecks" as the dictionary calls them — are well named. They are hard-working men who deal in tools seemingly made of pig iron. To a roughneck, a small wrench weighs 50 pounds and is about four feet long.

Then he ketches a chopper and drops that wrench on the floor while the pilot winces and the man who invented honeycomb construction feels an inexplicable stab of pain in the ribs.

The roughneck does not carry his tools in an attache case; he does not wear calfskin gloves nor delicate shoes. His callouses are thicker than some gloves and his shoes are heavier than some businessmen. It also seems sometimes that the steel safety toes in a roughneck's shoes are of thicker and stronger metal than that used to build helicopters.

This is the world of offshore support. It is the world operating more helicopters than any other activity on earth, excluding the military. To those in it, it is a hard-working world. To an outsider, it is a fascinating world.

4,000 PEOPLE EVERY DAY

There is as much money in supporting offshore oil and gas activity in the Gulf of Mexico as there is mud in Louisiana. Even replacing door handles on helicopters would be a lucrative sideline.

The scope and importance of the oil patch to the Louisana economy may be judged by the Yellow Pages. In the Greater Lafayette telephone book, for example, there are 295 pages of listings, 18 of which are headed by the word "oil." From oilfield equipment to oilwell services.

At least 4,000 people every day — or about as many as attend the HAA convention — are transported to and from oil rigs by helicopter. Every day. The total number of helicopters in the offshore support business in the Gulf has been estimated at about 300, and the number grows constantly. Distances involved are out to 130 miles (210 km) for regular rigs, and farther for new drillings being undertaken by specialized ships.

It is extremely difficult getting a handle on the Gulf of Mexico oil activity without getting into ponderous statistics. Like only two out of 100 oil wells are successful enough to be commercially profitable. And one-sixth of all U.S. oil comes from offshore wells, which cost about $600,000 each to drill — eight times as much as an onshore well.

But the helicopters, ahh, there's the link between all those statistics.

Equipment ranges from the Hughes 300 and Bell 47 (what else?) to the Aerospatiale Puma and even an occasional Sikorsky S-61 or S-62. Operators range in size from superbig Petroleum Helicopters Inc. (PHI) and Air Logistics, which together operate 300-plus machines, to small firms with one or two aircraft and perhaps a specialized knowledge of one company and its requirements.

The pilots of the great number of offshore-support helicopters are in their late 20s to about 40, with a few on either side of that range. Uniforms are a casual combination of shirt and trousers, but no insignia except a cloth patch in some cases.

The Gulf coast activities of the supporting operators are scattered along the coast from the Mississippi delta to the Mexican border. Of the 3,000 wells in the Gulf, most are in water less than 150 feet (45m) deep for the simple reason that's where the oil is found. The wells are all west of the Mississippi delta and are operated from about 500 platforms. Angular drilling has allowed the oil companies to drill from half a dozen to as many as 50 wells from the same platform.

PHI AND AIR LOG DOMINATE

The story of independent commercial offshore support in the Gulf area is fundamentally the story of two companies: Petroleum Helicopters and Air Logistics, both headquartered in Lafayette, La. Other operators are either "captives" of a solitary oil company or are quite small when compared with either PHI or Air Log. One oil company, Chevron, has an in-house operation under Part 91 that is bigger than most commercial operators. Chevron, based at New Orleans Lakefront Airport, has 35 helicopters including two S-62s, some S-58s, and a variety of other types.

In New Orleans, only six helicopter operators list themselves in the phone book: Air-Land Helicopters Inc., Air Logistics, Air Marine Inc., Helicopter Charter Inc., Petroleum Helicopters, and Spremich Enterprises, d/b/a/ Airborne Enterprises. The latter is a certified helicopter repair station also providing an air-taxi service.

In Lafayette there are only two helicopter listings: Air Logistics and Petroleum Helicopters. In Morgan City the only names that appear are those in the New Orleans pages, for a total of four. Far and away the dominant firms are PHI and Air Log, in that order.

The significance of this exercise is that FAA officials in the coastal area feel quite strongly that another helicopter operator who knew what he was doing could come into the business there and make it pay. The present operators there, as one might guess, do not agree. Not shown in the above are companies that are not based on the coast and for some reason do not maintain a separate local listing. The Chevron operation is one of these, as are other company flight departments. Air Oil is a new firm that also flies marine pilots out to ships entering the delta. Exxon has the services of Rotor-Aids Inc. almost exclusively.

HOW MANY HELICOPTERS?

The worldwide nature of the helicopter business is evident in the Gulf area, where most commercial operator personnel talk of jobs in any place imaginable, from West Africa to New Guinea. It is also the worldwide nature of the business that makes the number of helicopters engaged in the Gulf offshore business difficult to pin down. There are always dozens of helicopters engaged elsewhere. Between them, PHI and Air Log have somewhere around 325 helicopters, which is considerbly more than the 300 working on the Gulf at any one time. The rest are scattered all over the world in various numbers and places.

The decision late in 1976 to build two offshore oil ports off Grand Isle, La., and Freeport, Tex., is not expected to affect the overall nature of helicopter support even when such ports are fully operational, although some additional traffic might be generated. Bob Suggs, president of PHI, said the present effects are mostly political, including much resistance from environmental groups who want the projects stopped. A similar issue of the New Jersey coast promises to cloud the oil company activities there for some time, but helicopter firms are nonetheless planning to move operations there when the time is ripe.

That's the environment, a brief lineup of the companies and an introduction to the offshore oil business. It certainly isn't complete and doesn't pretend to be, but it paints almost all of the picture — unless, of course, you want encyclopedic reference book on the subject, in which case the Gulf certainly would provide the material.

How is it to handle that kind of business? The logical people to ask are the biggest operators, not because they have the only experience but because they have the most of it under one roof. I found it quite difficult to find the smaller operators in their offices, most of them being out of the country (they never seem to be out of town) or else in meetings.

Oil was first discovered in the Gulf of Mexico 40 years ago and Bob Suggs waited about a dozen years before starting Petroleum Helicopters to help out.

In all those years he's run through a passel of helicopters. It used to be said of Howard Hughes when he owned TWA that the airline owned at least one of every kind of airplane that had ever been built. The statement has a familiar ring about it when considering PHI. The company's current fleet of some 242 helicopters includes 19 Bell 47s, six Bell 205s, 144 Bell JetRangers, 10 Bell LongRangers, 20 MBB BO-105s, two Aerospatiale Pumas, 26 Bell 212s, 13 Hughes 500s, as well as a Rockwell Commander 680 and a Sabre 40.

Asked what was his favorite helicopter for his kind of work, Suggs replied: "The Sikorsky straight S-58, as far as living up to promises." As far as the S-76 is concerned, he says he's ordered 20 of them, but not on Sikorsky's terms. "Why should we finance United Aircraft?" he asks, "Their kind of contract is for speculators and we're not speculators. If they got out and studied the market they wouldn't need that kind of contract."

Even if/when he gets the S-76, Suggs does not feel it is the perfect aircraft. "We've been waiting 28 years for the helicopter that's going to solve all our problems and we're still waiting."

Ken Jones, president of Air Logistics, also leans heavily toward the S-76 because "there are a lot of good ideas coming down the line for the future and the S-76 represents a lot of them." Nonetheless, Jones says Air Log is interested in the Bell 222 except that information on it has been hard to find. "We briefly looked at the 222 but Bell has not had a presentation on it, although they were hoping to make it early this year. I'm not sure the 222 has the people capacity to fit the market we serve, but we'll take a close look at it."

Air Logistics' current fleet totals about 75, including five Bell 212s just delivered. The makeup is 47 Bell 206s, 13 Bell 212s, eight Bell 205s, four Hughes 500s, and three Hughes 300s. Jones notes the company is "very impressed with the S-76 and the way it's going to fit the market."

Concurring with PHI's Suggs on the S-76 contract arrangements, Jones said he has no intention of committing his company to such a sizeable liability so far in advance of delivery date. "It places far too large a commitment on the firm's balance sheet," he said.

One of the things often overlooked by people who view the offshore support business is the support needed by the supporters. By that I mean that each helicopter operator who has a widespread flight pattern in the gulf needs offshore refueling arrangements, maintenance contingencies away from home base, communications networks, and all the rest of it.

Air Logistics has $30 million invested in aircraft and equipment, which is one reason Ken Jones thinks FAA is dead wrong about the ability of another operator to enter the Gulf market. "It's been a tough road for us to travel, and it would be just as bad for anyone else who was not extremely well financed." Air Log is a division of Offshore Logistics,

Largest of the Gulf Coast operators, PHI has ordered three 18-place SA-330 Pumas.

which is the marine division of the two-part firm. For the first few years of the helicopter division's existence, the marine side of the operation carried the whole company, but now things are much more evenly balanced.

In the Gulf area, oil companies are reluctant to pay for IFR capability just for the sake of those magic initials. "They will pay for safety, but they won't pay for unnecessary equipment as they define 'necessary'," said Air Log's Ken Jones. "They *will* pay for safety, but even if we don't get paid specifically for IFR capability, we provide it as a safety factor on our own account. It is for this reason also that we use Ontrac location equipment to give us more pinpoint accuracy. We do quite a bit of night emergency flying and operate in an inclement environment. .and we generally regard it as important whether the oil companies pay for it or not."

ROOM FOR ONE MORE?

The safety and reliability of helicopter support in the Gulf is vital because there are more helicopters working out of the Gulf coast heliports than the rest of the United States combined. It stands to reason that if FAA has anybody knowledgeable in helicopters, they're here. At the New Orleans General Aviation District Office is Ron Steele, whose office at Lakefront Airport is papered with charts of the offshore areas.

"I do feel there is enough business here for more helicopter operators, and I don't mean necessarily just one," Steele said. "With enough capital and experience, a man will find an open market here. The present operators are turning people away and the bigger ones prefer business to each other."

He added that new operators are getting into the picture all the time with helicopters, but not always in the oil-support business. Some examples:

A new firm headed by Ben Seal out of Baton Rouge uses a 206B and specializes in collecting water samples for environmentalists.

Air Oil, mentioned previously, drops marine pilots on ships entering and leaving New Orleans, as well as other work.

Peterson Marine, in addition to handling marine pilots, also carries cleanup crews to handle occasional oil spills and recently handled a Mississippi River cleanup.

There are others, including those mentioned in the beginning of this account, but the point is that the work is varied, the area of geography is vast, and the capital required is substantial. Not to mention contacts, support facilities, and the other appurtenances. But obviously some entrepreneurs are finding those required elements. How well they will make out remains to be seen.

WHAT PRICE IFR?

Obviously, the big sales point in any offshore flight is that set of magic initials: IFR. How important is it? How often is it used?

Ken Jones has already been quoted to some extent on the subject but it does not simply end there. And that is where the whole question of safety, reliability, supporting facilities for helicopters, and Ron Steele's assessment of untapped business all come together.

Can a new business afford the venture? If IFR is not a strict operational requirement, but is a vital safety consideration as well as a loud and clear sales point, could a new entry afford not to have the finest and latest helicopters complete with IFR capability and all the rest of it?

On the other hand, if operators won't pay for unnecessary IFR capability, can the new operator afford to carry them anyway as a promotional "loss leader?" Well, perhaps he could if he had faith in FAA off-the-cuff estimates that PHI and Air Log have grown between, say 15 and 18 percent in the past year. That's an unscientific guess, but even half that is healthy.

There are no appropriate offshore regulations for the operation of helicopters, say the FAA people I spoke with, and that is something the agency will have to attend to soon. Such a move is precisely what the oil companies do not want. They do not fly IFR any more than necessary — not only because of excess cost — but also because of the door it would open for increased FAA control. That, above all, is what the entire Gulf partnership of oil companies and helicopter operators want no part of. Rotor-Aids is VFR-only for partly that reason. Chevron and its several dozen helicopters are VFR only for mostly that reason. Air Logistics operates IFR anywhere under FAR Part 29, while PHI operates IFR only under Special FAR 29, and with FAA approval.

(On the east coast it is another story where IFR is concerned. Oil companies insist on IFR because of the more dodgy weather conditions, and also want twin-engine aircraft. Thus, PHI has used a BO-105, a 212, and a Puma on the east coast for some predrilling surveys, while Air Logistics has followed suit with Bell 212 equipment.)

THE CORROSION STORY

The story of operations offshore in the Gulf or anywhere else is largely the story of equipment. Procedures are by now fairly well established but there is a constant battle to make the available equipment do the maximum work for the least expense, and expense equates with parts, service, and maintenance after the initial capital investment is dispensed with. The story of operations offshore, to sum it up, is the story of corrosion.

After flights to offshore rigs, engines are kept running while a garden hose with fresh water provides a washdown to eliminate salt from the internal engine components. Ditto the outside of the aircraft. Nevertheless, corrosion cannot be totally eliminated. It is found during every teardown, overhaul, inspection, and component replacement.

One solution is paint — applied liberally and thoroughly at the outset of an aircraft's service. I saw PHI maintenance specialists taking apart a brand-new aircraft so it could be dipped, sprayed, and generally slathered with appropriate amounts and types of paints inside and outside. I saw the flashy factory paint job disappear at the edge of a fairing, only to reveal nothing under the fairing but a single coat of primer. The PHI people were busy removing all fairings, decals and other impediments so they could properly paint their brand-new helicopter.

That seems an extreme step to take with a new aircraft, but I was shown the results of not doing it in the salt-laden corrosive environment of the humid Gulf coast. Chrome peels from door handles on helicopters with less than 100 hours flight time. And so on through the problem known to everyone who deals with machinery near the sea. Is there a solution? Certainly, but it starts at the factory with proper metal treatment and continues with good maintenance at the operator level.

The operators along the Gulf are proud of the maintenance job they've done with what they consider a hostile environment and aircraft that are not always as well protected from such hazards as they should be. PHI does as much of its maintenance as possible in the field because "customers find it hard to accept a breakdown on our part, even though they'll admit breakdowns are inevitable." The firm has an FAA-approved inspection program set up under Part 135 and it works on a calendar basis, not a 100-hour basis. This gives the customer more availability, although it is not the ideal use of manpower. PHI says that is acceptable.

Air Logistics — being a smaller operation — has an overhaul capability for everything up to the engine. It does not overhaul the Allison 250-C20 or Lycoming, but a shop is being set up for the C20. Jones said Air Log would like to do more work itself because it has confidence in its capability. "What it boils down to is that we do as much maintenance in house as the manufacturers will let us," he said.

As noted earlier, the Gulf offshore operations are varied and complex, and could easily be the subject of a sizeable book. Included would be (if my conversations are any indicator) some blistering comments on various manufacturers and their quality control — or lack there of — a few choice words on FAA in Washington, some pilot and mechanic evaluations on aircraft and engines, and a host of things that would not be quite appropriate in a general description of how the Gulf coast offshore industry is doing and hopes to do.

PUSHED TOO FAR

Reprinted by Permission
AVIATION CONSUMER (May 15, 1977)

D. Michael Coughlin, mild-mannered Clark Kent type from Acton, Mass., flies a Baron 300 to 400 hours a year, gets into scores of airports all over the country and thoroughly immerses himself in the work and fun of flying. As executive director of a company (Readak Educational Services) devoted to reading improvement programs for school children, he finds himself crossing between the two separate worlds of business and aviation. (The distinction is his.) The transition, it turns out, has not always been a pleasant one.

As the owner of two brand-new airplanes in the six years since he learned to fly — a Piper Arrow and his present modified 600-hp Beech Baron B55 — and collector of instrument and multi ratings and purchaser of much avionics equipment and many services of all kinds, he gets quite a bit of exposure to the world of general aviation.

As a businessman, he doesn't at all like what he sees. Chronicling his bizarre experiences with FBOs, airframe manufacturers and vendors of all types into a dossier, he found the sum total evoked a kind of shock. He decided to do something about it, and sat down and prepared a long, articulate, detailed account of all the strange and irrational aspects of general aviation he had encountered in just six short years.

He then mailed this off to some of the score or more of aviation publications he subscribes to and devours with gusto every month, and several magazines published his epic letter. As a result, he received some 50 replies from other businessmen and pilots who shared his sense of dismay, was contacted by a couple of manufacturers (Beech and Piper) who wanted to substantiate or absolve themselves from his beef, and was invited to join the National Business Aircraft Owners Association.

Coughlin's list of complaints contained items like the following: poor and delayed warranty service and backup, calloused or casual treatment at most airports, months-long delays in getting new avionics to function properly, regularly scheduled inspections finished way past deadline, aircraft salesmen who never follow up, an avionics manufacturer (King Radio) who took 10 weeks to answer his hot lead query for an expensive, new flight-phone; airplanes coming out of annuals with more problems than they went in with, his squawks during inspections returned unanswered, a new radar (RCA) that works seven months out of each year, seven trips to three shops to get his autopilot to work, logbooks not noted as requested during inspections, line personnel putting in the wrong octane, or the wrong oil . . . the list goes on and on.

Is Coughlin some kind of malcontent? "I believe in fair play," he told *The Aviation Consumer* in an interview near Boston. "And one day when I just got fed up with being shoved around, I said, 'Hey, I'm going to tell somebody about it, and I'm going to tell as many people as I can because I believe there are a lot out there who feel exactly the same way I do' — and I'm even more convinced of that now."

His problems started soon after he trotted into an airport in New Orleans with a Cessna learn-to-fly coupon in 1971.

"From the time I started renting, and working on my instrument rating, even then there were problems like dirty airplanes, no gas in the tanks, airplanes not back in time, sloppy practices."

In light of rigid federal laws concerning aviation, Coughlin sees the casual FBO performance and attitude posing a startling paradox. "The problem is a very curious anomaly since the FAA insists on very stringent rules which you as a pilot are expected to know and be held accountable for every time you fly an airplane."

In general, Coughlin's experience has been that while the piloting fraternity was friendly and pleasant, the FBO structure basically was not. "If he (an FBO) can sell you something, fine, but if not, well, I don't know. Coming up all the way from New Orleans to Boston I can count maybe three to four people that I can actually say I've gone back to time after time, because they're good people. The FBO in New Orleans that sold me the Piper — I never even got a letter of thank you. For that matter, I never got the warranty card for the airplane. I had a new King DME installed which didn't work for three straight months because they put the antenna this far from the transponder. It took me nine months to get the warranty card for the King DME, and every time I took it in to get it fixed because it didn't work, I got a hassle." And so on . . .

As Coughlin sees it, all too many FBOs are deficient in normal, effective business practices. "These are people operating a business enterprise on aviation desire. And the only thing going for the aviation industry is that there are a lot of people who love it so much they're willing to put up with it. It seems to me that as long as people put up with slipshod service, we're going to have it with us."

He gives low marks to the aircraft and avionics builders as well. "Why should the manufacturers do anything about it? With their record profits, why should they care? I've got $80,000 in the panel of that airplane (his Baron), and I don't know why it's wrong for me to expect it to work.

"When something is wrong in our business, everything stops and we solve the problem. We don't simply say, don't worry about it; sales will cover it. Don't worry about getting this money back to the unsatisfied customer. In our business we bust our tails serving these schools.

Does Coughlin, as a consumer, wish a plague on all their houses? "Hey, I don't want to wish a demise on them. I'm heavily into it."

Our sentiments also. But there's got to be a better way.

POWER LINES

by Gordon D. Jones

Reprinted by Permission THE AOPA PILOT (May 77)

I was instructing at a small airport in northwest Oregon after my discharge from the Air Force. The pay was short, the hours long, and my family was about to expand to three. To supplement my income, I decided to go into the crop-spraying and mosquito-control business on the side.

With the aid of a bank loan, a modified war surplus Stearman and a three-county contract I launched a new career. Before two seasons were finished, I found it necessary to take on a partner to help with the field work and drive the supply truck. Business mushroomed, and we soon had contracts with four neighboring counties for mosquito-control spraying.

This involved a great deal of field work, mostly done by the county crews, but there was one phase of it I insisted on doing personally — checking out the spray area from the ground. I'd drive to the area in my car in addition to overflying it and lay out an accurate map showing poles, wires, trees, buildings, fences, and any other hazards. Then I'd plot the approach, the runs and the turns, using prominent landmarks. Everything planned to the finest detail — there was not much margin for error in this business. I was a very careful pilot, and determined to maintain that status as a living example — not as statistic.

A note in my log book reminds me of that day Fate set a trap for me and my careful planning.

I had been spraying since dawn along the shores of the Columbia River, but was forced to suspend operations with the tank still half full of insecticide because of a strong westerly wind through the gorge. I returned to the field, gassed the plane, and was leaving the office when the phone rang. One of the county field men from Kelso, Wash., was on the line with an urgent plea for immediate spraying of a creek area north of the neighboring town

PIPER PAWNEE D

of Longview; also some floodwater pools along the nearby Columbia River where mosquito larvae were ready to hatch. Could I be there first thing Monday morning?

"Sure," I replied, "Be there before sunup."

As I drove home my mind juggled the details of the job, fitting the pieces into an efficient schedule: takeoff from Troutdale at 4:30 a.m., 30 minutes to fly the 50 miles to the spray area, arrive at daylight and start the run with the half tank of insecticide already in the plane. This would give my partner time to drive the supply truck to the Kelso-Longview airport and be ready for refilling after my first run.

I had sprayed this same area twice before and knew a half tank would cover it. Then two full tanks on the floodwater strips along the river bank and the day's work would be finished — all before 9:00 a.m. At least, this was the plan.

The next morning began like many other June days with ideal spraying weather — clear, calm and cool. I was airborne within five minutes of the scheduled departure time in the predawn grayness, and headed the Stearman downriver, slightly north of the sleepy lights of Portland. It was frosty cold when I levelled off for cruise at 1,000 feet, and I huddled forward in the cockpit to take full advantage of the windshield's protection against the prop's icy blast. Holding the stick between my knees left my hands free so I could cross my arms and tuck my hands under the armpits of my flight jacket for warmth.

There is an almost indescribable feeling of spiritual freedom when you're flying an open cockpit biplane in the murky dark of predawn. It makes you forget the cold and discomfort and the biting prop blast that digs its icy fingers under the corners of your goggles and makes your eyes run wet.

But reality has a way of asserting itself, even at rare times such as this, and soon I'm descending through the half light of departing night, down nearly to treetop level to start the spray run on the creek.

No more reverie, now, every nerve alert, senses straining. Eyes on a swivel, back and forth, relying on peripheral vision to probe the grayness. A firm but light grip on the stick, feet planted solidly on the pedals, everything under control. At treetop altitude in an open cockpit 90 mph becomes a fantastic speed.

Then lucid thoughts crowd out the fantasy and you find yourself wondering why you ever were so foolish as to get into this kind of profession — and for a lousy eight bucks an hour. But you clear your mind and go on, slipping and skidding along just inches over the trees, following the twisting course of the creek. An occasional glance backward to check the spray pattern. It unfolds as a wispy sheet from beneath the wing, swirling back beyond the tail until it's picked up by the wingtip vortices, churned out wide, then spiralled up, over and down, to finally plunge beneath the foliage and disappear.

But don't let it hypnotize you. Take it in short, quick glances. Eyes front, back, then front again. Keep that head on a swivel. And keep those turns shallow — you're too close to the trees to dip a wing very far. Let her skid!

A few minutes of this and the sky begins to pale into sunrise. Central vision is effective now, and the contrast is sharper on the horizon. "H-m-m-m, should be running out of goop any second now." A small bridge flashes by under my wings, the boundary of the control area. But the spray pattern is still full and flowing, so I continue the run, glancing rearward ever more frequently to catch the instant the lines go dry. The spray keeps coming, and I fly farther and farther up the creek, into an area I've never checked from the ground.

And that's when it happened. In the middle of a shallow right turn, my eyes sweep forward just as the spray quits. My left hand flicks the shutoff valve, then hits the throttle as I start to roll level to pull up and away. Something flickers in the corner of my eye, up there on a hilltop to my right — a quick flash of sunlight glistening off the bright aluminum cables of a Bonneville transmission tower.

My glance darts along those lines, all nine of them, three rows of three, — inch-and-a-half cables. I trace them down till they disappear in the treetops directly in front of the plane — and then I'm into them! Suddenly it's as though I'm looking over my own shoulder from behind, and that huddled, paralyzed figure in the cockpit belongs to someone else.

Fear can bring strange fantasies to the imagination. I wait for the explosion, the lightning bolts, the ripping apart and the mass of tangled wreckage. But it doesn't happen. And I stare in frozen fascination as those huge cables zap, zap, zap past my wings.

Then I awake to the realization that the engine is still there in front of me with its throaty roar full and strong, and the wings are still intact, and the plane is still flying. Is it possible?

In disbelief I raise my head and look back. All nine cables are hanging quiet, stretching from tower to tower — motionless.

I pull up then into a long, climbing turn, up through the shadow of the dark hills, finally breaking out into the full blaze of sunrise — and what a glorious sunrise!

As I point the nose of the plane south, toward the Kelso-Longview airport, a prayer of thanks is on my trembling lips and my lungs grab huge gulps of cold air. I try to keep my quivering feet from slipping off the pedals. I feel like jelly, bathed in icy sweat. But somehow I make it to the airport, through the landing, and cut the engine. I peel off my helmet and goggles and step out on the wing. My legs are rubber, and I sprawl flat on the wet grass, shivering.

I lay there for a long time.

PILOT

by Michael Jones-Kelley

Reprinted by Permission SOUTHERN WINGS (January 77)

They tried to tell Tom Kerr that he would never be a pilot, but he wasn't listening. The reason he wasn't listening is that Mr. Kerr is a deaf-mute, which is also the reason they told him he would never be a pilot.

Like many other pilots, Mr. Kerr's flying career began as the result of an errant notion that he would like to drive airplanes around the sky. So he mosied out to a flight school in Spartanburg and presented his request, in writing of course.

Before his flight training really got underway, he had to go for a physical. The medical examiner decreed unilaterally that Mr. Kerr could never be a pilot because, well, he couldn't hear or speak. And that, it seemed, was that. At least until Tom Kerr ran into the Moser family, owners of the FBO at St. Augustine.

Not long after moving to St. Augustine to take a job as a cold type instructor at the Florida School for the Deaf and Blind, Mr. Kerr took his children out to Moser's Aero Sport, Inc., to give them a pleasure ride in a small airplane. In the course of his visit, Mr. Kerr communicated his old frustration at not having been able to learn to fly to Ernie Moser, patriarch of the Aero Sport operation.

Col. Moser — a man not known for allowing bureauracy to get in the way of good sense — saw a potentially good pilot and told him to start over, that the people at Aero Sport would help him get his ticket.

At first it looked like defeat once again when yet another medical examiner refused to give Mr. Kerr his physical. But on a trip back to Spartanburg to sell his house (which he had built by himself, by the way), Mr. Kerr was told of a medical examiner in North Carolina who might be sympathetic.

The doctor in North Carolina simply picked up the phone and called some FAA official to see exactly how the regulations read. It turned out to be simple: Under FAA regs, the doctor *had* to give Mr. Kerr a physical, which he passed with no problem except a waiver for his deafness.

The instruction began, with Jim Moser acting as instructor and working with Mr. Kerr to develop a set of hand signals and signs (since Jim does not know deaf-mute sign language) to cover situations such as keeping up one's airspeed on final. It developed into a good working relationship ("fantastic instructor," Mr. Kerr says; "excellent student," Mr. Moser says) and some 45 hours later Thomas R. Kerr was a licensed pilot.

There are limitations, of course. Mr. Kerr cannot fly into controlled fields, although he sometimes will hop over to nearby Gainesville after having gotten a friend to let the tower know by phone what time Mr. Kerr will arive.

Mainly, he flies for fun.

Sometimes he will load his two children in the plane and just buzz around for a while. His son, Rudy, 17, is more interested in tennis than in flying (interested enough, in fact, to qualify for the world games for the deaf to be held in Rumania this year). Thirteen-year-old daughter Meme, on the other hand, is fascinated by airplanes and bombards her father with questions when they crawl through a new and unfamiliar model.

"I learned to fly for adventure," Mr. Kerr says, "not to prove anything to anybody."

So far the adventure has gotten him into a Cub, a 150, a Citabria, a Varga and a Great Lakes, his favorite. Right now, he's working on aerobatics in the Great Lakes with Jim Moser.

Even though he did not set out to prove anything, Mr. Kerr has encouraged other deaf-mutes to think about becoming pilots. One fellow, who had begun lessons in his native Maryland but had also met with problems, has already showed up at Aero Sport to have another go at it.

It may turn out that Ernie and Jim Moser became the country's leading experts on silent flight instruction.

THE TSO'd PENCIL

by Jim Weir

Reprinted by Permission THE AOPA PILOT (August 77)

"(expletive removed)! Broke the fool pencil again."

This chance phrase was spoken into an accidentally open microphone by a poor pilot who was unaware that he was the impetus behind the recent issuance of TSO c7734 (Pencil, Wood, IFR). You see, during a later playback of the ATC center tapes, it was determined that the pilot did, in fact, violate the FAR regarding airworthy equipment by having a non-approved pencil in use while operating under an IFR environment.

The FAA naturally concluded that pencils were the weak link in the multi-billion-dollar ATC structure, and forthwith issued a press release that, ". . .non-approved pencils may yet cause a serious crash between private aircraft and a jumbo jet."

In the ensuing hullabaloo, Congressman Pucksuckle (an aviation expert due to the fact that his father, once in 1922 and again in 1924, lost his lunch when a barnstormer looped the loop with old man Pucksuckle in the front seat) demanded that the airways be cleansed of such damn fool things as non-approved pencils. Hence, the rapid acceptance of TSO c7734.

Which is where I came in, friend, because I now own a full warehouse of TCO'd, PMA'd approved, blessed and tested pencils. You say you want to fly IFR? Good, because I've got the pencil for you, the only legal pencil to use, and all at the low pilot's discount price of only $13.85 (each, in packages of 10). And you can't go anywhere else for them, because between the FAA and me, we've got the only game in town.

When the TSO first came out, I estimated that I would buy pencils from the Hawk Pencil and Storm Door Co. of Horse Cave, Ky. for a nickel apiece and sell them as TSO'd for about a dime. This estimate was before I talked to the FAA about the tests and inspections required to launch such a critical-to-flight item as a pencil into the nation's navigable airspace. Did the FAA requirements add anything to the pencil's price? Oh, brother, did it.

First, Hawk Pencil would have to get its incoming inspection procedure up to snuff. No more of this old cut-down-a-pine-tree-and-make-a-pencil procedure. No siree. Each tree is now inspected by a licensed forestry engineer for dry rot, grain straightness, moisture content and compressability (woodpecker holes are cause for immediate rejection). The company saw is now calibrated to National Bureau of Standards specifications once a week by an FAA-licensed instrument shop, and perfectly good saw blades are replaced every 100 board feet or 10 days, whichever comes first.

The old saw blades are rebuilt and inspected by an FAA-approved repair shop at a cost only slightly over triple the cost of a good, new commercial blade. Do all these procedures cost money? Well, hardly at all when you consider the reliability of the "approved" product. Anyway, Hawk only adds $2.84 per pencil for these inspections.

Second, although Hawk has produced pencils for a hundred years, the FAA engineer at the local office handling our project (charming young fellow, Bob Gobbler) said that the "standard procedure" is to require a full-time inspector at the pencil factory—at my expense of course. Since these standard procedures are not to be trifled with, I hired a man willing to live in Horse Cave, Ky., (billed as Chicken Manure Capital of the World) for the pittance of $25,000 per year. Plus moving expenses, of course.

Good old Bob said that he personally didn't see any sense in this, but that since he had never really worked in industry, or had any real hands-on experience in flying, he couldn't bend standard procedures or his supervisor might get perturbed. Add another $1.63 per pencil for our resident inspector.

Now, about this pencil lead problem. Could Hawk prove traceability of the carbon used in making the lead back to the mine that the coal used to make the carbon was dug from? And exactly what procedures and approved measuring techniques was Hawk following to ensure that the diameter of the lead was within .00001 inch of the TSO specification, as required? Could these tests and reports be done? Of course, for money anything can be done. Add another $2.25 per pencil for pencil lead reports.

Well, finally we've got Hawk producing pencils to TSO standards. Well, almost, and all it has cost us so far is $6.72 for inspections plus a nickel for the original pencil.

Not quite so fast. We haven't done the engineering test procedure yet. You see, the FAA was concerned that the pencil might not work under any and all environmental conditions, or that the pencil may interfere with other systems on board the aircraft.

For instance, the color was all wrong. Yellow, they insisted, was reserved for a warning light or flag. A yellow pencil may be misinterpreted by the pilot as an unsafe condition and cause the pilot to take unnecessary precautions. Couldn't we paint the pencil pastel green or something? And how about making the paint fireproof? Add another 85¢.

How about temperature and altitude? Sigh. Put 100 pencils and me into an environmental chamber and freeze it to -60°, heat it to +180°, run it up to 25,000 feet and see if all the pencils would still write. The pencils all wrote. Me, I caught pneumonia. The price for all this testing and my hospital bills? Hardly a drop in the bucket. Add $3.16 per pencil.

Aren't we done yet? Not quite. You see, moving a conductor (pencil lead) across an insulator (paper) makes what engineering types call a self-induced charge. The FAA calls it static, and demanded a full-blown EMI (ElectroMagnetic Interference) test to prove that the pencil wouldn't interfere with the radios. Since the required FAA approved radio shop owner was a friend of mine, he barely charged me anything to do the tests. Well, hardly anything. Add $3.02 per pencil.

These pencils are now costing me $13.80 and I'm selling them for $13.85. After all this, I'm still making exactly what my original estimate called for — a profit of a nickel per pencil. By the way, out of this nickel profit, I've got to take the cost of printing a full set of installation and operating instructions for each pencil, plus a clearly printed warning that only an A&P mechanic is permitted to sharpen the pencil, and then only with an approved sharpener (see TSO c7735 for the sharpener).

By now you may be asking yourself why the FAA didn't require tests on the eraser. Oh, you mean you want an eraser? Well, that's nonstandard equipment and not covered under the TSO, but maybe for a few thousand bucks more we can get an STC. . .

1978 CHEYENNE III

AIR CARRIER

AVIATION YEARBOOK

INTRODUCTION

The airline industry had been holding its collective breath for years, hoping, praying that one of their worst fears would not become a reality. But it happened. . .this year. . .at Tenerife in the Canary Islands.

Two, fully-loaded Boeing 747's — the world's largest airliners — collided and most of those aboard perished in the burning cataclysm.

The probable cause? Something that has confounded man since he crawled from the cave: poor communications.

This issue of the *Aviation Yearbook* contains the Tenerife story. It is not enjoyable reading, but it — like other accident reports — contains valuable lessons. And this is why we have chosen to include such sobering material. We have no fascination for disaster, but as long as man chooses to remain mobile, he tacitly agrees to accept the implicit danger.

Another rare accident occurred this year when the crew of a Southern DC-9 encountered total power failure while skirting under severe thunderstorm activity. It is chilling to read about the pilot's heroic efforts to safely dead-stick the jetliner onto a highway.

Not all of this year's drama was accidental. Some was caused by the lunatic fringe of humanity.

Although there has been a marked decline in skyjackings during the past several years, 1977 is when this trend reversed. The first such major event of the year occurred when members of the so-called Japanese Red Army absconded with a Japan Air Lines' DC-8 carrying 141 passengers plus crew.

Unfortunately, the Japanese government blithely paid the hijackers' demand: six million dollars. Now that's a lot of *yen*. I say "unfortunately" because other idiots were encouraged by this payoff. And so, with visions of new-found wealth, these characters emerged from under a rock and hijacked a Lufthansa jetliner. The West Germans, however, were not so obliging. Instead of paying off with cash, the members of an elite, West German, anti-terrorist squad executed a brilliant, Entebbe-style rescue.

Politically, 1977 was the beginning of a new era. This was Jimmy Carter's first year in office. Like any new President, Carter began to activate those plans he had proffered during his campaign, including his program of deregulation, (one of the most potentially ominous, as far as the airline industry is concerned.)

With the consumer's welfare in mind, Carter wants to instigate a free-for-all between the scheduled airlines. To those with large cash reserves (such as United Air Lines, et. al.), deregulation is welcomed. After all, when you have enough shekels to weather the storm, adversity can be more easily withstood. Other carriers, however, might not survive if the storm of deregulation becomes too severe.

This issue of the JEPPESEN SANDERSON AVIATION YEARBOOK covers both sides of this hotly-discussed issue. Unfortunately, most of the rhetoric must be regarded as speculative because the ultimate shape of Jimmy Carter's reform package has yet to be determined.

Remember when Paul Revere warned the colonials? Everyone within earshot reacted to the message: "The British are coming; the British are coming!"

That was a couple of hundred years ago. During 1977, the British came again. This time no one heeded the warning and the colonials lost the battle without a shot being fired.

This confrontation was called "Bermuda II" and was simply a meeting to determine the rights of British versus U.S. air carriers. It's difficult to say why the Carter representatives allowed the British to walk away with the majority of the marbles, but they did. U.S. flag carrriers, notably TWA and Pan Am, lost ground while the rights of the British were substantially increased. The story is, of course, told in more detail on the pages that follow.

It is no wonder that the American flag carriers are somewhat apprehensive. Negotiations are about to begin with the Japanese. Will the Administration continue its give-away program and knuckle under to Japanese pressures? Hopefully not. It's difficult enough for the airlines to keep their heads above the international waters these days.

Enter Freddie Laker, the owner of Laker Airways. This was the year that Freddie's Skytrain began hauling passengers back and forth twixt New York and Londontown at cut-rate fares (on a standby basis).

Did this really bother TWA and Pan Am who carry most of the traffic from the U.S. to England? Nope, not at all. They were fully prepared to operate Skytrains of their own to compete head-on with Sir Laker. But did they? No way. The U.S. government wouldn't let them. "Too unfair to allow such corporate Goliaths to compete so unmercilessly against the small English firm."

Humm? On the one hand, the Administration wants to deregulate the airline industry to foster competition. But on the other hand, the Feds won't allow it. Hypocritical? Draw your own conclusions.

If the U.S. airlines survive the financial shocks of deregulation, they must then brace themselves for the need to replace their large fleet of DC-8's, 707's, and 727-100's. No, it's not that these aircraft have become antiquated. It's just that they make more noise than is permitted by Federal Aviation Regulation Part 36 which becomes fully effective in 1985. The Feds won't allow these magnificant aircraft to die of attrition. Instead, these aircraft have been mandated as obsolescent, prematurely destined to the graveyard.

But what will be done about the Concorde in 1985? It, too, violates these future noise standards. Will the U.S. government excempt the Concorde in 1985 while forcing its own airlines to comply? Will this, too become diplomatically expedient?

The years ahead are likely to be turbulent. And exciting, too, because whenever confronted by challenge and adversity, Yankee ingenuity somehow always rises to the occasion. (If and when it is *allowed* to rise, that is.)

Please, Mr. Carter.

Barry Schiff
December 1977

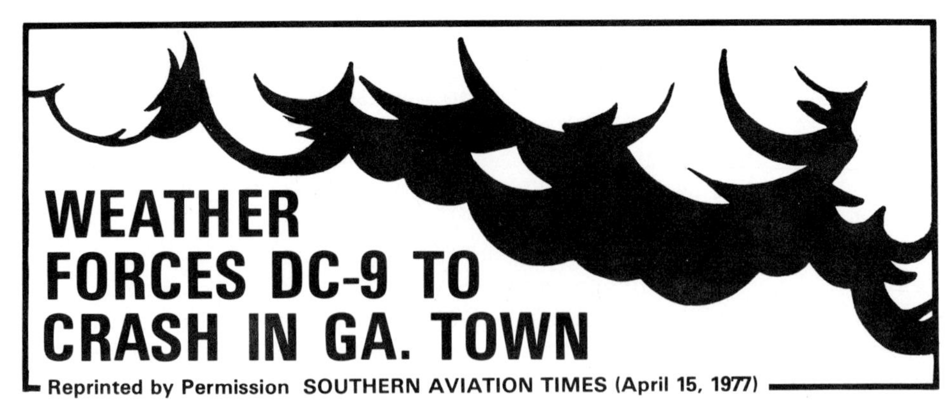

WEATHER FORCES DC-9 TO CRASH IN GA. TOWN

Reprinted by Permission SOUTHERN AVIATION TIMES (April 15, 1977)

Preliminary reports by National Transportation Safety Board (NTSB) and FAA investigators indicate hail and the ingestion of huge quantities of rain water into the engines of a Southern Airways DC-9 led to its crash in New Hope, Ga., on April 5.

The pilot, Capt. William McKenzie, 54, of La Place, La., First Officer Lyman W. Keele Jr., 34, of College Park, Ga., and 58 passengers were killed when the aircraft crashed and burned while attempting a forced landing on a street in New Hope. Seven persons in a car and another person standing in a yard were also killed.

The two stewardesses aboard the DC-9 and 13 passengers survived the crash but many sustained serious injuries and burns.

According to published reports, investigators are baffled as to why both of the DC-9's engines failed. Government and industry officials said there have been cases in which one engine of a twin-jet has flamed out because of weather conditions, but there has been no precedent for total power loss in a twin-jet induced by weather.

Officials of Pratt & Whitney, which makes the JT-8D engine, used extensively on DC-9's, 727s and 737s, said that no airliner using the engines had ever been lost because of ice or hail.

The aircraft was enroute from Muscle Shoals, Ala., to Atlanta, Ga., with a stop in Huntsville, Ala.

A survivor of the crash, pilot Don Foster of Decatur, Ala., recounted the flight to reporters:

"When we departed Huntsville, the flying was fairly smooth in spite of the severe weather for the first few minutes and then we encountered a family of quite severe thunderstorms," he said.

"We started to get hail. I was about three seats back from the rear of the airplane on the left side and I could look right into the left engine. And I could see what appeared to be serious hail damage to parts of the engine, inside the cowling.

"The right engine quit. It was lost completely. The left engine was operating at reduced power and it was missing and coughing, and it was obvious that it was going out," he said.

"When the left engine went, apparently the pilots made a few shallow turns to look for a place to land. And then as we got real low, they made a real steep turn and apparently tried to turn on to a small paved road," Foster said.

"And I guess they didn't quite make it, but they did a tremendous job of giving it a try anyway," he said.

A witness on the ground, J.M. Pickett, said, "He (the pilot) had it beamed dead center down the road. But it didn't work out. He clipped some trees and telephone poles on the left, tipped up with what looked like a gust of wind, sliced through Newman's grocery and began to spin and break up."

Investigators reportedly are also concerned with finding the reasons why the plane flew through such a violent storm. According to FAA's Jack Barker, sigmets were in effect indicating thunderheads reaching as high as 45,000 feet and a few possible tornados in the area when the aircraft took off from Huntsville for the relatively short hop to Atlanta.

The first indication the aircraft was in trouble came when the pilot reported a "busted windshield," apparently caused by hailstones, when the aircraft was at about 15,000 feet. The engines apparently failed shortly later when the DC-9 was at an altitude of about 10,000 feet.

NATIONAL TRANSPORTATION SAFETY BOARD TRANSCRIPTION OF COCKPIT VOICE RECORDER DATA SOUTHERN AIRWAYS FLIGHT 242 NEW HOPE, GEORGIA APRIL 4, 1977

Note: The reader is cautioned that the transcript of a CVR tape is not a precise science, but the best product of an NTSB group investigative effort. Conclusions or interpretations should not be made using the transcript as the sole source of information.

This transcript covers a period of approximately 25 minutes, beginning immediately after Southern Flight 242 took off from Huntsville, Alabama, and ending with impact. Both cockpit and ground tapes are used. Times are given in Greenwich Mean Time (GMT).

— JSAY

LEGEND

CAM	Cockpit area microphone voice or sound source
RDO	Radio transmission from accident aircraft
-1	Voice identified as Captain
-2	Voice identified as First Officer

Captain William W. McKenzie

First Officer Lyman W. Keele, Jr.

ST-A	Voice identified as Stewardess A (forward cabin)
ST-B	Voice identified as Stewardess B (rear cabin)
PA	Public address system in the aircraft
IC	Aircraft's intercom
HG	Huntsville Ground Control
HT	Huntsville Tower Control
CR	Company Radio
HD	Huntsville Departure Control
MC	Memphis Center
AC	Atlanta Center
AA	Atlanta Approach Control
UNK	UNK
*	Unintelligible word
#	Nonpertinent word
%	Break in continuity
( )	Questionable text
(( ))	Editorial Insertion
———	Pause

TIME & SOURCE	CONTENTS
2054:29 HD	Southern two forty two, Huntsville radar contact, uh, turn left heading one two zero, vector around restricted area, climb and maintain one seven thousand
2054:38 RDO-1	Okay, one seven thousand heading turn left to one two zero
2054:51 RDO	((Click of mike))
CAM-2	Flaps up Bill
CAM	((Clicking noise, trim noise))
2055:05 CAM-2	Slats up, climb check
CAM	((Trim noise))
2055:14 HD	Southern two forty two is clear restricted area, continue left turn resume own navigation direct to Rome
2055:21 RDO-1	Okay, direct Rome, two forty two
2055:31 CAM-1	I don't know what direction Rome is

2055:34
CAM-2 About one hundred and ten *

CAM ((Trim noise))

CAM-1 ((Sound of sneeze)) Excuse me ---

CAM-2 Bless you

2055:58
CAM-1 Well, the radar is full of it, take your pick

2056:00
HD Southern two forty two, I'm painting a line of weather which appears to be moderate to uh, possibly heavy precipitation starting about uh, five miles ahead and it's **

2056:14
RDO-1 Okay, uh, we're in the rain right now, uh - it doesn't look much heavier than what we're in, does it?

2056:22
HD Uh it's painting — I got weather cutting devices on which is cutting out the, uh, precip that you're in now, this, uh, showing up on radar, however it doesn't — it's not a solid mass it, uh, appears to be a little bit heavier than what you're in right now

2056:34
RDO-1 Okay, thank you

2056:35
RDO ((Sound of click))

2056:37
CAM-2 I can't read that, it just looks like rain Bill, what do you think? There's a hole

2056:40
CAM-1 There's a hole right here ((simultaneous with "There's a hole" above

CAM-1 That's all I see

CAM ((Trim noise))

2056:43
CAM-1 Then coming over we had pretty good radar

2056:48
CAM-1 I believe right straight ahead, uh, there the next few miles is about the best way we can go

2057:04
CAM-1 Rome's fifteen twenty

CAM-2 Yeah

2057:06
HD Southern two forty two squawk five six two three

2057:06
CAM-2 You can go ahead and put yours on Atlanta now if you like, cause I've already got mine on *

2057:15
RDO-1 Two forty two, roger

2057:34
CAM-1 If it gets rough how about hand flying

2057:36
HD Southern two forty two you're in what appears to be about the heaviest part of it now what are your flight conditions?

2057:42
RDO-1 Uh, we're getting a little light turbulence now and uh, I'd say moderate rain

2057:47
HD Okay, and uh, what I'm painting, it won't get any worse than that and uh, contact Memphis Center on one two zero point eight

2057:55
RDO-1 Twenty point eight good day now and thank you much

2057:58
RDO ((Sound of garbled acknowledgement))

2058:10
RDO-1 Memphis Center, Southern uh, two forty two is with you climbing to one seven thousand

2058:16
MC Southern two forty two Memphis Center, roger

2058:22
CAM-1 As long as it doesn't get any heavier, we'll be all right

CAM-2 Yeah, this is good

2058:26
MC Attention all aircraft, SIGMET, hazardous weather vicinity Tennessee, southeastern Louisiana, Mississippi, northern and western Alabama and adjacent coastal waters, monitor VOR broadcast within a hundred fifty miles radius of the SIGMET area

CAM-1 Oh #

2058:41
CAM-1 Southeast Louisiana

2058:44
CAM-1 Out of ten

2058:45
MC Southern two forty two, contact Atlanta Center one three four point zero five

2058:50
RDO-1 Thirty four zero five two forty two good day

2058:54
MC Good day

2059:00 CAM-1	Here we go * hold 'em cowboy
2059:06 RDO-1	Atlanta Center, Southern two forty two we're out of eleven for seventeen
2059:11 AC	Southern two forty two, Atlanta Center roger, expect Rome runway two six profile descent
2059:16 RDO-1	Expect Rome two six
2059:19 AC	TWA four eighty five expect Rome runway two six profile descent
2059:24 TWA584	Was that five eighty four?
2059:26 AC	(It was)
2059:30 AC	Five eighty four, let me know where you're proceeding direct Rome
2059:33 TWA584	Okay, we're heading one sixty five now, it'll be a little while later before we can go Rome
2059:37 AC	((Sound of mike key acknowledgement))
2059:46 SCAT 16	Southern Jax, this is Scat one six
2059:59 2100:00 RDO	((Two unidentifiable noises on radio channel))
2100:06 (SCAT 16)	Be there in ten minutes, need a fuel truck
2100:21 CAM-2	I can handle this all the way over *
2100:30 CAM	((Sound of rain))
2100:51 RDO-?	Three forty two Birmingham
CAM-2	One thirty three in your window uh partner
CAM-1	Thirty three
2100:54 342	Yeah, three forty two go ahead
2100:55 RDO-?	Can you just, uh, let the passengers stay on for right now?
2101:00 342	Yeah, it looks good for that, that's a pretty good little shower moving across the field right now

2101:04
TWA584 Uh, Center, TWA's five eighty four this — this is really not too good a corridor we're coming through here, it's too narrow between your limit and this line, uh, we're getting moderate uh, heavy moderate turbulence and quite a bit of precip in here

2101:26
AC Five eighty four, roger, it looks like uh, right now another fifteen miles to the south you should through the uh, southeastern edge of what I'm showing and, uh, maybe, a little better

2101:38
TWA584 Okay, it's good to have hope anyway

2101:40
AC Looks like you might have went through a little one right over there and uh, you ought to be out of it now, though

2101:48
TWA584 Yeah, we were painting a little one, but, uh, you know, you wouldn't let us go any further so we're sort of in a box

2101:55
AC You have another airplane on over there to your left hand side too, really couldn't go any other way

2102:01
TWA584 Yeah I know, it's just too narrow through here

2102:03
AC He'd be a lot harder than the cloud though

2102:31
DAL657 Atlanta, Delta six fifty one, uh, two eight for two seven zero

2102:35
AC Six fifty one, Atlanta Center, roger

2102:57
CAM-1 I think we'd better slow it up right here in this uh, #

2103:02
CAM-2 Got ya covered

2103:03
AC Southern two forty two, contact Atlanta Center one two one point three five

2103:07
RDO-1 Twenty one thirty five good day

2103:09
CAM ((Sound of click))

Time	Source	Content
2103:14	CAM	((Sound of light rain))
2103:15		
2103:17	CAM	((Two short garbled transmissions))
2103:20	RDO-1	Atlanta, Southern two forty two with you level seventeen
2103:24	AC	Southern two forty two Atlanta, roger altimeter two niner five six
2103:29	RDO-1	Roger, two nine five six
2103:30	AC	Eastern six eighty three, Atlanta altimeter two niner five six if I didn't give it to you
2103:35	EAL683	Okay, six eighty three
2103:48	CAM-1	Looks heavy, nothing's going through that
2103:54	CAM-1	See that
2103:56	CAM-2	That's a hole isn't it?
2103:57	CAM-1	It's not showing a hole, see it?
2104:01	EAL683	Uh, six eighty three's in the clear over here, expect it looks sort of dark there
2104:05	CAM	((Sound of rain))
2104:06	CAM-2	#
2104:08	CAM-2	Do you want to go around that right now?
2104:09	AC	* * ((ATC to Eastern 683 garbled transmission — frequency change))
2104:18	EAL683	Eastern six eighty three, good day
2104:19	CAM-1	Hand fly at about two eighty five knots
	CAM-2	Two eight five
2104:20	AC	Thank you much
	RDO	((Sound of static on radio channel))
2104:30	CAM	((Sound of hail and rain))
2104:42	TWA584	Atlanta, TWA five eighty four, one nine zero

2104:46
AC Atlanta, roger

2104:50
RDO-1 Southern two forty two, we're slowing it up here a little bit

2104:53
AC Two forty two, roger

2105:03
AC TWA five eighty four, would you like to go on and descend?

2105:06
TWA584 Yes sir we would — five eighty four

2105:45
TWA584 TWA five eighty four will take that lower altitude whenever you're ready

2105:49
AC TWA five eighty four, roger, descend and maintain one four thousand cross the forty mile fix north of Atlanta VOR at one four thousand, the al-imeter Atlanta two niner five six, twenty nine fifty six

2105:53
CAM-2 Which way do we go cross here or go out --- I don't know how we get through there, Bill

CAM-1 I know you're just gonna have to go out**

CAM-2 Yeah, right across that band

2105:59
TWA584 Fouteen thousand two niner five six cross the fix northeast of, uh, northwest of Atlanta, and uh, one four thousand and say again the fix

2106:01
CAM-1 All clear left ap-proximately right now, I think we can cut across there now

2106:06
AC Dallas intersection on the Rome arrival

2106:09
TWA584 Okay, Dallas at fourteen, five eighty four

2106:12
CAM-2 All right, here we go

2106:18
AC TWA five eighty four let me turn - know when you turn toward Rome

2106:20
TWA584 Five eighty four roger, looks like about --- that's about it for now, we're headed uh, to intercept, uh, the Atlanta three thirteen, that's about the best we can do for awhile

2106:25 CAM-2	We're picking up some ice, Bill
2106:29 CAM-1	We are above ten degrees
CAM-2	Right at ten
CAM-1	Yeah
CAM	((Sound of two clicks))
2106:30 AC	I show the weather up northwest of that position north of Rome, just on the edge of it, I tell you what, maintain one five thousand
2106:38 TWA584	Maintain one five thousand, we paint pretty good weather one or two o'clock
2106:41 CAM-2	He's got to be right through that hole about now
2106:42 AC	Southern two forty two descend and maintain one four thousand at this time
2106:46 CAM-1	Who's that?
CAM-2	TWA
2106:48 AC	Southern two forty two, descend and maintain one four thousand
2106:53 RDO-1	Two forty two down to fourteen
2106:55 AC	Affirmative
CAM	((Heavy hail or rain sound starts and continues until power interruption))
2107:00 AC	Southern two forty two Atlanta altimeter two niner five six and cross forty miles northwest of Atlanta two five zero knots
CAM	((Sound similar to electrical disturbance))
2107:21 AC	TWA five eighty four what's your speed?
2107:22 PA/ST-B	Keep your seatbelts on and securely fastened, there's nothing to be alarmed about, relax we should be out of it shortly
2107:24 TWA584	We're doing about two seventy five right now
2107:26 AC	Roger, you can reduce to two five zero, if unable, advise
2107:29 TWA584	We can, that's okay, back to two fifty

2107:31 AC	Southern two forty two, what's your speed now?
CAM	((Sound similar to electrical disturbance))
2107:39 AC	Southern two forty two Atlanta, what's your speed?
2107:49 AC	TWA five eighty four uh, descend and maintain one four thousand
2107:53 TWA584	Okay, one four thousand, five eighty four
2107:55 AC	Yes, expedite to one four please
2107:57 CAM	((Power interruption for 36 seconds))
2108:33 CAM	((Power restored))
CAM	((Sound of rain continues for 40 seconds))
2108:34 AC	Southern two forty two Atlanta
2108:37 CAM-2	Got it, got it back Bill, got it back, got it back
2108:38 PA/ST-B	* * check to see that all carry-on baggage is stowed completely underneath the seat in front of you, all carry-on baggage * * put all carry-on baggage underneath the seat in front of you, in the unlikely event that there is a need for an emergency landing we do ask that you please grab your ankles, I will scream from the rear of the aircraft, there is nothing to be alarmed but we have lost temporary APU power at times, so in the event there is any unlikely need for an emergency you do hear us holler, please grab your ankles, thank you for your cooperation and just relax, these are precautionary measures only
2108:42 RDO-1	Uh, two forty two, stand by
2108:46 AC	Say again
2108:48 RDO-1	Stand by
2108:49 AC	Roger, maintain one five thousand if you understand me, maintain one five thousand, Southern two forty two

2108:55 RDO-1	We're trying to get it up there
2108:57 AC	Roger
2108:59 TWA584	TWA five eighty four's in the clear for awhile
2109:05 AC	Uh, TWA five eighty four report out of one five thousand
2109:09 TWA584	We're out of fifteen in the clear
2109:11 RDO	((Mike keyed))
2109:15 RDO-1	Okay, uh, two forty two uh, we just got our windshield busted and uh, we'll try to get it back up to fifteen, we're fourteen
2109:24 CAM-2	Fifteen thousand
2109:25 AC	Southern two forty two you say you're at fourteen now?
2109:27 RDO-1	Yea - uh - couldn't help it
2109:30 AC	That's okay, uh, are you squawking five six two three?
2109:36 CAM-2	Left engine won't spool
2109:37 RDO-1	Our left engine just cut out
2109:42 AC	Southern two forty two roger, and uh lost your transponder squawk five six two three
2109:43 CAM-2	I am squawking five six two three, tell him I'm level fourteen
2109:49 RDO-1	Five six two three, we're squawking
2109:53 AC	Say you lost an engine and uh busted a windshield?
2109:56 RDO-1	Yes sir
2109:59 CAM-1	Auto pilot's off
CAM-2	I've got it, I'll hand fly it
2110:00 AC	Southern two forty two, you can descend and maintain one three thousand now, that'll get you down a little lower
2110:02 IC	Sandy ---

2110:04 CAM-2	My # the other engine's going too, #
2110:0- RDO-1	Got the other engine going too
2110:08 AC	Southern two forty two, say again
2110:10 RDO-1	Stand by - we lost both engines
2110:14 CAM-2	All right Bill get us a vector to a clear area
2110:16 RDO-1	Get us a vector to a clear area Atlanta
2110:20 AC	Uh, continue present southeastern bound heading, TWA's off to your left about fourteen miles at fourteen thousand and says he's in the clear.
2110:25 RDO-1	Okay
2110:27 RDO-1	Want us to turn left?
2110:30 AC	Southern two forty two, contact approach control one two six point nine and they'll try to get you straight into Dobbins
2110:35 RDO-1	One two -
2110:36 CAM-2	Give me - I'm familiar with Dobbins, tell them to give me a vector to Dobbins if they're clear
2110:38 RDO-1	Give me, uh, vector to Dobbins if they're clear
2110:41 AC	Southern two forty two, one twenty six point nine, they'll give you a vector to Dobbins
2110:45 RDO-1	Twenty six nine, okay
RDO	((Sound of click on radio channel))
2110:49 EAL683	Eastern six eighty three
2110:50 CAM-2	Ignition override, it's gotta work by # — #
2110:52 AA	Learjet triple nine mike reduce speed to one seven zero knots
2110:56 CAM	((Power interruption for 2 minutes and 4 seconds))
2113:00 CAM	((Power restored))

2113:03 CAM-1	There we go
2113:03.5 CAM-2	Get us a vector to Dobbins
2113:04 RDO-1	Uh, Atlanta, you read Southern two forty two
2113:08 AA	Southern two forty two Atlanta approach control uh, go ahead
2113:11 RDO-1	Uh, we've lost both engines - how about giving us a vector to the nearest place we're at seven thousand feet
2113:17 PA/ST-B	Ladies and gentlemen, please check that your seatbelts are securely again across your pelvis area on your hips
2113:17 AA	Southern two forty two roger, turn right heading one zero zero, will be vectors to
2113:18	Dobbins for a straight-in approach runway one one, altimeter two niner five two, your position is fifteen, correction twenty miles west of Dobbins at this time
CAM-2	What's Dobbins weather, Bill? How far is it? How far is it?
2113:31 RDO-1	Okay, uh one forty heading and twenty miles
2113:35 AA	Ah, make a heading of one two zero Southern two forty two, right turn to one two zero
CAM-2	Declare an emergency, Bill
2113:40 RDO-1	Okay, right turn to one two zero and, uh, you get us our squawk haven't you on emergency?
2113:45 AA	Uh, I'm not receiving it but radar contact your position is twenty miles west of Dobbins
2113:50 RDO-1	Okay
2113:51 AA	Delta seven fifty nine, contact approach control one two seven point two five now
2113:56 DAL759	((Sound of mike click))
2113:58 AA	Eastern six eighty three, contact approach control on one two seven point two five

2114:02
EAL683 Eastern six eighty three

2114:03
CAM-2 Get those engines (*---)

2114:04
AA Eastern one forty three reduce speed to one seven zero knots

2114:07
EAL143 Roger

2114:09
AA Eastern six eleven reduce speed to two one zero knots

2114:12
RDO *

2114:14
AA TWA five eighty four descend and maintain one one thousand, you can expect an ILS runway two six, and that altimeter two nine five two, localizer frequency one zero eight point seven

2114:24
RDO-2 All right, listen, we've lost both engines, and, uh, I can't, uh, tell you the implications of this uh, we uh, only got two engines and how far is Dobbins now?

2114:34
AA Southern, uh, two forty two, uh, nineteen miles

2114:40
RDO-1 Okay, we're out of, uh, fifty eight hundred, two hundred knots

2114:44
CAM-2 What's our speed, let's see what's our weight Bill, get me a bug speed.

2114:45
AA Southern two forty two, do you have one engine running now?

2114:47
CAM-2 No

2114:48
RDO-1 Negative, no engines

2114:50
AA Roger

2114:53
AA Eastern one forty three fly heading one nine zero

2114:56
EAL143 Roger

2114:59
CAM-1 One twenty six

CAM-2 One twenty six

CAM ((Sound of trim noise))

2115:04
CAM-1 Just don't stall this thing out

CAM-2 No I won't

CAM-1 Get your wing flaps

CAM ((Sound of lever movement))

2115:11
CAM-2 Got it, got hydraulics so we got

CAM-1 We got hydraulics

2115:13
AA Eastern six eleven, reduce speed to one seven zero knots

2115:17
EAL611 Roger

2115:17
CAM-2 What's the Dobbins weather?

2115:18
RDO-1 What's your Dobbins weather?

2115:22
AA Stand by

2115:25
CAM-2 Get Dobbins on the approach plate

2115:28
AA TWA five eighty four reduce speed to one seven zero knots

2115:32
TWA584 One seventy, five eighty four

2115:42
CAM-1 I can't find Dobbins

CAM-1 Tell me where's it at? Atlanta?

CAM-2 Yes

2115:46
AA Southern two forty two Dobbins weather is two thousand scattered, estimated ceiling three thousand broken, seven thousand overcast, visibility seven miles

2115:57
RDO-1 Okay, we're down to forty six hundred now

2115:59
CAM-2 How far is it? How far is it?

2116:00
AA Roger, and you're approximately uh, seventeen miles west of Dobbins at this time

2116:02
CAM ((Sound of windshield wipers coming on))

2116:05
RDO-1 I don't know whether we can make that or not

2116:07
AA Roger

2116:09
AA — Eastern one forty three, contact approach control one two seven point two five

2116:11
CAM-2 — Ah, ask him if there is anything between here and Dobbins?

CAM-1 — What?

2116:13
EAL143 — Roger

CAM-2 — Ask him if there is anything between here and Dobbins

2116:15
AA — Eastern six eleven reduce speed to one seven zero knots

2116:18
EAL611 — We're doing it six eleven, what is he a Martin or con — uh, nine ?

2116:22
AA — DC - nine

2116:25
RDO-1 — Uh, is there any airport between our position and Dobbins, uh

2116:28
IC — ((Sound of three chimes))

ST-A — Sandy

ST-B — Yea

ST-A — They would not talk to me --- when I looked in the whole front windshield is cracked

ST-A — Okay so what do we do

ST-B — Ah, have they said anything

ST-A — Ah he screamed at me when I opened the door just sit down so I didn't ask him a thing, I don't know the results or anything, I'm sure we decompressed

ST-B — Ah yes we've lost an engine ---

ST-A — I thought so

ST-B — Okay Katty, have you briefed all your passengers in the front?

ST-A — Yes, I told them I checked the cockpit and help me take the door down

ST-B — Have you removed your shoes?

ST-A — No I haven't

ST-B — Take off your shoes, be sure to to stow them somewhere right down in the galley in a compartment in there with the napkins or something

ST-A — I got them behind the seat, so that's no good

ST-B	It might keep the seat down now
ST-A	Okay
ST-B	Right down in one of those closets, I took off my socks so I'd have more ground pull with my toes, okay?
ST-A	You'd have what?
ST-B	So I took off my socks so I wouldn't be sliding
ST-A	Yea
ST-B	Okay
ST-A	That's a good idea too
ST-B	Okay
ST-A	Thank you, bye bye
2116:29 AA	Southern two forty two uh, no sir, uh, closest airport is Dobbins
2116:34 RDO-1	I doubt we're going to make it, but we're trying everything to get something started
2116:38 AA	Roger, well there is Cartersville, you're approximately ten miles south of Cartersville, fifteen miles west of Dobbins
2116:44 CAM-2	We'll take a vector to that yes, we'll have to go there
2116:45 RDO-1	Can you give us a vector to Cartersville?
2116:47 AA	All right, turn left, heading of three six zero be directly, uh, direct vector to Cartersville
2116:52 RDO-1	Three six zero, roger
CAM-2	What runways? What's the heading on the runway?
2116:53 RDO-1	What's the runway heading?
2116:58 AA	Stand by
2116:59 RDO-1	And how long is it?
2117:00 AA	Stand by
2117:02 AA	Eastern one forty three, contact approach control one two seven point two five
2117:08 CAM-1	Like we are, I'm picking out a clear field

2117:12 CAM-2	Bill, you've got to find me a highway
CAM-1	Let's get the next clear open field
CAM-2	No * (---)
2117:17 AA	TWA five eighty four turn left heading one one zero
2117:21 TWA584	One hundred ten degrees five eighty four
2117:25 AA	Eastern six eleven, uh, reduce speed to one seven zero knots and contact approach control one two seven point two five now
2117:35 CAM-1	See a highway over - no cars
CAM-2	Right there, is that straight?
2117:39 CAM-1	No
2117:44 AA	Southern two forty two the runway configuration
2117:44 CAM-2	We'll have to take it
2117:55 AA	At Cartersville is uh, three six zero and running north and south and the elevation is seven hundred fifty six feet and, uh, trying to get the length of now — it's three thousand two hundred feet long
2117:58 CAM	((Beep on gear horn))
CAM	((Gear horn steady for 4 seconds))
2118:02 RDO-1	Uh, we're putting it on the highway we're down to nothing
2118:07 CAM-2	Flaps
CAM-1	There at fifty
CAM-2	Oh # Bill, I hope we can do it
2118:14 CAM-2	I've got it, I got it
2118:15 CAM-2	I'm going to land right over that guy
2118:20 CAM	* (---)
CAM-1	There's a car ahead
2118:25 CAM-2	I got it Bill, I've got it now, I got it

CAM-1 Okay

2118:30
CAM-1 Don't stall it

CAM-2 I gotta bug

2118:31
CAM-2 We're going to do it right here

2118:32
TWA584 Eleven for five, five eighty four

2118:33
CAM-? ((Woman's voice)) Bend down and grab your ankles

2118:34
CAM-2 I got it

2118:36
CAM ((Sound of breakup))

2118:38
CAM-? * ((#))

2118:39
CAM ((More breakup sounds))

2118:43 End of tape

NTSB SAFETY BULLETIN

FLIGHT 242 NOT TOLD OF NEARBY AIRPORT

The following safety recommendation (A-77-67) was forwarded to the FAA Administrator on September 28, 1977.

On April 4, 1977, Southern Airways, Inc., Flight 242, a DC-9-31, crashed at New Hope, Georgia, as its crew attempted an emergency landing on a highway. Following failure of both engines and as the flight descended through 7,000 feet, the crew requested. . ."a vector to the nearest place." The Atlanta Approach Controller told the flight to turn right, issued a vector for Dobbins Air Force Base, and advised the flight that it was 20 miles west of Dobbins. Before this turn instruction was issued, the aircraft was 6 miles from Cornelius Moore Airport and headed in that direction. The airport has a published instrument approach procedure.

Investigation revealed that the airport was not included on the video map of the radar display and that the Atlanta Approach Controller did not know that the Cornelius Moore Airport existed. The Board recognizes that the Cornelius Moore Airport is located in Atlanta Air Route Traffic Control Center's (ARTCC) airspace near the boundary of the adjacent Atlanta Approach Control's airspace, and that air traffic control facilities generally do not depict areas outside their jurisdiction in as much detail as areas within their jurisdiction.

The Board also recognizes that although the Cornelius Moore Airport may not have been suitable for DC-9 type aircraft in this situation, the Board nevertheless believes that had that airport been depicted on the radar display of Atlanta Approach Control, it would have been available for immediate consideration by the controllers. Also, in an emergency situation involving a smaller aircraft, the depiction of similar adjacent airports would provide controllers greater latitude in assessing their options and facilitate coordination with adjacent facilities.

Since a portion of adjacent airspace is normally displayed on a facility's radarscope, the Board believes that the video mapping for this adjacent airspace should then contain the same airport information as the adjacent facility thus increasing the air traffic controller's capability to assist aircraft during emergencies. Specifically, those airports outside a facility's boundary, but within the area in which handoffs normally are accomplished, should be included on the video display.

Therefore, the National Transportation Safety Board recommends that the Federal Aviation Administration:

Require that each air traffic control facility depict on the map portion of its radar displays, those airports immediately outside of that facility's jurisdiction to the extent that adjacent facilities depict those airports on their displays. (A-77-67) (Class II, Priority Followup).

BAILEY, Acting Chairman, McADAMS, HOGUE, and HALEY, Members, concurred in the above recommendation.

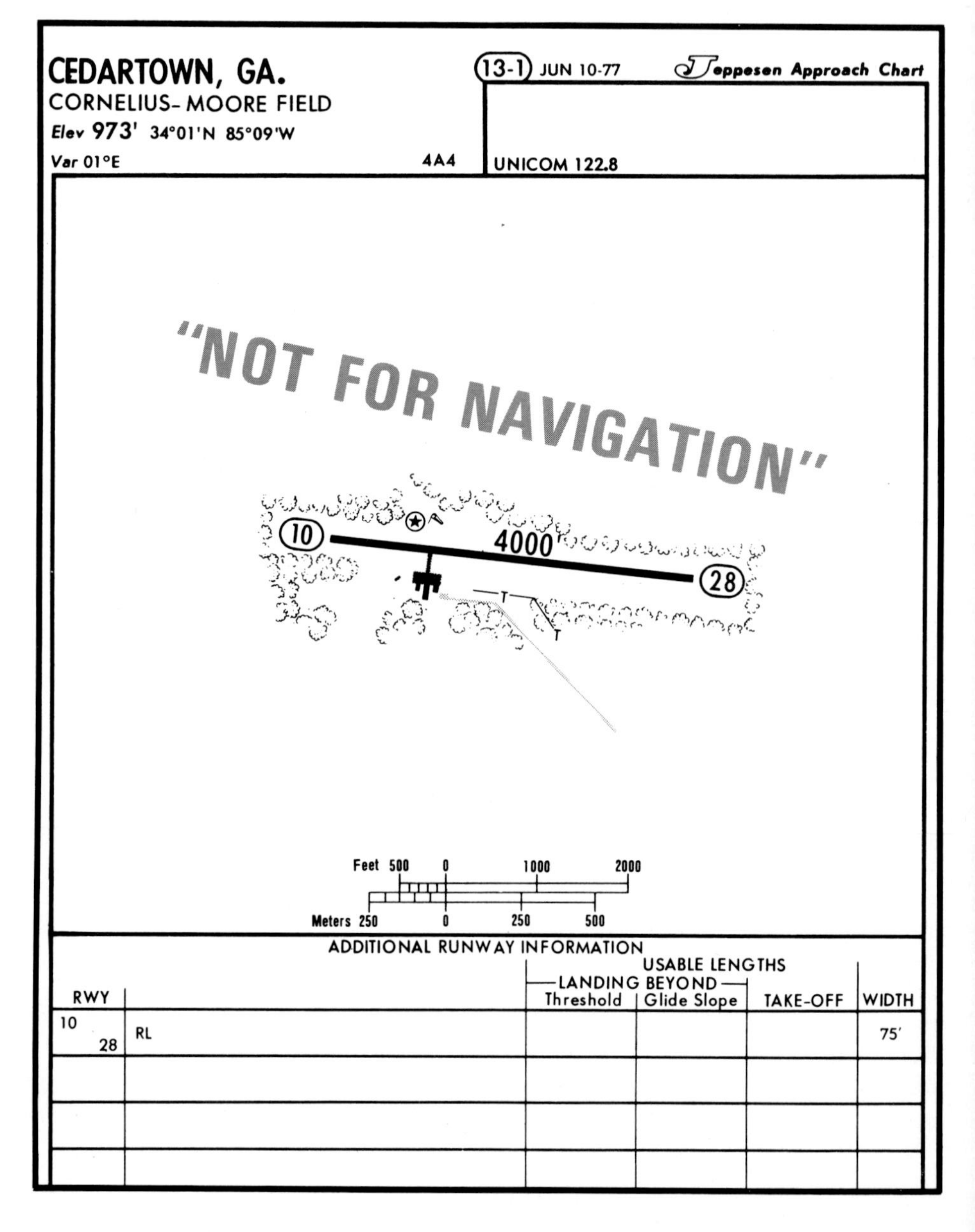

FATAL COMMUNICATION FAILURE

Reprinted by Permission
FLIGHT OPERATIONS (May 1977)

by Richard H. Jones

In aviation, words not spoken can often be as fatally misleading as any that are. A recent Florida aircraft accident offered dramatic proof of that.

Three crew members of a Douglas DC-7 were killed when their plane went down in heavy thunderstorms. A simple lack of communication between the tower and the pilot led to the tragedy. In a suit arising from the crash, a Federal district judge found that the controller had failed to advise the pilot of severe thunderstorm activity in the area and neglected to ascertain the crew's extent of knowledge of the prevailing weather conditions. The judge also decided that the pilot was negligent in failing to obtain an adequate weather briefing prior to takeoff and in not seeking any updated weather information from the controller despite having an opportunity to do so. Looking at the combined wrongs, the court stated that 30% of the blame for the accident could be placed on the pilot.

That flight was scheduled to leave Miami International Airport very early (shortly after 4 A.M.) on a June morning. Destination was to be Bimini. As takeoff time approached, weather conditions west and southwest of the airport gradually worsened. Severe thunderstorms moved into an area about four miles west of the airport. The crew could see lightning flashes from these storms, but the real extent of the storms could not be visually determined in the darkness.

The pilot had received a weather briefing about an hour before takeoff when he had filed his flight plan. At no time between then and takeoff, however, did he obtain any revised or more current weather information. The pilot violated good operating practice by not getting an update, according to the judge, particularly since there was known thunderstorm activity in the area.

Shortly before the DC-7 was to depart, the controllers changed the operations into and out of Miami International from an east operation to a west operation. The first plane to take off under the changed operation was a Braniff flight. The tower warned that pilot that there was rough, severe weather west of the airport. Additionaly, prior to leaving, the Braniff pilot was informed that he could make a right or left turn after takeoff to steer clear of the storms.

Still, before the DC-7 taxied into position, the controllers were informed by an Eastern jet trainer inbound from Dade-Collier Jetport that its radar indicated a line of thunderstorms west of Miami International. The jet trainer elected to fly around the storms rather than attempt to penetrate them.

Despite this report from the trainer and the weather information that was available to the controllers from the radar system in the tower, no updated briefing was made to the DC-7.

In his key findings, the judge stated that the controller knew, or in the exercise of reasonable care should have known, that the Douglas pilot did not have the extensive knowledge of the severity of the thunderstorms that the controller had. The judge also believed, for a number of reasons, that the controller knew or should have known that the pilot could not appreciate the dangerous proximity of the storm activity. Weather to the west of the airport had not been covered in the pilot's enroute briefing. The pilot gave no indication to the tower that he was aware of the intensity of the storm activity.

CONTROLLER NEGLIGENCE

The language used by the judge that the controllers "knew, or in exercise of reasonable care should have known," is quite common in personal injury cases. I made note of this in a previous article, but it bears repeating. In these types of cases the law is not concerned very often with exactly what a person did or did not do. Rather, behavior is measured by a standard of "reasonableness." What would a reasonable controller have done under these circumstances? If the person's conduct comports with what the community would regard as reasonable action, then it is not negligent. But the key point is that it is not the actual person's (or controller's in this case) perception of whether his act is reasonable that is determinative. It is this somewhat flexible and unstructured reasonable-man standard that is the measuring stick for negligence.

The DC-7 was cleared for takeoff in a westerly direction on a heading of 270 degrees toward and into the storms. Seven minutes after takeoff it crashed in the Everglades, about nine miles west of Miami International.

The judge found that the controllers failed to issue immediate turn instructions to the crew, although they knew they were flying straight into the thunderstorms. Indeed, the departure radar controller failed to recognize the danger, and instead gave routine instructions to another aircraft. Thereupon the judge held that this litany of acts and omissions by the controllers constituted negligent performance of their duties and was the proximate cause of the accident.

"Proximate cause" is another legal term that pops up all the time in negligence cases. Basically, it can be defined as the act or omission that was directly or principally responsible for the accident sued upon.

Another interesting item from a legal standpoint about this case is the judge's division of responsibility for the accident. As mentioned earlier, he found that the controller's actions were 70% responsible for the crash and the pilot's failure to get an updated weather report 30% responsible. This is an example of what is called "comparative negligence."

Proximate cause and comparative negligence are only legal sidelights in this case, however. The paramount point is that here was truly an avoidable accident, and the tragedy could have been averted if only there had been some timely communication between the pilot and the controllers.

THE CONTROLLER AND WEATHER

Reprinted by Permission
INTERCEPTOR (June 1977)
Originally printed in
THE MAC FLYER.

by David D. Thomas
Consultant to GAMA and Pan American Airlines

A basic principle of the air traffic control system is stated in the Federal Aviation Regulations as the "pilot in command of an aircraft is directly responsible for, and is the final authority as to, the operation of that aircraft." Another historic premise is that air traffic control clearance is limited to the purpose of preventing collision between known aircraft and allowing an aircraft to proceed under specified conditions within controlled airspace.

The entire concept of pilot-controller responsibility is tied up in a neat package in the Airman's Information Manual: "If ATC issues a clearance that would cause a pilot to deviate from a rule or regulation, or in the pilot's opinion, would place the aircraft in jeopardy, it is the pilot's responsibility to request an amended clearance." The manual goes on to say that "the pilot in command of the aircraft shall not deviate from the provisions of an ATC clearance unless an amended clearance is obtained."

The historic picture is clear. Controllers are responsible only for the separation of known aircraft and pilots are responsible for all other phases of flight — including severe weather avoidance. The current situation is not that crisp and clear, and the controller finds his plea of not having any responsibility for assisting pilots in avoiding terrain, windshear and thunderstorms falling upon deaf ears as accident investigation dramas are unfolded to the public eyes by the media.

A FRESH LOOK

It is not my purpose to argue what the current rules say, or fail to say, or what any controller or pilot did, or failed to do, in any specific accident. Rather, I hope we can take a fresh look at the controller, the pilot, and the weather with the objective of formulating a system that will increase flight safety by making better use of available tools, skills and information.

The most productive area to examine is weather. According to the National Transportation Safety Board, weather is the most frequently cited causal factor in general aviation fatal accidents and has been for several decades. We all know that weather plays an equally important role in air carrier accidents, as even the best professional pilots sometimes encounter weather conditions beyond their capabilities.

The subject of "weather" is much too large for a short presentation. Therefore let's concentrate on "severe weather" for the moment. While there are no universally agreed upon definitions of "severe weather" the principal concerns to most pilots are:

— any thunderstorms in the path of his airplane, especially in a terminal area.
— low level windshear.
— heavy precipitation.
— icing.
— strong cross winds and gusts.
— wake vortices.

Controllers may say that all this is interesting but that FAR 91.5 requires each pilot in command to familiarize himself with weather reports and forecasts before beginning a flight and that ATC is responsible only for separating aircraft.

HISTORIC RELATIONSHIP

Let us consider changing the historic pilot-controller relationship to one where the pilot is responsible for all the things he can control, or knows about, and the controller is responsible for assisting and advising the pilot for things the controller knows about but pilot does not. Specifically, let's bring the controller into the weather loop in a positive, programmed and official basis.

The controller has long been in the weather loop to varying degrees. In airport traffic control, wind direction and velocity are predominate factors in selecting active runways. Visibility and cloud heights have long determined whether operations could be conducted under visual flight rules.

The spacing between aircraft in terminal areas is more dependent upon the possibility of wake vortices than on runway occupancy times. En route controllers have provided excellent service in providing vectors to steer aircraft around threatening weather cells.

When radar was first introduced into air route traffic control, the radar presentation was a welcome relief from "manual separations standards" used whenever radar was not available. Controllers shuddered when the thunderstorms rolled in and they watched the scopes develop weird blotches which obliterated aircraft tracks. Many of us vowed that we would clean up the radar displays by eliminating weather returns.

CLEAN SCOPE RADAR

As time passed, more and more radar data became "processed" with "raw" radar becoming less useful until finally the day of the "clear scope" arrived and the controller watched only radar beacon returns with neat alphanumeric tags attached.

Pilots and controllers alike want the controller to have a display that will enable the controller to see all air traffic. Both want the controller to know where severe weather is located so that the controller may assist in weather avoidance.

Previous attempts to outline severe weather on clean scopes containing alphanumeric information have not been too successful. The controllers should have at their fingertips weather information from radars whose sole purpose is to detect and depict weather. It is suggested that a useful exercise would be to consolidate the dislays from National Weather Service radars and display the weather information in a form convenient to the controller. This would allow him to refer to it without detracting from the advantages of his clear scope free of weather returns. Providing such a consolidated display in each sector or terminal areas does not appear to be a major technical or economic problem if digitized information is used.

Another useful bit is information on the tops of thunderstorms as well as their geographial extent. While there has been little definition of thunderstorm intensity in terms of height, it is apparent that a thunderstorm more than 50,000 feet high contains much more energy than one which is 20,000 feet high. Additionally, the 20,000-foot junior size can be overflown by many airplanes where as the 50,000-foot variety must be circumnavigated by all civil aircraft.

WINDSHEAR AND WAKE VORTICES

The most talked about weather subject in aviation today is "windshear." Unfortunately, there is no operational capability to read and report windshear except by aircraft experiencing it. If a pilot experiences too much of it too low on the final approach course, the final point of impact may not be the runway.

At the present time, acoustical radar, such as that used to measure wake vortices, is probably the most effective tool for measuring windshear provided that precipitation is not too heavy. But windshear does occur in heavy precipitation. The answer is either to improve acoustic doppler radar to overcome the noise of the rain sufficiently to detect windshear or to develop other radar techniques which do the job. Regardless of the answer to the technical question, a windhsear detection system must be developed and should become part of the air traffic control system.

Another important element is that weather created by aircraft called wake vortices. It has been demonstrated that wake vortices can be detected by acoustical radar but there is no program for installing the radar. Acoustical radar is expensive but so are large and unnecessary separation standards. The present separation standards are based upon a possibility that wake vortices may exist rather than on certain knowledge that they do. If wake vortex detectors were available, the controller could use reduced separation standards on those days when the vortices dissipate rapidly and increase standards on days when he could see that vortices continue to dance in the approach path.

The controller needs a precise tool

at the major airports to determine the existence of wake vortices so that he can truly arrange for the safe, orderly, and expeditious flow of traffic.

In summary, let's make weather information available to controllers in a useful form so that he may assist pilots in avoiding severe weather. Some work in these areas is now underway, such as acoustic radar test, controller and flight service specialist access to computer-stored weather data, and stationing of meteorologists in air route traffic control centers. The planned collocation of consolidated FSS facilities with ARTC centers is a big step forward. All of these are good and excellent measures, but they are the mere beginnings of a truly effective system.

I am making no implication that the air traffic control system is not safe, as indeed it has performed its assigned mission well. Every controller can feel proud of his part in making the U.S. Air Traffic Control System in the best in the world. I personally feel very proud of any small role that I may have played in the development of the system and I make no apologies for it. My purpose is merely to indicate that times and techniques are changing and that we need more tools to make the system safer in severe weather. These tools, when given to the controller, can and will be used to make the honored creed of "safe, orderly and expeditious flow of traffic" more meaningful with regard to safety.

Presented to the Air Transport Association of America operations conference at Atlanta, Georgia.

ABOUT THE AUTHOR

Mr. Thomas served with the CAA and FAA for many years and is presently a consultant to the General Aviation Manufacturing Association and Pan American Airlines. Formerly president of the Flight Safety Foundation, he is known as one of the most influential protagonists on air traffic control in the United States. Among the awards he has received for his work are the Department of Commerce Meritorious Service Award, the Medalion Award of the Air Traffic Control Association, the Laura Taber Barbour Air Safety Award, the President's Award for Distinguished Civilian Service, the Civil Service League Award, the Rockefeller Public Service Award, and the Monsanto Air Safety Award.

ALPA PRESIDENT QUESTIONS FAA INFORMATION TRANSFER IN SOUTHERN CRASH

Excerpted from speech by J.J. O'Donnell. President Air Line Pilots Association, at Aviation/Space Writers Association San Fransciso, May 4, 1977

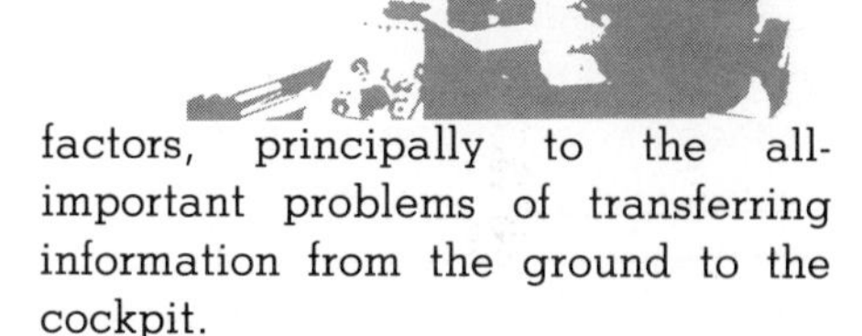

. . .What really concerns us is that while FAA is constrained by economics in aircraft certification, it also has decided its next-generation air traffic control system is to be based on ground control and ground data rather than data to be made available to pilots in flight.

In this arena, too, FAA is giving insufficient attention to human factors, principally to the all-important problems of transferring information from the ground to the cockpit.

In removing the pilot further from the decision-making loop, con-

Captain John J. O'Donnell, President, Air Line Pilots Association

centrating reliance on ground control, safety of the system is downgraded accordingly.

Yet, while the FAA moves ahead in this area, work on two important elements in the airborne equation lags behind. That's in head-up display and airborne traffic display to assist in avoiding collisions.

It's common knowledge the most critical phases of flight are the terminal approach and the landing.

Head-up display has been a significant element in military low-level operations for more than a decade. Boiler-plate airborne traffic displays for commercial aircraft have been built and they work. The time slot system is being flown by NASA at this very time.

But FAA, says more work needs to be done on both. Thus, as FAA generates a new environment for its controllers, it fails to come to grips with the problem of making certain that pilots have essential information that is being made available to controllers, who will make judgements what to do with it, or may become too busy to transfer it to the pilot who needs it.

A recent accident raises this question again, what weather was available, who had it, was it transferred to the pilots in a timely manner, or transmitted at all?

Once again, problems encountered in transferring timely adequate information raises its ugly head.

Listening to the tower tape released by the FAA on the Southern tragedy, we can only imagine the crew's workload during their descent. Two pilots, one flying the airplane, the other desperately maintaining communications. Both at the same time trying to restart their engines, to communicate with passengers as well as the ground, their windshield smashed.

All this in a period of about eight minutes and the FAA in ATC publicly critcizing the crew for allegedly not declaring an emergency. The tower tape clearly showed the crew did request confirmation of the 7700 Squawk.

Under normal circumstances, the human mind copes well with day-to-day conditions of flight. With bad weather, short runways, equipment malfunction, out-of-service navaids and other emergencies, at times the mind becomes overloaded beyond acceptable limits. Uncertainty caused by knowing that information on which to make decisions has been too frequently screened out by the inexperienced on the ground detracts considerably from human performance

Yet, the FAA would remove the pilot even further from the information loop. It continues to place more authority in the hands of ground-based people and their mechanical systems while doing little to reduce the pilot's workload or enhance the safety of this flight.

Removal of the pilot from the information and decision-making process causes too many slippages in the system. Accordingly, we feel the level of safety is reduced in proportion, and to a level we feel is unacceptable in terms of today's technology.

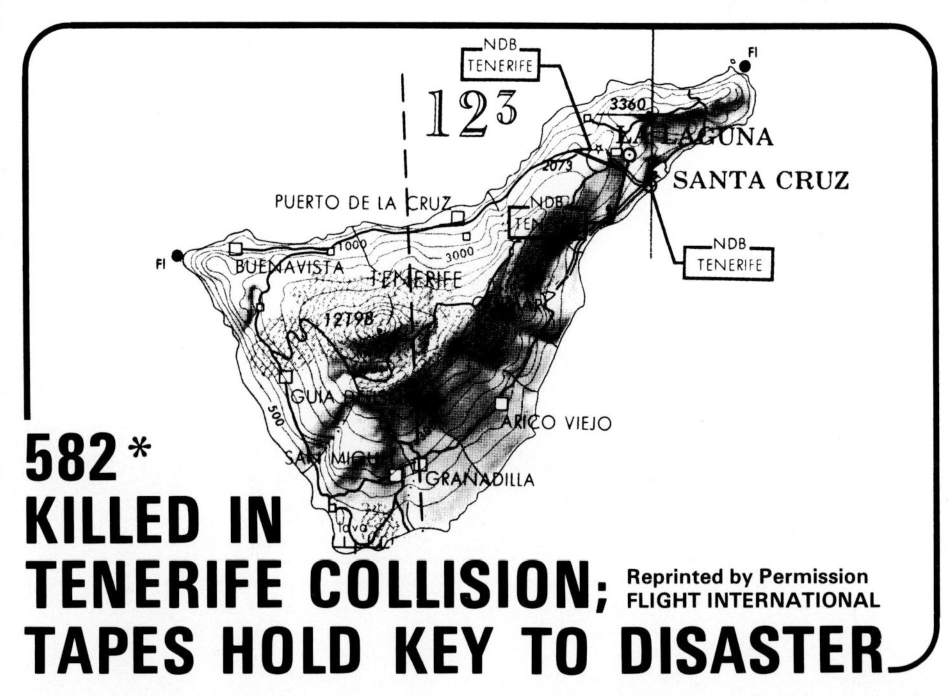

582* KILLED IN TENERIFE COLLISION; TAPES HOLD KEY TO DISASTER

Reprinted by Permission
FLIGHT INTERNATIONAL

This article was compiled from three articles published April 2, April 9, April 16, 1977.

Two Boeing 747s, belonging to Pan American Airways and KLM Royal Dutch Airlines, collided on the runway at Tenerife Santa Cruz airport Sunday, March 27, at 1714 hr GMT.

Final casualty figures in the Tenerife disaster are still not clear as *Flight* closes for press. It is certain that there were no survivors among the 235 passengers and 14 crew of the KLM 747. Only 15 of the 68 survivors from the Pan American aircraft are believed to have escaped unhurt. It is reported that of the injured, 11 are seriously and six very seriously hurt. Total death toll as we go to press is 579.

A USAF C-130 landed on the undamaged section of the 11,155ft runway on March 29 and flew out 55 people, one of whom died from burns en route to the United States on board a USAF C-141. Two more have died in hospital in Tenerife.

The Pan American 747 had left Los Angeles on Saturday, March 26, with 364 passengers on a charter flight to join a Mediterranean cruise leaving the Canaries. Fourteen more people joined at New York, and two airline staff boarded at Tenerife. With a crew of 16, a total of 396 people were on board the Pan American 747 when the collision took place. The survivors include Capt Victor Grubbs, commander, his co-pilot and the flight engineer.

According to the Spanish civil governor, six aircraft had been diverted to Santa Cruz Airport after a bomb was set off in a Las Palmas terminal flower shop, allegedly by a Marxist separatist group. According

*Correct January 1978.

Photo: Henri Bureau/SYGMA

to another report, there were eight aircraft parked on the tarmac, making access to the taxiway difficult.

The Pan Am aircraft had been on the ground three hours when Las Palmas re-opened and both 747s prepared to leave. The Pan-Am aircraft at least had refueled. At 1700hr visibility was reported as 2,500m with a variable 14kt wind. By 1800hr, after the accident, it had dropped to 100m with continuous drizzle. Over the same period patches of cloud were drifting across the airfield surface, and coverage by 1800hr was eight-eighths on the ground. By 1900hr visibility was zero. Visibility at the time of the accident was probably patchy, variable and worsening. Tenerife has no surface-movement radar.

The KLM aircraft had waited three and half hours at Tenerife after being diverted from Las Palmas. It is reported to have been carrying 9,000lit of JP-1 kerosene. Both airlines have denied that either aircraft was carrying any Jet B fuel.

According to one survivor: "It was just like a car crash at first. Then . . . everything caught fire."

Both CVRs were flown to Washington on April 3. The Pan Am recorder is already described as "very legible," but the KLM CVR is not of American manufacture and has been sent to Sundstrand for transcription.

The (tapes) were to be heard by the

NTSB experts in the presence of Spanish, Dutch, KLM and Pan American representatives. According to KLM's Karel Ledeboer, the US National Transportation Safety Board (NTSB) is the only authority capable of playing back tapes which have been subjected to impact and fire. Both CVRs remain in the custody of the Spanish authorities. All transcripts and the readouts from the flight-data recorders will be sent to the Spanish Government.

According to the deputy director of Tenerife Airport, Senor Juan Linares, the Dutch aircraft was told to hold. "The order was issued in clear English by the control tower," Linares is reported to have said. "The word used was 'standby'." His statement was based on the tower tapes, which were listened to also by the Dutch team.

According to the Dutch, "the request (to take off) was not approved and (KLM captain) was not given the vital clearance to take off."

According to Tenerife sources the air traffic control tower tape went like this:

> KLM to tower: "KLM ready for take-off."
> Tower to KLM: "Maintain holding position."
> Tower to Pan Am: "Have you left the runway?"
> Pan Am: "No."
> Tower: "Do so, and tell me when you have done it."
> Tower to KLM: "Stand by. I will call you for take-off."

According to Dutch sources there is a 15-to-20-second "blank" here. Then:

> KLM to tower: "We are taking off."
> Pan Am: "We are still on the runway." Then: "He's crazy — he'll kill us all . . ."

The US and Dutch teams, representing the States of Registry which have rights to participate in the inquiry and to have access to all evidence, arrived by helicopter in Tenerife on March 28. Head of the Dutch team Frans van Reijsen, confirms that the Pan American aircraft attempted to turn off on to the grass verge. "What seems to have happened," he is reported to have said, "is that the KLM aircraft received preliminary control clearance giving all the necessary information for the first part of the flight. The KLM commander repeated his instructions. He should have asked for take-off clearance, which he did not."

Mr van Reijesen claimed that the Pan American aircraft had missed its turn-off to the parallel taxiway. He said that the control tower message passed to KLM could have given rise to misunderstanding. There was a possibility that part of the message had been garbled. According to Reijsen, visibility at the time of the accident was 300 yards and the KLM captain could not have seen the Pan American aircraft when he began his take-off run.

The Pan American commander, Capt Grubbs, is reported to have confirmed that he had been authorised to back-track along the main runway, and to turn off on to the taxiway. He is reported to have said that he heard the KLM captain tell the control tower "We're going," and that he tried to break into the conversation while simultaneously turning his aircraft on the grass to avoid a collision.

According to the Pan Ameican co-pilot in a television interview, his aircraft had been cleared to enter the runway and to turn off on to the taxiway. One dispute is over whether or not the Pan American aircraft had missed its turn-off from the main runway. Mr Reijsen is reported to have said: '"We presume there was a misunderstanding over the position of the Pan American aircraft on the runway." He said that the Pan American aircraft had been told to leave the runway at Exit 3 and that instead it had been carrying on to Exit 4. The suggestion that Pan American had overshot its turn-off was rejected by vice-president operations Bill Walltrip. "It is my opinion that the Pan American aircraft was operating correctly," he said.

The control tower tapes seem to indicate confusion about the turn-off taxiway which the Pan Am aircraft was supposed to take in order to clear the runway and allow the KLM aircraft to take off. The Pan Am crew appear to have been intending to take turn-off C4, although they apparently requested to clear the runway at C3; the collision took place beyond C3, between C3 and C4. the actual exchanges were reported as follows:

Tower to Pan Am: Taxi into the runway and leave the runway third, third to your left, third.

Pan Am: Third to the left, OK.

Tower: The third one to your left. (Pause)

KLM: Would you confirm that you want Pan Am to turn left the third?

Tower confirmed, but it seems that the Pan Am crew considered one of the taxiways to be inactive and therefore did not count it. Certainly it led back to a crowded apron, but it is not believed to have been physically blocked. Pan Am was following KLM down the runway; when KLM reached the end of the runway the captain asked for airways clearance.

KLM: The KLM 4805 is now ready for take-off and we are waiting for our ATC clearance.

Tower: You are cleared to the Papa Beacon, climb to and maintain FL90, right turn after take-off, proceed on heading 040 until intercepting the 325 radial from Las Palmas VOR.

KLM read the message back.

Tower: OK. Stand by for take-off. I will call you.

Tower than asked the Pan Am captain to report when the runway was cleared, and no further exchanges are reported between the KLM captain and the tower. The last Pan Am transmission was: "We'll report runway cleared." It was conjectured last week that the KLM captain heard only the words "runway cleared" and assumed that it was safe to take off. Confirmation of the exchanges will await transcription of the CVR tapes.

A Pan American spokesman said that the early note issued by the Spanish authorities appearing to clear the control tower of blame was "entirely premature and uncalled-for." It is known also to have greatly angered the Dutch authorities. But it

Photo: Henri Bureau/SYGMA

is probable that the report of the statement was taken out of context in the heat of the moment. According to one report, the original statement by the Spanish authorities referred to three possibilities, including "an inherent fault in the equipment or methods of the airport." The accident could also have been due, the statement added, to a mistaken order from traffic control or to a failure on the part of one or other pilot to follow instructions. The offending passage said: "One can discard. . .lack of communication on the part of the control tower or its staff."

The accident, the worst in air transport history, is likely to be the most expensive in terms of hull insurance and passenger liability. The KLM 747 hull (PH-BUF) is believed to be covered for $40 million and that of the Pan American 747 (N736PA) for $23 million. Current price of a new 747 is about $45 million, compared with about half

Photo: Henri Bureau/SYGMA

that value for the early models delivered in 1970 and 1971. Of the Pan American hull's $23 million, 45 per cent is placed with Lloyd's and London companies, 15 percent in Europe, and 20 per cent in the USA. There is also reinsurance in London. According to London insurance brokers Willis, Faber and Dumas, the eventual insurance bill could be at least $100 million and possibly $200 million. Two-thirds might be carried directly or indirectly on the London market. C.T. Bowring of London immediately paid KLM $40 million for the 747 hull. Sedgwick Forbes of London has paid Pan American some of the $10-35 million as the London market's share of the airline's 747 insurance.

Passenger liability, assuming limited damages, could be $75,000 per passenger (including $17,000 legal fees). The Warsaw limit of $20,000 is raised in the case of US passengers to $75,000 and to the same amount by KLM, which is one of the European airlines which has voluntarily accepted the higher ("Montreal") limit of $75,000. Thus the minimum passenger liability would be over $40 million. But there is personal insurance to add, and the insurance market points out that the Pan American aircraft was carrying passengers originating in California, a state which has a reputation for being one of the most lawsuit-conscious.

• Damages totalling nearly $2,000 million have been claimed on behalf of 306 Californians killed in the Tenerife disaster. A suit has been filed in the San Francisco Federal Court on behalf of the sons of two victims, and it is being presented as a "class action" which will apply to all those killed.

It claims $1.5 million per victim in general and special damages, and $5 million each in punitive and exemplary damages. Pan Am, KLM and Boeing have all been named as defendants and are alleged to have been negligent in "managing, taxiing, operating and maintaining" the aircraft.

Photos: Henry Bureau/SYGMA

AIRPORT GROUND MOVEMENT CONTROL

Reprinted by Permission FLIGHT INTERNATIONAL (April 16, 1977) Edited for Space

Contrary to the layman's belief, an airliner is under positive control not only from take-off to touchdown, but in fact whenever it is moving.

Ground control at a busy commercial field is usually conducted by specially qualified staff who relay start-up, push-back and taxiing instructions — and sometimes route clearances — for outbound aircraft, and taxiing and parking advice for arriving aircraft. Ground controllers have to co-ordinate with other air traffic staff, especially if there is more than one runway in use, and the notion of installing both operators in a high building — the universally accepted control tower — still holds good. Ground-movement and aerodrome controllers can share the top floor, an arrangement which simplifies co-ordination; but at busy airports the two functions are often split between two floors. Heathrow ground-movement control (GMC), for example, is located at the top of the control tower, above approach control.

At a simple airport, in clear weather and with a steady traffic flow, GMC can be handled with few instructions. It is when airport layouts become complex, at night and in low visibility, or when aircraft movements increase suddenly, that the ground-movement task becomes especially demanding.

Airport layout has a large bearing on ground-control workload, irrespective of all other conditions. At the perfect airport every runway would have a taxiway leading to each threshold so that departing aircraft would enter the runway only when ready for take-off. Many older airports have a perimeter track, which satisfies this need, but when the runway is lengthened there is often insufficient traffic to justify extending the perimeter track. If traffic is light it is acceptable for a departing aircraft to backtrack for half a mile and line up, but extra vigilance and awareness is required of ATC. Manoeuvring on the runway usually comes under the brief of the aerodrome controller, not the ground controller, so that all arriving and departing traffic can be coordinated.

An aircraft back-tracking in poor visibility, without radar monitoring, has to follow strict procedures. Regulations, although not standard, normally allow not more than one departing aircraft to enter the runway at any time. Nor is incoming traffic allowed within about five miles while there is a departing aircraft on the runway. In very low visibility, when the approaching aircraft might need to execute an automatic landing, backtracking and departing aircraft can upset the instrument- landing- system signals, and such ground manoeuvring in bad conditions is avoided completely. To minimise these limitations, well equipped airports have a parallel taxiway along the full runway length with entry points at each threshold — preferably with overtaking bays — and high-speed, angled turn-offs. The taxiway system should always be on the same side as the terminal building and apron — an obvious requirement, but one which the planners of luxurious new terminals tend sometimes to forget.

The sea of lights bewilders many a newcomer to an airport, and when at strange fields most crews taxi slowly, concerned at taking a wrong turning or passing a stop position. A good GMC operator will be aware of an unsure driver, but there is no guarantee at any airport that all false movements will be seen, and if visibility is poor the onus for accurate taxiing is almost entirely on the crew.

An effective aid is the "stop-bar," a string of red lights across the taxiway which shows how far aircraft can move. Not only is this unambiguous, but it can reduce the number of instructions needed to control a given situation. Stop-bars are usually associated with taxiway "blocks." With such an arrangement the GMC operator illuminates centreline lighting through the specified movement blocks and terminates the track with a stop-bar. It is expensive, and almost impossible to install without severely disrupting a busy airport. But for these limitations, the block and stop-bar system would probably be more popular.

By no means the final solution, but the most effective monitor available to date, is the airfield surface movement indicator (ASMI) — a very short-range, high-resolution radar. Since 1956 one has been operational at Heathrow, where parallel runway movements are accompanied by a circulating taxiway flow around the central area. With stop-bars and taxiway blocks, this is an almost ideal GMC system. The ASMI should be little more than a monitor, although it has proved valuable not only in ensuring that aircraft obey instructions — no short-cuts or overruns — but also in providing immediate feedback of traffic flow. Re-clearances can thus be issued as aircraft approach stop-bars.

The Heathrow ASMI, made by Decca, can show detail out to a range of 4,600m. Discrimination is sufficient to see a $3m^2$ target at 2,700m, although in practice considerably better performance is achieved (at Heathrow a dead hare has been detected at 700m range). The radar can therefore monitor not only aircraft and vehicles, but also potential obstructions — debris, dislodged maintenance-pit covers and over-eager reggie-spotters. And airport security can be tightened by aiming a sensitive radar at the boundary fence.

ASMI picture resolution is remarkable: smooth surfaces appear dark (as radar energy is reflected away from the receiver) while grass areas appear speckled. Aircraft shadows thrown across grass areas show tail fins and even such distinctive features as the 747's upper-deck bump. Not all aircraft are distinctive, but the canny operator can differentiate among many types.

Heathrow's radar display, although virtually flicker-free at the refresh rate achieved, has recently been replaced by a conventional TV display. This is less affected by ambient light and is more restful to the eye than a radar display. The number of monitors showing the same ASMI picture is unlimited, a high-quality camera channel being used to convert the basic radar display to TV presentation. A typical ASMI costs £300,000.

Decca has installed ASMI at Heathrow, Paris Orly and Rome Fiumicino — two independent systems at the Italian airport. The company also has orders for Turin and Milan. Philips produced an ASMI for Amsterdam Schiphol several years ago, and is currently adding a bright display.

GROUND RADAR A RARITY

Although a number of major American airports have ground-movement radars (the US dubs them ASDE, airport surface detection equipment), there are relatively few fields worldwide with ground radar. This is because many airport authorities still see movement radar as not worth the investment. To them the main asset is improved safety, and as always safety is difficult to price. If GMC operators use the same control techniques, with or without an ASMI, the radar's value is limited to the detection of aircraft which fail to comply with movement instructions. If one aircraft taxis into the path of another, the carved-up aircraft soon alerts the controller. Aircraft crossing runways constitute the greatest risk, and radar, although unble to see a landing aircraft until it is almost on the runway, can be valuable for detecting such conflicts even when the visibility is poor. The occasions when an ASMI proves its worth as a primary control device are therefore infrequent, even at a busy airport with a complex runway plan. Its value, in general, has to reside in the provision of a streamlined ground-control service — and someone somewhere might be able to prove that timely re-clearances can save a worthwhile amount of brake wear and engine time.

One criticism often levelled at ASMI is its inability to associate an aircraft with its callsign. Labelled secondary surveillance radar (SSR) facilities would assist GMC operators, but conventional SSR cannot differentiate between targets less than 1.6 n.m. apart. Systems are under development which will produce interrogation pulses from independent aerials, thus triggering SSR transponders over only a small area at any given time. Bendix has proposed a system which operates on this principle, not only on the airfield but also in the terminal area. An experimental airfield-only system, being evaluated at the National Aviation Facility Experimental Centre, has successfully differentiated between targets only 15 feet apart.

Such systems are not improbable in the future, but meanwhile the ASMI is available only to selected airports and conflict detection depends on aircraft crews and GMC operators' eyeballs.

U.S. BLOCKS AIRLINE COMPENSATION AGREEMENTS

Reprinted by Permission NEW ZEALAND & AVIATION DIGEST (No. 63)

The Tenerife jumbo jet collision will focus world attention on acceptance of an international insurance agreement. Up to now American opposition or, more properly, the opposition of American lawyers has held up its introduction.

Known as "the New Zealand package", the agreement, if adopted, would limit airline liability to passengers killed or injured in aircraft accidents. World acceptance of limited liability would make it quicker, easier and less expensive for a passenger or his estate to recover compensation from an airline in the event of accident.

The principles underlying "the New Zealand package" are similar to those governing much of New Zealand's insurance legislation which aims to insure that claims for accident compensation are met quickly and with a minimum of cost.

The principle of limited liability for loss or damage in aircraft accidents was first established by the Warsaw Convention in 1929. At the time it was a completely new concept in accident compensation.

Primarily creation of European nations when international aviation was in its infancy, the Warsaw Convention aimed to standardise the law which varied from country to country in its relation to aircraft accidents, and to set the rates of compensation payable to all.

The convention was ratified by the nations and was signed by the United States in 1934 when airline liability was limited to $8300.

Following World War 2 and the advent of galloping inflation, the convention's provisions were revised upwards on a number of occasions till the United States finally renounced it in 1965 because it considered that the level of liability had become completely inadequate.

Finally, at international meetings in Guatemala in 1971, New Zealand's representative, Mr Peter W Graham, suggested a way out of the impasse by establishing a maximum compensation figure of $US100,000 reviewable every five years. Hence the term "the New Zealand package".

It was agreed that the $US100,000 limit would be unbreakable in all cases but that any country could provide a system of supplementing the compensation awarded under the Guatemala Protocol.

In the case of the United States, work was begun on investigating the feasibility of providing another $US200,000 in compensation by arranging an interline agreement between American carriers subject to the approval of the US Civil Aeronautics Board.

Before the Guatemala formula could be written into international law, however, it had to be ratified by

30 states five of which, between them, provided air services catering for 40 per cent of the world's total scheduled international traffic.

Now, six years after the Guatemala meetings, the protocol still remains unratified with the result that compensation limits have been held down to $US75,000.

The failure of the United States to ratify the protocol has been the major reason for withholding its passage into international law. Other states have preferred to wait for the United States — western world's largest provider of air services — to act before taking their own action.

Aviation Digest suspects that American delay results from the pressures put on its legislators by leading US lawyers who would see ratification as eating into their respective incomes.

Objections won't be put like that, of course. Legal opposition will be wrapped up in a host of high sounding arguments around the theme that ratification would interfere with the freedom and individual rights of American citizens.

As stated in Aviation Digest 59, the evil of the American attitude — and it will affect world aviation as a whole — is that it must be leading to a blockage in the free flow of accident information between manufacturers which formerly helped promote air safety.

American lawyers apparently look only at maximum awards, ignoring the fact that, concerning each individual's circumstances, lesser awards may well be adequate for the majority of airline passengers or their estates.

The American legal viewpoint is illustrated in the following comments taken from a paper presented to a symposium on the Guatemala Protocol by Lee S Kreindler, an American lawyer who specialises in handling air crash cases:

"Under the Guatemala Protocol, if it ever comes to be,. . .if I represent passengers in (an) accident, ... and I can only recover $100,000 or $300,000 . . . . and I have a $600,000 case . . . I am going to look for other people who may be legally obliged to pay.

"A little bit of negligence is all you need. Any negligence is enough to establish liability.

" . . . No matter how responsible, no matter how faulty the airline is, and no matter how limited the fault of the manufacturer is, the manufacturer is going to end up paying practically everything."

Mr Kreindler conveniently ignores the fact that — at the time at which he made his speech — a CAB survey

showed that about one half of all recoveries were $100,000 or less and that, with a limit of $300,000, more than 80 per cent of all cases had full recovery.

Further, says Mr Robert P Boyle, an international aviation consultant with the US Department of Transportation, "if you take into account the value of the money to the claimant in six months to a year from the date of the loss versus a larger sum received five to seven years later, and after expensive litigation, it may be that even these few not fully compensated cases become even fewer".

Mr Boyle said it was not easy to achieve agreement at Guatemala. To many countries the $US100,000 limit set would be an "unthinkable" amount for their nationals to recover in their courts.

Under a system where limited liability did not exist a claimant might eventually recover $500,000. That would normally be after expensive litigation and ensuing delays which, in the United States, averaged about six years.

"Out of the $500,000," says Mr. Boyle, "will come all court costs, attorney's fees, expenses of investigation, expert witnesses and other fees.

"Assuming the usual one third contigent fee and substantial —but not unreasonable—investigative and other expenses, this means recovery by the claimant (after six years) will probably be of the order of $300,000, since $200,000 will go for attorney's fees and other litigation costs.

"Under the Guatemala Protocol with a supplemental system providing a minimum of an additional $100,000 the same claimant would get $200,000. After deduction of expenses he would be left with a net $180,000.

"The award could be made within a year of the date of the accident. If this sum was invested for the next five years at a modest six per cent, he would arrive, on the date when the unlimited liability claimant was awarded $500,000, with $240,000.

"Thus the real difference to the claimant is not the difference between $200,000 and $500,000 which seems to be the effect of a limitation system. The real difference is about $60,000 and the advantage of having it quickly without waiting six years may be well worth the price."

There is some hope that the United States may yet endorse the Guatemala Protocol — an action which would almost certainly result in its ultimate reproduction in international law.

One of the last acts of President Ford before he handed over to Jimmy Carter was to ask the senate to ratify the document.

There is one snag: American lawyers have built up a $40 million litigation industry from aviation's mishaps.

Thus it is to be hoped that the rich pickings they will see in prospect following the Tenerife accident do not prompt them to bombard the American senate with the high sounding but specious arguments that could deter it from giving the protocol its approval.

NYA PAN-AM BUILDING/CRASH CLOSES HELIPORT

Reprinted by Permission
AIR LINE PILOT (July 77)

The commuter helicopter accident, which took the lives of five persons in New York City recently, also led to the grounding of the carrier's entire fleet of Sikorsky S-61 helicopters and the revocation of the landing certificate for the heliport atop the Pan American building in mid-town Manhattan where the accident occurred.

New York Airways was boarding passengers through its rooftop terminal on the 59-story building when the right landing strut of the idling copter collapsed, apparently when the upper landing gear mount broke.

The helicopter tipped over causing the blades to hit the roof and break apart. Impact of the rotor blades with the roof sent pieces of the blades flying into the waiting line of passengers killing four and injuring several others. A woman on a street two blocks from the building was also killed when pieces of metal fell on her. No one in the helicopter was killed.

NYA, who only recently started the service from the three New York area airports, Kennedy, LaGuardia and Newark, to the Pan Am building and optimistically reported what were called "dramatic increases" in traffic of about 30%.

The carrier voluntarily suspended its other 100 daily flights between the three airports but was able to resume the service within less than a week after all of its helicopters were thoroughly checked.

Service to the Pan Am building however must wait for a decision from New York's mayor.

PILOT SUSPENDED FOR ATC NON-COMPLIANCE

by Richard H. Jones

Reprinted by Permission
FLIGHT OPERATIONS (April 77)

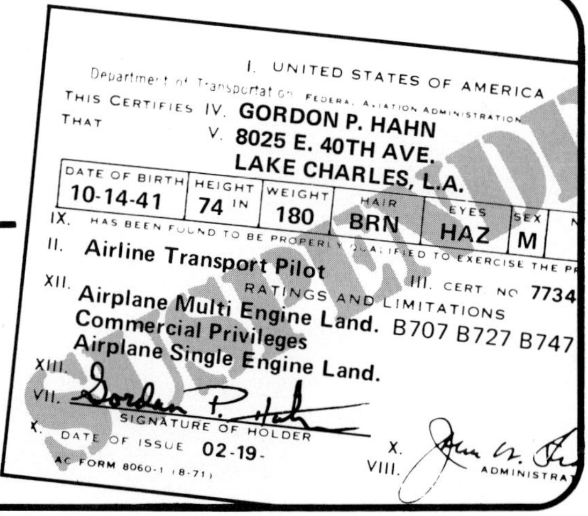
I. UNITED STATES OF AMERICA
Department of Transportation FEDERAL AVIATION ADMINISTRATION
THIS CERTIFIES THAT IV. GORDON P. HAHN
V. 8025 E. 40TH AVE.
LAKE CHARLES, L.A.
DATE OF BIRTH 10-14-41 | HEIGHT 74 IN | WEIGHT 180 | HAIR BRN | EYES HAZ | SEX M
IX. HAS BEEN FOUND TO BE PROPERLY QUALIFIED TO EXERCISE THE PR
II. Airline Transport Pilot III. CERT. NO 7734
XII. RATINGS AND LIMITATIONS
Airplane Multi Engine Land. B707 B727 B747
Commercial Privileges
Airplane Single Engine Land.
XIII.
VII. SIGNATURE OF HOLDER
X. DATE OF ISSUE 02-19-
AC FORM 8060-1 (8-71)
X. VIII. ADMINISTRA
SUSPENDED

Arguing with an air traffic controller over the latter's authority to order a reduction in flight speed has cost a transport pilot a stiff suspension. In a decision by the National Transportation Safety Board handed down earlier this year, an air carrier pilot received a 60-day suspension (1) for not complying with a Chicago O'Hare International controller's request to reduce speed, (2) for then refusing to accept clearance to proceed to a holding pattern, and (3) for arguing with the controller while in flight over the controller's authority to make these orders. The Board classified that conduct as operating the plane in a careless manner so as to endanger the life and property of others, in violation of Part 91.9 of the Federal Aviation Regulations (FAR).

The administrator had originally slapped the pilot with a 60-day suspension. This term was reduced to 20 days by the law judge, but was reinstated in full by the Board. What the Board characterized as the willful nature of the pilot's violation and the safety hazards inherent in such conduct warranted the longer enforced layoff, in the Board's opinion.

The incident took place during a passenger flight from Fort Wayne, Indiana to O'Hare. As the plane prepared to enter the approach traffic to Chicago the controller radioed the pilot to reduce his speed from 200 to 180 knots. The planes were being slowed to that speed in order to achieve the proper spacing and sequencing for the approach.

At the time this information was relayed to the cockpit, that plane was descending from about 11,000 feet at a speed of about 355 knots. The pilot thereupon told the controller that the plane's minimum speed then would be 200 knots. The controller told the pilot that, if he would not slow down to 180 knots, he was cleared to proceed to the Chicago Heights VOR to a holding pattern. After refusing to move into the holding pattern, the pilot engaged in what the Board termed a "protracted discussion" questioning the authority of the controller to issue the previous orders.

PILOT'S EXPLANATION

In defending himself against charges that he operated the aircraft in a careless manner, the pilot argued that his plane at no time came within the minimal separation from another one of five miles laterally or 1,000 feet vertically. He also contended that none of the evidence produced showed that the controller's preoccupation with his flight prevented him from maintaining safe control over other flights. For those reasons, the pilot stated that he was being punished for the manner in which he exercised his right not to accept a clearance or to request an alternate clearance. To the pilot, this case was basically a question of style, not of safety. The pilot also said he was concerned with saving fuel.

The Board, in affirming the administrator's opinion, put little stock in the pilot's contentions. What was particularly objectionable was the pilot's persistent "badgering" (the Board's word) of the controller.

The Board also pointed out that, while the pilot was correct in stating that no actual safety problems did result in this incident, 91.9 was still violated. All that need be shown, they said, to establish a breach of that section was that, in the judgement of the Board, potential danger was unnecessarily presented. The policy behind the regulation is to prevent inherently dangerous practices.

The Board summarily rejected the law judge's reasons for lowering the pilot's suspension from 60 to 20 days. In his opinion, the law judge reduced the sanction because of the pilot's prior exemplary record; the fact that the pilot's actions were not deliberate, willful or reckless; the pilot's concern for fuel consumption; his belief that, under the circumstances, he could not legally reduce the aircraft speed below 200 knots; and the judge's own feeling that the pilot had learned from this experience and would react differently in the future.

These reasons failed to sway the Board. As far as their flight records were concerned, aircraft crew members are expected to resemble Caesar's wife, and a "prior exemplary record" should be considered the norm, not the exception.

In reaching its decision in this case, the Board was obviously influenced by what it characterized as the "willful" or "deliberate" nature of the pilot's conduct. That a stiff suspension followed the Board's finding that the pilot's acts were intentional is not surprising. There are innumerable examples throughout the law where wrongs intentionally done are more severely punished than those inadvertently committed. This is especially true in the penal law, where intent is very often an essential element of the most serious crimes. Similarly, in civil cases, wrongfully injured parties generally can collect larger damage awards when the party at fault acted deliberately. Courts have little sympathy for the person who intentionally injures another, either physically or in some other respect. The law reflects society's value judgments (however imperfectly at times) and deals more harshly with a person who acts in such a manner.

One other interesting point about this case: It never really addressed the pilot's contention that the controller did not have the authority to order the speed reduction. The Board did find that the reduction requested by the controller was reasonable under the circumstances. But this appears to be the type of hindsight analysis it refused to allow the pilot to engage in, while claiming that his acts had not created a present danger.

CAN CONTROLLERS PENALIZE PILOTS?

by William B. Cotton, Chairman Airline Pilots Association ATC Committee Washington, D.C.

Reprinted by Permission
FLIGHT OPERATIONS
(August 77)

In your April issue, Richard Jones's Aviation Law only tells part of the story of that airline captain's ATC incident. His certificate suspension was not for noncompliance with ATC (FAR 91.75) as the headline implies, but for careless operation (FAR 91.9) with respect to the discussion on the frequency. The pilot's right to refuse a clearance has always been upheld by the courts in FAA violation cases. What really is at issue here, and which was never examined by the law judge or the NTSB, is the controller's ability to penalize a pilot who does not accept clearance.

When flying at 355 knots, it is not unreasonable for a pilot to require 200 rather than 180 knots while still 48nm from the airport. On a Boeing 727, flaps and leading edges must be extended while flying at less than 200 knots, resulting in a 30% increase in fuel flow in level flight. In fact, the controller's handbook specifies 250 knots minimum for speed requests at that distance and altitude. Therefore, the 180-knot request was improper for all the arrivals to O'Hare from that position, not just the one at issue. But even given that closer flights had reduced to 180 knots, a slightly altered heading would have resulted in the proper spacing on final approach by causing the intercept of the approach course to occur further out. The airspace was available, without any further coordination, and the controller would have used this technique for a military aircraft *unable* to slow to 180 knots.

So, when the controller responded to the request for an amended speed of 200 by issuing a penalizing hold, the pilot questioned his authority to do that. After all, if one is penalized for exercising a right, that right is just imagined. The controller's handbook clearly states that speed adjustment requests, beyond those in the guidelines, must be made with pilot concurrence. One should not expect that nonconcurrence will result in arbitrary and unnecessary penalty.

This was the issue which triggered the discussion on the frequency. As stated in Jones's column, it did not cause any danger. It was prolonged by the controller's inept vectoring (issuing a left turn to a heading when he meant right) and by his transmissions to another aircraft during the period in question. The only "wrong" found by the Board was "unnecessarily presenting potential danger." Since the pilot did not feel he was presenting any danger, and as none resulted, his judgement was correct. How can he be accused of a "wrong intentionally done"? Especially in view of FAA's repeated exhortation to question and clarify any clearance not understood *on the frequency,* how can a pilot be held as careless for doing just that?

As Jones closes, he raises one of the important remaining issues. What can a pilot do when a controller gives an instruction beyond his authorization? And beyond this, what can a pilot do when, after refusing a clearance, he is penalized beyond any requirement of traffic separation because of his refusal? The Board has not addressed this question, but its answer is crucial to the rights and responsibilities of all pilots.

DAVID B. PEAT: PILOT WITH A CAUSE

by Jeanette Mabry Reddish

Reprinted by Permission
FLIGHT OPERATIONS (August 77)

When the head of the air safety committee for a major airline becomes embroiled in a dispute with flight authorities, it's likely that he believes there are some important issues at stake. Such is the case with Captain David B. Peat, chairman of ALPA's central air safety committee at United Air Lines.

A January 1975 disagreement with Chicago ATC gave rise to a case in which Peat says his object was to define the extent of the pilot's authority in relation to the air controller. Peat lost his case and he failed to achieve his objective. He declares gloomily, "The outcome is a giant step forward for the authority of the air traffic controller and perhaps the kiss of death to the authority of the pilot, be he flying a Cessna 150 or a Boeing 747, because of the precedent that has been established."

These words are not spoken lightly by the 46-year-old Peat. Flying is the very core of his existence. Peat's father was one of the early barnstormers. In 1924 he teamed up with Clifford Ball to construct an airport in Pennsylvania; they founded an airline company which was awarded the nation's first airmail contract by the Government. The airline, only one of many flying-related undertakings of the senior Peat, eventually became Capital Airlines and was ultimately merged into United. Peat's mother, a sprightly septuagenarian, is also a pilot. When she soloed in 1929 she was among the first women to dare venture aloft.

The younger Peat was fiddling with the controls of aircraft when he was six or seven. He had his license by the time he was 18 and he's flown for United for 22 years. Unsurprisingly, he married a stewardess.

Disappointed by the NTSB ruling, Peat continues to be concerned by the sometimes uneasy arrangement between pilots and controllers. Since air carrier jets are subject to IFR clearance most of the time, observes Peat, "The controller is flying the airplane from the ground. The captain is not really in command anymore.

"The controller is better equipped than we are to tell where we are and what the traffic is. However, he's not in a position to actually handle our plane because he doesn't know what the flight conditions are, the weather. And that's where our problems lie. Weather is our major enemy up there," explains Peat. "So you get into situations where the pilot says he doesn't want to move somewhere because of the weather and yet the controller says that he has no choice because there's another plane in the way."

Controllers are not at fault for the situation, says Peat. "It's the system's fault. We got this arrangement simply because the ground equipment has been improved faster than ours."

The best solution is to give the pilot up in the air access to a cockpit display showing the traffic situation. The technology — not available when the current system began evolving — is now readily obtainable, suggests Peat. It was the original lack of this aircraft technology that prompted the division of information which now exists, he opines.

Peat is aware that the technology he would like to see installed in airplanes is costly and that airlines are reluctant to spend unless the cost benefits are well defined. Ironically, it may be the lawsuit liability of airlines in accidents such as the one in the Canary Islands that might make the installation of the type of equipment Peat would like to have feasible. That equipment would have permitted the pilots of those two ill-fated aircraft to have been aware of each other's whereabouts at all times despite the dense fog.

Peat believes some of the complicated computer systems currently envisioned for air traffic control by the FAA are unnecessarily complex and inappropriate. He says, "We pilots need a way to know what the traffic is and where the weather is and we don't need all that elaborate stuff."

CO-PILOT LIABILITIES

by Arnold W. Scott

Reprinted by Permission PROFESSIONAL PILOT (May 1977)

Remember Gerald W. Knouff? He was the copilot on the Air East (Allegheny Commuter) Beech 99A that crashed on approach to the Johnstown-Cambria County Airport on Jan. 6, 1974. Of the 17 persons on board, all perished except Knouff and four passengers who escaped with fractures, head and internal injuries.

According to news reports, the United States has filed a countersuit against this hapless copilot to recover any and all awards the government may have to pay for actions brought against it as a result of the accident, plus court costs. That's no drop in the bucket, Jack. At last audit Uncle was being sued "either directly or for contribution or indemnity for damages claimed to be $5,135,867.68." Add to that a reasonable sum for legal fees and you're talking in the neighborhood of five and a half big ones. Pretty steep on a copilot's take-home.

It all began Dec. 22, 1975 when Knouff filed a claim to recover medical expenses related to the accident. It was denied Jan. 16, 1976. Knouff then filed suit July 14, 1976 to the tune of about $12,000, representing medical expenses, lost earnings and reimbursement for pain and suffering. Apparently, this angered Uncle Sam and he retaliated.

Papers filed in U.S. District Court for Western Pennsylvania by the Justice Department allege that Knouff, as an employee of Air East "had operated in the past as well as on the accident flight in conformity with company procedures, which he knew or should have known violated the approved Air East operations manual, Federal Aviation Regulations and good and safe operating practices; he knew or should have known that adherence to such company procedures created a risk of extreme danger to himself and others, and his adherence thereto constituted negligence." Further, as copilot of the flight, Knouff was "vicariously liable for the negligence of other members of the crew of said aircraft, and their negligence is imputable to him.

"As a direct and proximate result of the aforesaid negligence of Gerald W. Knouff, both actual and imputed, N125AE crashed on approach to Johnstown-Cambria County Airport before reaching the runway."

Quite a stink was kicked up in the commuter airline industry as a result of this accident. Former Air East pilots testified before NTSB that after being personally checked out by Allan McKinney, the company's vice president for operations, they were allowed and expected to bust approach minimums on a routine basis in favor of "company minimums" with MDAs as low as 200 ft AGL. Reason: to achieve a highly competitive completion factor which, undeniably, is a strong bargaining tool when the time comes to renew or renegotiate service contracts.

It was also revealed that "it was a regular company practice to enter low fuel weights on the load manifests when a maximum load of (15) passengers were aboard. The low fuel weights were entered to show that the aircraft was within weight and balance limits. (Former Air East Pilots) also stated that passenger seats were never restricted from use to keep the aircraft within weight and balance limits. It was an unwritten policy to accept additional passengers and to fly the aircraft overweight and out of c.g. limits, if necessary."

Knouff was also negligent, says Uncle, for allowing the captain to make the ill-fated flight from Pittsburgh and for "failing to inform the company of the captain's lack of fitness on or about Jan. 6, 1974 when he knew, or in the exercise of reasonable care should have known, prior to the flight, that said pilot was ill, reckless and unable to discharge the duties of an aircraft pilot." This charge apparently stemmed from a deposition Knouff made to investigators that Captain Daniel W. Brannon had not eaten dinner before leaving Pittsburgh, as was his usual practice, and that he had allowed the aircraft to deviate slightly on the taxiway before taking off.

Besides Knouff, Jeff Wilkinson, the captain of another Air East Allegheny Commuter that departed Pittsburgh for Altoona directly behind the doomed Johnstown flight, discounts this contention. He said he asked Brannon to have dinner with him and two other pilots but that he refused and elected to remain with his aircraft. Wilkinson said that he observed the Johnstown flight's

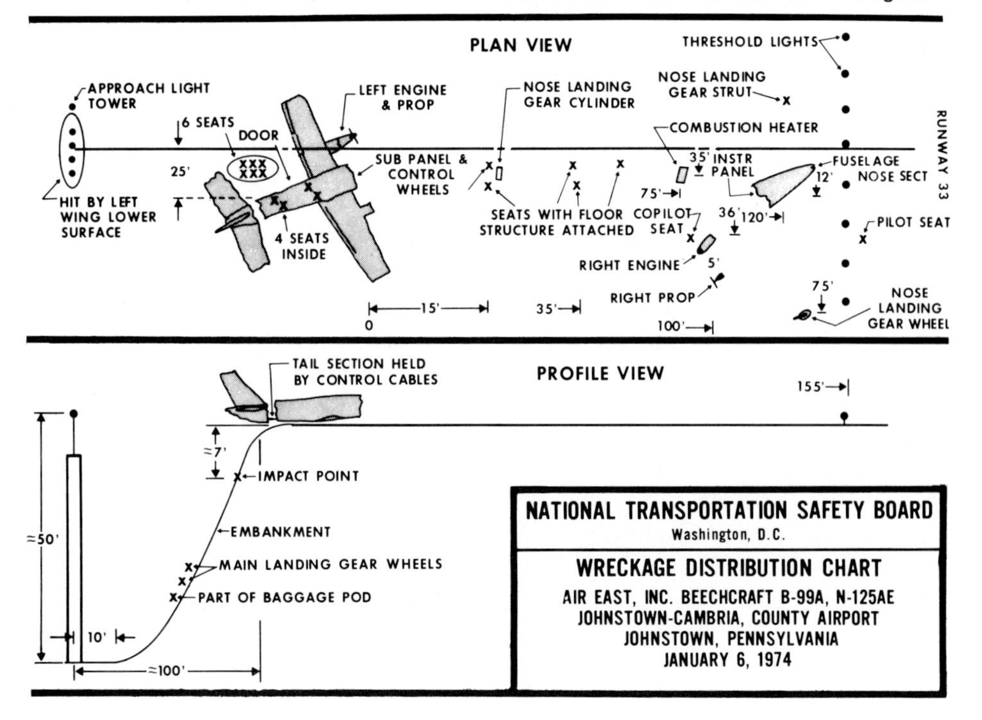

takeoff and saw nothing out of the ordinary that would cast suspicion or doubt on the captain's physical or emotional condition.

Two months after the crash, FAA stripped Air East of its FAR Part 135 certificate.

James L. Weisman, Knouff's Pittsburgh attorney, says that the government's countersuit and allegations are highly unusual. He believes the U.S. has no intention of recouping its losses from the copilot and therefore has filed a petition for dismissal.

NTSB has concluded that the probable cause of the accident was "a premature descent below a safe approach slope followed by . . .[accelerated] stall and loss of aircraft control. The reason for the premature descent could not be determined, but it was probably the result of: (1) a deliberate descent below the published minimum descent altitude to establish reference with the approach lights and make the landing, (2) a visual impairment or optical illusion created by the runway approach lighting systems, and (3) downdrafts near the approach end of the runway." Survivors' attorneys disagree, claiming that the approach light system was improperly installed by FAA. The government, however, by filing this countersuit, says it was the copilot's fault. Incredible.

Even though it has denied all liability for damages in the accident, the government has nevertheless made several substantial settlement offers totaling as much as $1.25 million.

What lessons are to be learned from this tragedy? Copilots, beware. Don't let anyone, including yourself, take you for granted. Remove all the legal mumbo-jumbo and your purpose for being in the cockpit extends far beyond raising and lowering the landing gear, reading off checklists or operating the radios. Indeed, it involves more than just sitting on your butt and picking your nose while building time toward that "dream job."

As copilot, you are second in command, possessing the same authority and responsibility as the captain and entrusted with the same degree of care and safety when required to assume command. Be prepared, however, to justify your action. Be especially prepared to defend your inaction, because failure to act may imply consent or, in the case of Gerald W. Knouff, "impute negligence."

And when should you assume command? Only *you* can make that decision. It brings back memories of the Allegheny Convair crack-up at Tweed-New Haven (CT) Airport, June 7, 1971. Asked why he didn't assume command when he saw that an accident was inevitable, First Officer James A. Walker replied: ". . .There was a thought in my mind. . ., It's better (to have) one man flying the airplane in perfect control, than (to have) two men fighting over it. . . . Had he been incapacitated in any manner, I mean, I would have, because that is the only time that I can take an airplane away from a captain."

Where are the principals to this accident and what are they doing now? Happily, Jerry Knouff has fully recovered from his injuries and flies an MU-2 and a Bell 206B JetRanger for a Johnstown corporation. He admits it took him a good year to settle down and feel comfortable in an airplane again. During the two weeks he lay in intensive care lingering near death, doctors told his pregnant wife, Heidi, that she could write him off as a basket case. This may have had something to do with their daughter being stillborn. Gerald Jr., now six months old, has given them a new hope and reason for living.

Air East is still the fixed base operator at JST and has reapplied to FAA for another Part 135 certificate, this time under a different name.

Air East President Mac McKinney, has split the scene and has reportedly gone into the water purification business. Son Allan, his vice president for operations, and former Chief Pilot Jim Tallant, lost their licenses but have since been reexamined and are now flying a Jet Commander for a Pittsburgh firm based out of JST. Former Check Pilot Merle Luther is now the assistant chief pilot for Henson Aviation in Hagerstown, MD, flying between DCA, BAL and HGR in Henson's Beech 99s for Henson's segment of the Allegheny commuter system.

Former Captain Jeff Wilkinson, the Air East pilot who found considerable disfavor with FAA when he testified in behalf of the late Captain Brannon (and who refused to admit he had busted minimums routinely or flown overweight and out of c.g. limits), lost his license too — but he said to hell with it. It wasn't worth the hassle to pick up where he had left off. He is now a disc jockey for a Johnstown radio station.

And what about the four surviving passengers? One, a medical doctor, retired suddenly from practice shortly after the accident. Another passenger, a 12 year old girl who sat directly behind Copilot Knouff, escaped with serious injuries, but lost her twin sister, aunt and cousin in the crash. The other two passengers, a woman and a 19 year old boy, received critical injuries and are reportedly still under professional care.

Captain David W. Brannon and eleven of his passengers aboard Air East Flight 317 lie buried in cemeteries located throughout the Keystone State.

Remains of Air East Flight 317. (*Tribune-Democrat* photo by Bob Donaldson)

AIRLINE REGULATORY REFORM: PRO

The question of regulatory reform or "deregulation" created much controversy in 1977. The two articles following present the "for" side of the argument.

A SUMMARY OF UNITED AIRLINES POSITION ON REGULATORY REFORM

Reprinted from material supplied by Public Affairs Division United Air Lines

BACKGROUND

United Airlines supports a new federal aviation law to increase competition gradually and reduce government regulation of the air-transport industry. A new law that gradually increases market influence and reduces government influence will benefit both passengers and the air-transport system

For the passenger, regulatory change would mean continuation of a sound air-transport system and the benefits of innovation and efficiency that the marketplace demands of sellers. It would mean service even more closely tailored to market demand — better tailored regarding price, frequency of flights, and variety and quality of service.

For airlines, regulatory change would permit better matching of prices and costs for a reasonable profit. Under the current regulatory framework, the airline industry's 3.6-per cent return on total capital over a five-year period put it last among 30 industries — including the railroads — in Forbes magazine's latest survey. Unless that return improves, airlines will not attract enough private capital to replace obsolescent aircraft and accommodate growth. In that case, either the air-transport system will deteriorate or government capital will be required.

Gradual transition to a market system would avoid the possibility of wholesale disruption of the air-transport system. A gradual change would help ensure continued stable employment in the industry.

In short, United favors regulatory change in pricing, market entry and market exit. Here are the specifics of our proposal:

PRICING

Carriers should be permitted to adjust their fares within a zone of 10 per cent above and below the price in effect a year earlier. Adjustments would be made market-by-market, in contrast with the current practice of adjusting fares on a system wide basis. Fare changes within the zone would be considered lawful and wouldn't be subject to suspension by the Civil Aeronautics Board. Fare adjustments outside the zone would still be subject to the CAB's review.

COMPETITION

To protect the public from unreasonably high fares as the CAB's control over price is reduced, more reliance should be placed on competition. Therefore, the new law should include a pro-competition policy statement.

United also proposes that the new law include a two-tiered entry provision that would increase the ability of both new and existing carriers to enter new markets.

New airlines should meet the same standards that existing carriers have already met. New carriers would be required to file an application for authority to serve a specific route, go through a CAB hearing as they would under existing law, and demonstrate that they are fit, willing and financially able to provide the service they propose. They would also be required to demonstrate that the service is required by the public convenience and necessity, as they would under existing law. To permit a gradual transition to a market-oriented system, United endorses Senator Kennedy's suggestion made in 1976 that the CAB be allowed to certify no more than two new scheduled interstate carriers each year.

These procedures would apply to carriers which are not already certificated to provide scheduled interstate passenger service such as World Airways, Midway Airlines, The Flying Tiger Line, Southwest Airlines and PSA. The procedures would also apply to newly formed companies.

Existing scheduled carriers would be allowed to expand their route systems within prescribed limits at their own discretion every two years beginning in 1980. United's proposal calls for dividing airlines into three groups. Each airline would be allowed to increase its route system by a given percentage every two years.

Under the formula, the largest carriers, such as United, would be allowed to expand proportionally less than smaller carriers in their group.

After 18 months of continuous service on a new route, a carrier's authority to fly that route would become permanent.

Existing carriers could still seek new authority beyond the limits prescribed for automatic route expansion. Carriers would follow existing certification procedures by filing an application with the CAB.

MARKET EXIT

Along with increased competition should be greater ability for carriers to terminate unprofitable service. Freedom of exit must be tempered by provisions to protect small communities from being shut off from the national air-transport system. For example, proposals have been made that the government directly subsidize continued service to small communities. It's been proposed that the goverment award contracts for such service based on bids from carriers. It's also been proposed that a new class of carriers — limited air carriers — be created to specialize in serving small markets. Some solutions along these lines can and should be found.

United, for one, does not contemplate wholesale abandonment of small markets under reformed regulation. United, which serves 113 cities across the country, is one of the nation's largest local-service carriers. But we don't plan to eliminate service at the vast majority of the small cities served because they're contributing to the airline's financial results.

ANTITRUST IMMUNITY

To preserve the nation's integrated air network, it's imperative that the

CAB's power to grant antitrust immunity be retained. Hundreds of intercarrier agreements now let passengers travel on several airlines with just one ticket, permit baggage transfers from one carrier to another, and provide compensation for lost or damaged baggage without requiring that passengers determine which airline on their trip was responsible.

These agreements are clearly in the public interest. But without antitrust immunity, airlines could be forced to abandon such agreements rather than open themselves to the threat of extensive litigation.

THE SIX GREAT MYTHS

Former President Ford and incumbent President Carter, conservative and liberal Senators and Congressmen, economists of all persuasions, and newspapers ranging from *The Wall Street Journal* to *The Washington Post* have all called for reform of airline regulation.

United Airlines is supporting reform, greater competition, and more decision-making by the marketplace rather than by governmental decree.

In the midst of the debate about this issue six great myths have been circulated by opponents of reform.

MYTH NUMBER ONE: Bills pending in Congress, and proposals advanced by United Airlines would "deregulate" air transportation.

THE TRUTH: *No one* in Congress or the Administration is proposing total economic deregulation, nor is United Airlines. Every proposal is for *reform* of the system. The Civil Aeronautics Board would continue to regulate where necessary, but more decisions would be made by the marketplace reflecting the needs and desires of passengers and shippers. More competition would gradually develop, and some flexibility in pricing — downward and upward — would be permitted, in a narrow range without government interference.

The result of reform would be better-tailored service at reasonable prices for the public, and an opportunity for carriers to suit their products to their markets and earn a reasonable return.

MYTH NUMBER TWO: United Airlines supports change because it is the biggest airline, it has strong financial reserves and has a large fleet of airplanes.

THE TRUTH: United supports change because less regulation will allow the airline industry to have sufficient earnings to replace their aircraft fleets while offering the public continued good service at reasonable prices.

The airline industry has the worst earnings record of the top 30 industrial groups in the United States. Unless it can improve that record it will go the way of the Northeast passenger railroads as equipment and service begin to decay. We are not the way we were. We must plan for tomorrow's needs today if we are to preserve the world's finest transport system. That means change.

While United is big and strong it is also a high cost airline and more vulnerable to competition than most other airlines. But freer competition is preferable to decay under out-of-date and burdensome government regulation.

MYTH NUMBER THREE: Regulatory change will result in lowering safety standards.

THE TRUTH: This statement is a red herring. The Federal Aviation Administration regulates airline safety. Its powers would not be changed. Furthermore, strict standards of fitness and ability would be part of any new law regulating airline economics. There are simply no grounds for believing that airline safety standards would be lowered.

MYTH NUMBER FOUR: Regulatory change would result in loss of service at small communities.

THE TRUTH: The proposals for change include *greater* safeguards for continued small community service than exist under present law. There has been considerable abandonment of small communities in recent years by existing airlines. Proposed new laws would insure against the loss of that service, and United Airlines, which serves more small communities than any other airline, has testified publicly that it has no plans for wholesale abandonment of small communities.

MYTH NUMBER FIVE: 300,000 airline jobs would be in jeopardy under regulatory reform.

THE TRUTH: This is a fanciful story that simply is not supported by any evidence. United supports gradual change in economic regulation of the airlines with build-in safeguards to protect service to communities presently receiving it. Given that there will not be disruptions in air service, what reason is there to believe there would be any disruptions in airline employment? Answer: None.

In fact, United believes the form of regulatory changes it supports will bring about expansion of the present air transportation system. This will create additional opportunities for advancement for present employees and new job opportunities for potential newcomers to the industry.

Under the present tight regulatory framework, employment in the airline industry has actually declined during the 1970s. In 1975 alone air carriers reduced employment by nearly 10,000.

Not only would more airline jobs open up under regulatory reform, but there would also be more jobs available in the related aerospace and engine manufacturing industries.

MYTH NUMBER SIX: Airline fares would increase under regulatory reform.

THE TRUTH: Under existing regulation airlines are penalized whenever they offer discount fares to the public. Some fares, such as youth and senior citizen fares, have been outlawed by the CAB. Regulatory reform could open up a full range of those fares, according to the needs of the market. Management and passengers, not government, will decide where and how to offer these fares.

In the long run inflation may drive up the general fare level, whether or not there is reform. But reform will permit new efficiencies to pass through to passengers and shippers, and will also permit the offering of a full — and easily understood — range of discounts.

AIRLINE REGULATORY REFORM: CON

Arguments against regulatory reform or "deregulation" are contained in the following articles.

THE CRUEL HOAX OF DEREGULATION

an AIR LINE PILOT editorial

Reprinted by Permission
AIR LINE PILOT (May 1977)

Last February, the General Accounting Office (GAO) issued a report entitled "Lower Airline Costs Per Passenger Are Possible In The United States And Could Result In Lower Fares." Those touting deregulation as the answer to the airlines' woes promptly claimed that the GAO supports deregulation. Sen. Edward M. Kennedy (D-Mass.), who had requested the report in the first place, jumped at the chance to headline the report by saying that airline passengers will save millions once airline regulation is lifted.

The 188-page report is not what Senator Kennedy and the other deregulation proponents would have you believe. First of all, it is not a GAO report in the usual sense. It is a GAO analysis of a 1972 study by Dr. Theodore E. Keeler, an economics professor at the University of California at Berkeley.

Dr. Keeler estimated that unregulated airline fares were lower than those regulated by CAB. He attributed this to lack of price competition and too much service competition — mostly in the form of frequent flights.

The GAO updated the Keeler study to cover the six years from 1969 to 1974. Assuming that a deregulated industry would mean easier entrance and exit to particular markets and that airlines could raise or lower fares as they chose, the conclusion was that passengers would have saved \$1.4 billion to \$1.8 billion annually.

What Senator Kennedy and those favoring deregulation have not bothered to tell is that the GAO report represents only the GAO/Keeler views and assumptions and stems primarily from analysis of the California and Texas intrastate airline situations. On page 39 of the report, the agency carefully hedged by saying ". . .but this report does not answer a number of questions about what might happen if the form of airline regulation were changed or if regulation were abandoned completely."

What Senator Kennedy should remember is that the GAO recommended "that CAB continue to work toward improving airline efficiency under its existing legislation by emphasizing higher load factors, higher seating densities, and the other factors identified in this report and by increasing its reliance on competition to determine service and prices."

The GAO also recommended that Congress provide CAB with "legislative guidance defining current national objectives for air transportation, and the extent to which increased competition should be used to achieve those objectives."

Before those who make the laws go

off and legislate a vital industry out of existence, perhaps it would be wise for them to read the GAO report in its entirety. The GAO actually supports the airlines in what is already known to be necessary to restore earnings to the industry. It does not support deregulation as many contend.

What is really lacking in all this palaver is a sound national air transportation policy which will guide the industry and its regulators for the rest of the century. Congress should give guidance as GAO suggests. It should not dismantle an industry under the guise of promising lower fares. This is a cruel hoax which must not be perpetuated any further.

U.S. AIRLINES AND THE FREE MARKET SYSTEM

by Sam I. Aldock, President,
Systems Analysis and Research Corp.

Reprinted by Permission
AIR TRANSPORT WORLD (July 77)

The recent statements, testimony, articles and letters emanating from opposing sides of the continuing U.S. airline regulation/deregulation issue suggest that both groups are overlooking the fundamentals of providing a sound air transport system.

The debate has for all practical purposes evolved into a confrontation between the airline industry and the U.S. government. Unfortunately, the government is not an objective third party in this debate, as is normally the case when the public is involved. It supports deregulation, although some of its proposals apparently would increase regulatory staffs. It does so with studies which were designed not to analyze the issues, but rather to "sell" its position.

This is surprising, since there has been no public outcry from communities or travelers for a radical change. Is it possible that travel needs are being met by the present system at reasonable cost?

There may be reasons for the absence of dissatisfaction. Today, 23 airlines serve more than 600 U.S. cities, ranging in population size from under 5000 to millions. Superimposition of carrier routes on a map resembles a closely woven spider web. It is probably correct to say that in the vast continental U.S., one can move from almost any point to any other point, regardless of distance or isolation, within a fraction of a day. Air fares are lower in the U.S. than in any other part of the world, and generally are rising at a lesser rate than the cost of living. Technically, the U.S. air transportation system is the most advanced in the world.

The apparent basis for the deregulation effort lies primarily in 1) the unique and rather spectacular results achieved by intrastate carriers in California and Texas, and 2) a deep-seated, honest belief in the "free market" theory.

However, the suggested extrapolation of the California and Texas experiences to a national system is fallacious. Lower fares and increased market growth might well occur, but only in limited situations and possibly for limited periods. Other parts of the system would deteriorate.

What has become obscured in the current philosophical debate is the fact that the maintenance of a sound national air transportation system requires elements of free competition *and* government regulation.

Naturally, a basic conflict is inherent in this prescription. A competitive system with maximum use of the free market mechanism is required if the benefits of competitive innovation are to be optimized. On the other hand, if these benefits are to be other than temporary, and if it is desired to maintain a "transportation network," the government must monitor entry into the system. Complete freedom of entry without price regulation inevitably will result in a "system" comprising a few large carriers serving considerably fewer markets than are served today. Additionally, it will create price differentials difficult for the traveling public to understand.

A basic policy determination or reaffirmation is necessary to a resolution of the regulation/deregulation debate. The U.S. government must decide if it desires to continue to maintain a widespread network of airline services with required standards of service to cities of all sizes.

If this is the goal, it necessarily will require regulation and some route protection. Alternatively, if the national policy decrees that the decisions of the marketplace shall determine the scope and shape of the "system," then complete elimination of regulation is the answer.

Unfortunately, because of the stance adopted by the government agencies involved, the facts for making such a decision have not been assembled or analyzed so that a layman can make an intelligent judgment.

Under the assumption that, as a matter of policy, the U.S. opts for a national air transport system which binds the cities of this country together, it must be recognized that regulation cannot be avoided. At the same time, within the regulatory framework, competition and market mechanisms must be utilized as much as possible to insure the benefits of innovative competition. The current Federal Aviation Act in its present form is adequate for this purpose.

What is needed are regulators who recognize the need for regulation but who, because of their distrust of regulation, will take advantage of every opportunity to effectively use competition and the market to enhance the system. In other words, we need regulators who have the wisdom to blend both elements, without permitting one to obliterate the other.

A good illustration of this philosophy is the present program for subsidizing air service to small communities. Under this program, with minimum regulation, carriers are given all the incentives deemed by government to be essential for developing service to small communities, whether by reducing costs, increasing revenues or innovating with new equipment that produces better results.

This program contrasts sharply with the subsidy program in existence prior to 1961 when class rate subsidy was introduced. The old system lacked incentives, and the carriers were subjected to a considerable degree of bureaucratic control.

The cornerstone of the class rate subsidy program involves prescription of the necessary parameters to insure control but, once this is done, the allowance of complete freedom of action within these parameters. The government maintains control of the subsidy level by first deciding what cities are eligible for subsidy and then by limiting the payment to two daily roundtrips in any given market.

A more important constraint, however, operates without the need for government regulation. It is imposed by forces of the marketplace, because a significant amount of commercial revenue must be generated by the carrier in addition to the subsidy payment if the operation is to be maintained at a profit.

Thus the service provided to a community is responsive to its needs: if the commercial revenues fall below the amount needed to produce a profit between a pair of points, an incentive is provided to the airline to correct the situation. It may alter the service pattern, or the equipment, or the timing, or it may petition the Civil Aeronautics Board to discontinue service entirely if no need is demonstrated.

The marketplace, by its response, in effect votes on the carrier's performance rather than on an inflexible plan approved by the regulators. If the carrier develops more commercial revenue than was estimated, the excess is shared by the carrier and the government — insuring that the airline will not be penalized for its success.

The program requires minimum regulation and yet achieves what the government intends. It provides a city with the pattern of service it needs, rather than simply plugging the city into the national transportation system by a connection with the nearest hub. It provides carriers with incentives to develop traffic, without infringing on managerial initiative. The class rate program is a splendid example of government regulation and the market mechanism working together to produce maximum benefit to all.

It is ironic that the CAB proponents of deregulation, in a presentation to the Senate aviation subcommittee on March 21, 1977, have now proposed a new program which among other things would prescribe the number of flights, the types of equipment, the precise markets to be served and the time of day for each flight. A separate contract would be required for *each* city served by an air carrier.

The net result of this system would be to leave no room for market forces or for carrier management innovation. In addition, such a system could be accomplished only with a large bureaucracy. Thus, rather than decreasing the regulatory burden, the subsidy program proposed by the CAB would be a major step backward.

This is a good example of using detailed regulation, which is neither essential nor desirable, when the market mechanism will produce better results for the public, the carriers and the government.

THE SPECTRE OF DEREGULATION

Reprinted by Permission AIR TRANSPORT WORLD (May 77)

A guest editorial by Morten S. Beyer, president of Avmark Inc. and one of the industry's most astute appraisers of airline costs

The spectre of deregulation has cast a pall of doubt, uncertainty and fear over the U.S. airline industry.

Investor confidence has been seriously undermined. Financial institutions are cold to new debt or equity funding for modern aircraft. The aggressive and confident spirit that sparked the growth of the airline industry has been all but snuffed out. New high-technology aircraft promising greater economy, lower energy consumption and a quieter environment languish on manufacturers' drawing boards.

Promising the public a pie-in-the-sky 25% to 50% reduction in air fares, the deregulators (or reformers if one prefers to call them that) simply don't know what they are talking about. The basic premise they advance is that deregulation will bring lower fares to the public and higher load factors (and profits) to the airlines.

They repeatedly cite the example of Texas and California intrastate airlines which operate outside Civil Aeronautics Board control and which offer their passengers fares that are 35% to 50% lower than those of the CAB-regulated trunks and local service carriers. On the surface their case makes sense: If PSA, Air California and Southwest can do it, why can't TWA, United or Allegheny?

This simplistic comparison has been eagerly and unquestioningly accepted as gospel by the advocates of deregulation. It has been rejected equally emphatically as economic heresy by the airline industry.

No one has bothered to determine the answers to two basic questions:

- How do the unregulated intrastate carriers do it?
- Can the trunks and locals

achieve the same fares and profits if they are deregulated?

Avmark Inc. has just completed a comparison of the operations of two of the major unregulated intrastate carriers, Air California and Southwest, with those of the regulated local and Hawaiian carriers.

The operations of Southwest and Air California are not dissimilar to those of the local and Hawaiian airlines. True, the former do not have the responsibility of serving small communities, but the locals receive some $75 million in subsidy annually to finance this service.

Both groups of carriers use the same aircraft types — Boeing 737s and McDonnell Douglas DC-9s. Fuel, maintenance and financing costs are comparable, as are expenses for landing fees and airport charges. Both groups provide ticketing and reservations, and offer minimal inflight service. Air California is strongly unionized. So why the big discrepancy in operating costs and fares?

There are the three major areas of difference between the two groups of carriers:

- The intrastate operators install 15-20 more seats than the locals in similar aircraft.
- The intrastates have average load factors of 68.75% compared to 53.45% for the locals.
- Employee wages on the intrastates are 19.3% lower while productivity is 40% higher.

These factors combine to give the intrastates a seat-mile cost of 4.33¢ compared to 6.47¢ for the locals. Revenue passenger-mile cost differs even more dramatically — 6.29¢ for the intrastates vs. 12.10¢ for the locals.

Significantly, aircraft-mile cost for the two groups is substantially the same: $4.92 for the intrastates, $5.30 for the locals. By what alchemy will deregulation bring about major reductions in costs for presently regulated operators?

Seating Capacity

The trunks and locals could, if they wanted, install as many seats in their aircraft as the intrastates — or more. Aloha and Hawaiian have 115 seats in their DC-9-30s and 737-200s, the same number as Air California and Southwest. The locals, on the other hand, configure their DC-9s and 737s with between 94 and 104 seats. Piedmont even uses only five-abreast seating in its 737-200s — and probably is the only one in the world to do so.

The locals feel these low densities are necessary to meet trunkline competition. Their reasoning is condoned by CAB's economists in subsidy and ratemaking computations. The locals aggressively promote their "Wide Ride" and "First-Class Legroom" in their advertising.

But DC-9-30s and 737-200s are certificated for up to 125 or 130 passengers, and are operated in this configuration in Europe and other parts of the world.

First-Class Must Go!

Similarly, the trunks generally utilize space that would accommodate 30 or more seats on a Boeing 727-200 (and up to 120-150 seats on a 747, DC-10 or L-1011) for such amenities as first-class sections, coatrooms, carry-on baggage racks, galleys and so forth. To match intrastate seating densities, these amenities would have to go — particularly first-class sections, which contributed more than $1.5 billion to trunk revenues in 1976.

But to the extent that the regulated carriers can see their way clear to increase seating densities, costs per seat-mile can be comparably reduced.

Load Factor

The reformers seem to believe that the magic wand of deregulation will somehow result in a 25% jump in load factor. But nothing could be further from the truth in the highly competitive markets of the locals and trunks. History has shown that increased competition brings *lower* load factors, not higher.

The highest load factors in the regulated part of the airline industry are achieved by the Hawaiian carriers. These load factors, ranging in the mid-60s, are achieved because Aloha and Hawaiian share their market and practice an oligopolistic restraint in their scheduling. Neither seeks to put the other out of business.

In the California and Texas intrastate markets, PSA, Air California and Southwest have what amounts to monopolies, carrying virtually all of the local traffic in the markets they serve. The California Public Utilities Commission, which regulates that state's intrastate carriers, has been careful to protect them from each other, and trunks and locals have long since abandoned more than token competition.

Prior to CAB sanction of capacity limitation agreements, the trunks drove load factors in the major multicompetitor transcontinental markets below 40%. Today we are witnessing the start of a new low-fare battle among American, United and TWA aimed at filling excessive numbers of empty seats on these same routes.

To the degree that deregulation will create more competition and bring new entrants into the marketplace, history tells us to expect lower load factors — at least until the weakest carriers are driven out and schedule restraint (if not monopoly) takes over.

Wages and Productivity

Wages and related costs comprise 40% of the scheduled airlines' expenses. These carriers are all heavily unionized, and the individual wages of airline employees are higher than those in any other major industry in the U.S.

Larger, more efficient aircraft, coupled with fare increases, have blunted the impact of rising wage scales and increasingly restrictive union work rules.

But the fact remains that the major economic advantage enjoyed by the relatively new intrastates arises from their lower wage scales and the higher productivity of their employees. While locals as a group pay their employees an average of $18,533 a year, the intrastates pay only $14,882. A pilot for Southwest flies 775 hours a year, while his counterpart at Allegheny flies only 579 hours. A Hughes Airwest flight attendant serves 6805 passengers a year while an Air California attendant serves 12,554. All of these figures are as of mid-1976.

In summary, Avmark's analysis indicates that intrastate airline employees produced 88% more revenue ton-miles, handled 101.7% more passengers and serviced 56.8% more weighted departures than their local service counterparts.

Some of this higher productivity is due to route structure and equipment as well as management effectiveness. The intrastates *must* operate at this level of efficiency in order to survive.

However, the bulk of the difference comes from labor contract conditions.

The only trunk whose labor costs approach those of the intrastates is Northwest Airlines. In 1975, 29.71% of Northwest's total operating expense was for labor, a figure which compares favorably with the 28.10% estimated for Southwest. If the other trunks emulated Northwest's example, they would have approximately 67,000 fewer employees and their annual wage costs would be reduced by $1.4 billion.

How To Reduce Costs

What would happen to the costs of the local service airlines if they were to match the seating density, load factors and wages of Air California or Southwest?

Assuming that revenue passenger-miles stayed the same and that aircraft miles, employment and wages were reduced in order to bring about desired improvements in costs, Avmark has concluded that:

- Increased seating density would reduce local service costs 10%, from 12.10¢ per passenger-mile to 10.89¢.
- An increase to intrastate levels in load factors would cut costs per passenger-mile 13.3%, to 10.49¢.
- Higher density seating and higher load factors together would reduce operating costs by a total of $298 million for the local industry, dropping passenger-mile cost to 9.62¢.
- A 40% improvement in employee productivity among the locals would eliminate 8541 of the present 29,975 jobs.
- Reducing wages to intrastate levels would cut average salary and related costs by $4381 per employee. Total cost reduction for the local industry would be $276.4 million, and passenger-mile costs would drop 20%.

The lower wage costs which give the intrastate carriers such a large economic advantage really have nothing to do with regulation or its absence.

Additional seats and/or higher load factors must assume either a sharp curtailment of scheduled miles flown or a substantial increase in passengers to fill the added capacity.

Most of this won't happen. Higher seating density can reduce seat-mile costs, and lower fares will probably stimulate additional traffic. But some increase in airline income is needed just to stay ahead of inflation, higher wages and increased fuel costs.

Improved load factors can be realized only through more capacity limitations and more monopoly markets — an obvious anathema even in today's regulated industry.

Finally, it is grossly unrealistic to expect deregulation to turn back the clock. Airline labor unions, their members, and managements alike would be loathe to give up hard-won wage benefits and work rules to satisfy reformers' dreams of a deregulated industry. Not surprisingly, labor is solidly opposed to deregulation.

Although today the risk seems minimal that knowledgeable investors will put large amounts of financing at the disposal of new entrants into the airline business, it is possible that the country will again be swept by the "airline fever" that emerged after World War II. In this case, new deregulated airlines could spring up like mushrooms to challenge incumbents with low-cost, non-union operations.

Half of the fare cuts promised by reformers would be possible only if wages were to be rolled back. The other half is dependent on wildly unrealistic dreams of higher load factors, as well as on the increased seating density which already is needed just to absorb continually rising costs.

It is nonsense to think that deregulation will work.

DOES BERMUDA II DELIVER?

Reprinted by Permission
FLIGHT INTERNATIONAL (July 23,1977)

"In return for a few significant gains, British air transport will be substantially worse off because some of the broad groundrules for international air transport, enshrined in Bermuda 1, are undermined. Because of the importance of Bermuda 2 to every nation's air services, and most of all to our own, there ought to be no misunderstandings or glossing over of the factual position,"writes Sir Peter Masefield in a letter to *The Times* .

Sir Peter sees the more important gains of Bermuda 2 as the opening up of new North Atlantic routes; some rationalisation of fifth-freedom rights into Europe within five years; a move towards a form of capacity control; and new ways of agreeing fares and rates. On the minus side, he says, there is a disturbing retreat from the principle of fair and equal opportunity for the two nation's carriers which will disadvantage Britain and the many small nations likely to use Bermuda 2 as a guide.

He cites four causes for concern:

- Change-of-gauge rights (whereby a US airline can, if the destination country approves, carry US-originating traffic onwards from London in smaller aircraft) will allow more US services into London than UK operators can justify to the US.
- The value of the three-year UK-exclusive non-stop London-Houston B.CAL route is devalued by a parallel US all-cargo service and the existence of a US-exclusive non-stop service from nearby Dallas/Fort Worth. In addition, a US carrier will be granted a perpetual monopoly service to London from an as yet unnamed US gateway.
- Skytrain's economics will be prejudiced if the US carriers flying out of London to New York are allowed to offer low-cost standby/fill-up fares on scheduled services.
- Notable by its absence is any significant reference to charter services, one of the causes of airline losses on the North Atlantic.

ON TO JAPAN: U.K. BILATERAL

by Joan M. Feldman

Reprinted by Permission
AIR TRANSPORT WORLD (September 77)

By all accounts, the U.S. came out on the short end of the so-called competitive stick in the recent bilateral negotiations with Britain.

Behind the resulting recriminations and posturings lie two questions: 1) How much of what one government official called "an organized retreat" was inevitable; and 2) what the "retreat" portends for the future of U.S. international air policy.

U.S. devotion to free enterprise internationally, whether in aviation or in other foreign trade areas, has not been what it's cracked up to be for years. As far as the U.K. goes, "we haven't had pure Bermuda in three or four years," said one agency official, referring to the National Airlines capacity agreements on the Miami-London route, for example. Elsewhere in the world, such as in Mexico and Japan, capacity determinations have hardly been unfettered. There are side agreements to those respective bilaterals to prove that.

Moreover, the International Air Transportation Policy Statement issued last autumn, although a compromise and hardly a gem of clarity in expression, put down in black and white, for the first time, the fact that the U.S. just might consider — if pushed — such hitherto unmentionables as capacity controls and a hold on multiple designation.

Well the U.S. was pushed by the U.K. — and it went over the edge. For the first time in the main body of a bilateral, a formula for capacity control and the ultimate right to designate multiple flag carriers in a market were conceded.

Protectionism, of course, is hardly revolutionary for the United States. The OPEC cartel participation by U.S. oil companies, cargo preference, and orderly marketing agreements for various goods such as steel and textiles are only the more recent examples of protective stances taken by the U.S. government.

DID U.S. GIVE TOO MUCH?

And the U.S. delegation went into the bilateral talks knowing it would have to give something. Not, of course, the 50-50 balance of benefits, total elimination of multiple designation and total predetermination that the U.K. initially threw out on the table. Those demands are considered to be too ludicrous even for comparison with what the U.K. eventually received. But some of the fifth freedom rights were bound to go, as were some of the multiple designation points, one member of the government team explained.

What sets critics on edge is how *much* the U.S. gave: most of its fifth freedom rights out of London, all but two multiple designation points, all-cargo beyond rights in the Pacific, a written formula for capacity review, a concession to "work toward" a multilateral agreement on charters with European nations, formidable

Canadian and Latin American rights to the British. And all this was without adequate compensation.

The portent thus is for difficult times in the future if every U.S. bilateral partner wants the equivalent for itself — plus whatever special circumstances might dictate.

Apparently, the fault does not lie in the lack of a strategy. There was one, at least in the initial stages of the talks following the British denunciation of the Bermuda agreement in June 1976. But whatever strategy existed prior to special ambassador Alan S. Boyd's accession to the throne of chief negotiator was tossed out over the Atlantic in the shuttle of trips between March, when Boyd came aboard, and June 22, when the draft pact was initialed. What happened?

Boyd insists that the ultimate threat, withdrawal of U.K. airline permits for U.S. service and probable retaliation, just would not have produced what he thinks he got: "a fair agreement" which only the future will prove, as he told a Washington luncheon audience.

For once though, most of the U.S. airlines were ready to go to the mat if need be on cessation of service. "We would have lost tons," said one Pan Am representative on the scene, but the carrier had already started moving equipment and personnel to Amsterdam in case a switch was required.

Cessation, apparently, was something that Boyd just could not bring himself to impose upon his British counterpart, Patrick Shovelton, given all the possible consequences for England's already teetering economy and government. Although U.S. delegation members, both government and industry, had heard a lot of advance billing on Boyd's talents as a negotiator, some apparently feel he bent over backward to avoid embarrassment to Shovelton.

Thus the great advantage for the U.S. that Boyd was supposed to bring — avoiding the need to negotiate with U.K.'s former top representative and the ability to speak through one voice, rather than through a myriad of agency and airline voices — were more than counterbalanced by either Boyd's concern for Shovelton's future or instructions from President Carter to get an agreement no matter what in order not to add to England's problems.

But there are those who feel that while England may indeed be "a sovereign nation" (one of the excuses used for the U.S.'s inability to drive a harder bargain), the U.S. is no less so, and one with a lot more bargaining chips (traffic, markets) that were used in London.

JAPANESE TALKS ARE NARROWER

There is still time to use those chips, however. In Japan, for example, where that government is ready to cash in on the U.K. winnings at the roulette table.

The Japanese talks, which get down to the nitty-gritty in October with an anticipated completion by the end of the year, are much more narrow than were the British. The latter denounced their treaty with the U.S. and the delegates started from scratch. (That does not mean the new U.K. treaty should be considered "Bermuda II." The end result is not seen as either the lasting product its predecessor was, or as worthy of serving as a model for U.S. air rights talks the world over.) The Japanese have not denounced their civil air agreement, but rather are seeking a "revision" of something they see as a remnant of the post-World War II occupation.

In the initial sparring, the U.S. held its own with the Japanese. It had demanded two things from Japan prior to getting down to revision talks: 1) acceptance of Continental Airlines in the Saipan-Tokyo market; 2) release to the U.S. supplementals of all the slots at Haneda Airport previously agreed to but not granted.

So Continental will operate daily service Saipan-Tokyo, but at *ATW* presstime World Airways and the Japanese were sitting down informally to work out slot assignments for congested Haneda.

The supplementals were not so sure they hadn't, once again, been sacrificed at the aviation altar as in London, since the slots weren't nailed down prior to the beginning of revision talks. But the U.S. State Department insists that this bilateral will, without doubt, include supplementals as part of the package, unlike the U.K. pact where supplemental services are being handled in a separate charter bilateral. That, in fact, is "the big issue" in Japan, *ATW* was told.

In return, the Japanese will want plenty. For example, U.S. officials figure Japan regards the U.K. capacity formulas as "too liberal." They want "as much capacity control as they can get away with," one official said, although he added, this is something "they effectively have now" with the congestion problem at Haneda. Japan also is seeking numerous beyond rights to Mexico and South America, and probably will chop down the U.S.'s current liberal beyond service to get them.

The State Department is trying to figure out a way to do something about the Pacific fare situation during the talks. The U.S. Transportation Secretary's comments notwithstanding, any talk of low fares resulting from the British act is "a bunch of crock," said one who ought to know.

In Japan, things are different. The U.S. would like lower fares, particularly if the Japanese insist on a very restrictive capacity stance. How the U.S. would accomplish this "if we don't get rid of IATA" is something else again.

Such an end run might become a negotiating tactic for the Japanese talks for another reason. For there is a firm conviction in the U.S. that that country must get more in future bilateral trades than it did in the U.K.

This is not to say that the U.S. is not prepared to give up certain items. It knows that the post-World War II drive by other nations worldwide to create sophisticated air service systems has now succeeded. But, it is felt, the U.S. flag system would be decimated if the U.K. approach — meaning no *quid* for each *quo* — is continued.

Significantly, the seeming death wish of the Carter Administration to appoint yet another "special ambassador" has been sidetracked. Instead a high State Department official and CAB will once again lead the U.S. delegation. Most importantly, the U.S.'s top aviation negotiator, Michael Styles, will be back in harness.

Someone told *ATW* in August that the U.K. treaty "is not the end of the world." There are plusses, such as blind sector rights in Europe, all-cargo flexibility on the Atlantic, and Caribbean rights. But, if the U.S. lets much more of the accommodating attitude observed in London slip through, the trend to protectionism and restriction certainly will be accelerated.

TOKYO: THE BUSTLE . . . AND THE SILENCE

by Warren Goodman

While 21 million passengers a year crowd and jostle their way through Tokyo/Haneda's inadequate facilities, the $1,000 million Narita Airport stands empty and unused six years after completion. Japanese officials say it will open this year — but as we go to press fresh protest riots break out.

Reprinted by Permission AIRPORTS INTERNATIONAL (June/July 77)

Milling crowds at Haneda — and the emptiness of Narita. Congestion is a daily scene at the one Tokyo Airport, while the other lies ready and waiting — as it has done for six years.

"It was the best of times, it was the worst of times, it was the age of wisdom, it was the age of foolishness, it was the epoch of belief, it was the epoch of incredulity."

THOSE WORDS describing London and Paris in 1775 seem to summarize the airport situation in Tokyo in 1977. It is the best of times because air traffic has doubled in six years, growing from 10.7 million passengers in 1970 to 21 million in 1976 (6.8) million domestic and 14.2 million international).

It is the worst of times because of Tokyo's air traffic must be handled at Haneda (officially called Tokyo International Airport). The airport has a capacity of 460 movements a day or 167,900 a year, and has handled more than that every year since 1971. It is probably the world's most congested airport.

Tokyo has another airport at Narita (officially called New Tokyo International Airport). It was completed in 1972 and has not yet been opened, giving it the undisputed title of world's most under-utilized airport

CONGESTION AT HANEDA

In 1975, Haneda was sixth among the world's airports in number of passengers, ranking just behind Kennedy International Airport in New York. In 1976, Haneda's 21,008,000 pasengers put it almost even with Kennedy which handled 21,033,000.

But Haneda has only two runways, Kennedy, has four; Haneda has only two passenger terminals, Kennedy has eleven; Haneda has a 2300 hr to 0600 curfew, Kennedy operates 24 hours a day; and Haneda has a total area of just 427 hectares (1,055 acres), where Kennedy is more than five times as large.

It is therefore hardly surprising that Haneda is so congested. Arriving by air, one immediately notices that almost all the available tarmac between the passenger terminals and runway 15-33 is occupied by aircraft and an incredible array of ramp vehicles.

There are 72 "spots" for loading an unloading passenger planes, but only 22 of them are adjacent to the passenger terminal complex. The other 50 are arranged in two straight lines along former runway 15R-33L, deactivated in 1970.

Many of the aircraft parked at these remote spots are wide-bodies, so there is a constant procession of buses (up to six per large aircraft), crew cars, baggage and cargo carts, caterers' vans, fuelers and other service vehicles shuttling and forth. The sign of the centre section of the International Arrivals Terminal says "Tokyo International Airport", but the cluster of vehicles on the ramp side makes it look more like a bus terminal than an airport.

Inside the passenger terminals, there seem to be people occupying every available square foot of floor space and every available seat. Arriving pasengers are crowded eight and ten deep around the baggage claim counters. Departing passengers form queues from every ticket counter position to the opposite wall.

Departing international passengers form a six-abreast queue to go through immigration and Customs clearances at a single point. The queue often extends for 200 feet or more and a passenger is lucky to get through in 20 or 30 minutes. Some airlines advise passengers to allow at least an hour.

About the only part of Haneda which does not look unusually crowded to one familiar wth US and European airports is the road side of the terminal, despite the fact that there is only a single-level roadway. One reason for this is the availability of a monorail from the passenger terminal to downtown Tokyo. It runs on five-to seven minute headways and the fare is 280 yen (about $1) as against 700 yen ($2.50) for the bus service to and from major hotels or about 2,500 yen ($9) for a taxi. Recent surveys show that about three quarters of Haneda's terminating domestic passengers use the monorail.

Another reason for the comparative lack of roadway and kerbside congestion is the small number Tokyo passengers who use private cars. Many of the Japanese nationals who fly, especially on international flights, go in large groups and tend to use buses. Very few foreign nationals drive because of the difficulty of coping with Japanese directional signing.

As a result, the number of rental cars and public parking spaces is extremely low compared with airports in the US and Europe. Haneda has space for only 1,400 cars (about a tenth the space provided at Kennedy) and parking charges are low (free parking for the first 30 minutes and 50 yen — less than 20 cents — for each 30 minutes thereafter).

Signs of protest: riot police (top) form up in the deserted car park for yet another confrontation with protesters . . . while a farmer steadfastly goes on growing carrots (his plot is behind the tower, above) where the airport access road should run.

OPERATION OF HANEDA

Haneda Airport was constructed in 1931, but was operated and greatly expanded by the US Occupation Forces from 1945 to 1952. It returned to public use in July 1952 and became the responsibility of the Civil Aviation Bureau of the Ministry of Transport.

A passenger terminal was urgently needed but the financial position of the country at that time and the comparatively low priority of air travel in the minds of most people made it financially and politically impossible to appropriate the necessary public funds. The financing, construction and operation of the passenger terminal was therefore delegated to a private company, Japan Airport Terminal Co Ltd.

The terminal was opened in 1955 and expanded to its present configuration, with separate international and domestic wings, in 1971. JATCO continues to own and operate the cargo terminals and the car park as well as the passenger terminals — but not the aircraft gate positions adjacent to the building or any other portion of the aircraft side.

JATCO receives its revenues from rentals for ticket counters, airline offices and other leased areas and from user charges on the common-use areas. The charge for international departures is 11,000 yen (about $40) per aircraft.

By agreement among the airlines, charges for international arrivals and for all domestic movements are based on actual cost and apportioned among the airlines on a "10-45-45" formula. Ten per cent is shared equally among the airlines, 45 per cent is apportioned by aircraft movements, and 45 per cent is apportioned by number of passengers carried.

NO EXPANSION FEASIBLE

Shigehiro Kusu of JATCO told *Airports International* that the terminal buildings have reached the maximum size feasible within the present boundaries of Haneda. He said that JATCO had considered double-decking the terminal roadway and adding additional levels to the car park, but felt that the additional congestion could not be tolerated as long as Narita remains closed and Haneda is so crowded. Paradoxically, after Narita opens and the construction becomes possible, it may not be needed.

Ichiro Sakamoto, Chief of the Air Safety Division of the CAB's Tokyo International Airport Administration Office, told *Airports International* that the 440 daily slots available for scheduled flights have been allocated since 1970.

The capacity of Haneda is 460 movements a day, with a further limit of 86 movements for any three-hour period or 34 movements for any hour. The Ministry of Transport reserves 20 of the slots each day for check flights, VIP flights (such as visiting heads of state), and non-scheduled flights. The remaining slots have been divided between 148 international and 292 domestic flights a day.

Therefore recent increases in passenger traffic, and any future increases, can be attained only by using larger aircraft or increasing load factors.

The only way to expand the capacity of Haneda would be to enlarge the airport beyond its present boundaries and build additional runways in Tokyo Bay, which borders the airport on almost three sides. The Ministry of Transport has considered that possibility but has shelved all such planning or discussion so as not to generate additional controversy while the battle over the opening of Narita still rages.

Asked what Haneda can do to cope with rising traffic until Narita opens, or if Narita never opens, Mr Sakamoto says: "The CAB and the Terminal Company are already doing all that can be done, so things can't get any better. But Haneda is operating at 100 per cent of capacity and we will not permit any additional flights, so things can't get any worse."

Despite Mr Sakamoto's analysis, it is hard to see how conditions at Haneda can fail to get worse if Narita does not open. Tokyo's traffic demand is expected to grow from 21 million passengers in 1976 to 35.6 million in 1980 and 61 million in 1985. Air cargo, which has more than doubled since 1970 is expected to grow from 447,700 tonnes in 1976 to over a million tonnes in 1980 and 2.2 million in 1985.

NARITA IS THE ANSWER

The need for a new airport to serve Tokyo has therefore long been obvious. The Japanese legislature designated the site about 60 km (38 miles) west of Tokyo, in 1966, and created the New Tokyo Airport Corporation (since renamed New Tokyo Airport Authority) to plan and construct it.

The first phase of the airport includes a 4,000 m (13,000 ft) runway, a 32-gate passenger terminal, a large hangar, cargo buildings, an operations administration building, a commissary, and a 150-room "guest house" for flight crews and in-transit passengers. All of these facilities were completed in time to open the airport in 1972.

The airport did not open as scheduled, and has not opened yet, because of a number of problems created by an anti-airport coalition composed of farmers and other landowners in the Narita area, university students, and environmentalists.

The major obstacle to the opening of the airport arose when officials in Chiba City on Tokyo Bay refused to allow the construction of a vital portion of the 44 km (28 mile) pipeline which the Airport Authority had planned to bring fuel to the airport.

At the end of 1975, the Authority began working on a plan to bring the fuel in through the Pacific Ocean port of Kashima. That plan was also blocked by local opposition and the Authority is now working on a third plan which would require bringing the fuel to Chiba and Kashima by tanker or barge, transferring it to dockside storage tanks, transferring it later to rail tank cars, rather than pipeline, to take it to the airport boundary, and then transferring it to the underground distribution system already in place on the airport.

An Airport Authority official told *Airports International* that he did not know exactly what all this handling

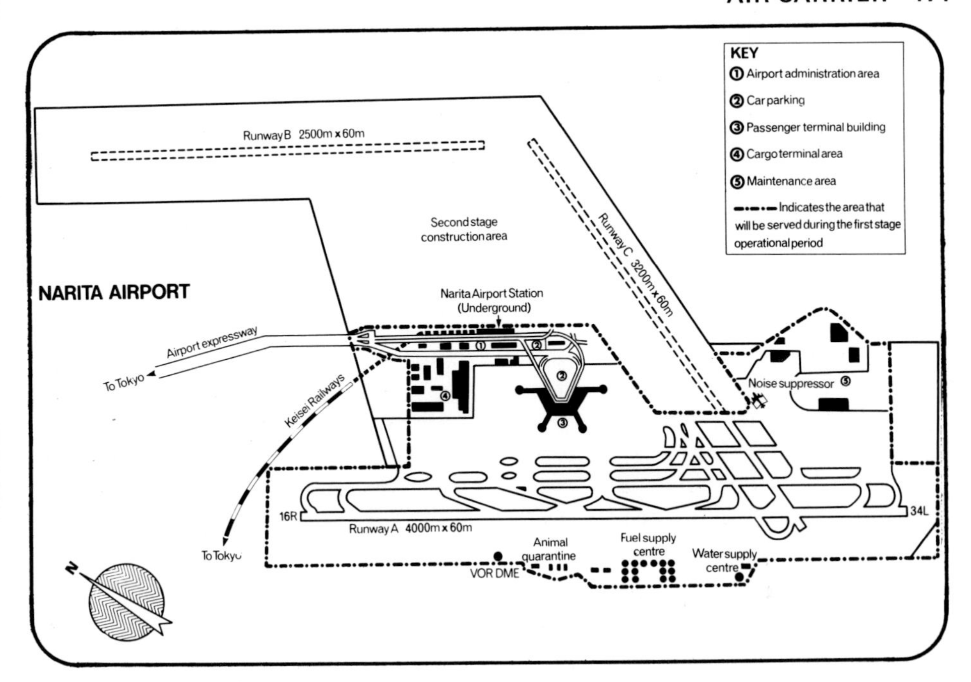

would cost or what it would mean in terms of charges to the airlines, but said: "It will be a very high price."

Meanwhile, the Airport Authority has until recently been wrestling with another problem which would have prevented the opening of the airport even if the fuel were available. In 1971, the airport's immediate neighbours erected a steel tower 32 m (105 ft) high on private property directly in line with Narita's only runway and only 1,100 m (3,600 ft) southeast of the threshold. A year later they put up a second tower 60 m (200 ft) high and only 700m (2,300 ft) from the runway.

The towers were constantly occupied by volunteer sentries who hurled epithets at the police or other goverment officials over loudspeakers and sounded an alarm to summon others to block the way with their bodies if any attempt was made to dismantle the towers.

Another problem which might not prevent the opening of the airport, but would certainly interfere with its access, has been created by a farmer who owns a carrot patch and has refused to give it up. The carrot patch blocks the otherwise-complete four-lane divided roadway which is the main entrance to Narita. The two left lanes are paved up to the patch and beyond it, leaving a gap of about 200 feet.

NARITA IS STILL A QUESTION

The Airport Authority has invested about 270,000 million yen (about $1,000 million) in Narita, about 20 per cent of which came from the Federal Goverment. The remainder was borrowed by the Authority, which is now borrowing additional funds to pay the interest on the outstanding loans and to meet the mounting costs of maintaining the unused airport.

The Authority has also found it necessary to make changes as airport technology and procedures have changed. Two years ago the number of gates at the passenger terminal was reduced from 32 to 28 in order to have 20 wide-body gates instead of only 12. Early this year, the Authority began to make changes in the departure halls to provide for the security inspections which were not required when the airport was designed.

Airlines and other tenants have invested an additional 100,000 million yen. Japan Air Lines, the largest investor other than the Authority itself, is now spending about 270 million yen a month on maintenance. Its facilities which would employ about 2,000 people when fully operational, include a larger hangar, a hotel, and one of the world's largest and most automated cargo terminals.

The only part of this which JAL can use until aircraft land at Narita is the cargo control computer for its entire system. It has been in operation since 1976, employs about 100 people, and is the only thing at Narita which is in operation.

To return to the words of Dickens which opened this article, it might well be called "the age of wisdom" when the decision was made in 1966 to build Narita. But it might well be considered "the age of foolishness" to spend close to 380,000 million yen to plan and construct an airport which could not be opened five years after it was completed.

It certainly seemed "The epoch of belief" when officials of the New Tokyo Airport Authority and of the Ministry of Transport told *Airports International* in April 1977 that the problems of getting fuel to the airport would be solved in three months, that they would go to court and get permission to take the steel towers down in a few months, and that Narita Airport would be opened by the end of the year.

It was "the epoch of incredulity" for me because I had been told the same thing by some of the same officials a year and a half earlier.

But two weeks later, as this article was going to press, the court order was issued. Acting with unusual speed the authorities surprised the protesters by moving in at 0300 the next morning.

Cranes, demolition crews and police took the two steel towers down by 0800. About 200 people who tried to obstruct the demolition were routed with tear gas and 14 of the protesters were arrested.

The next day 800 protesters stormed Narita, threw firebombs and rocks, and burned fuel-soaked tyres. Another 25 protesters were arrested, but the airport runway received its first aircraft since its construction five years earlier — and the Goverment announced its firm intention to open Narita before the end of the year.

PILOT'S VIEW: NOISE ABATEMENT TUG OF WAR

Has flight safety been compromised in the interest of noise reduction? The author urges a "get-clean-quick" method as an alternative to the "get-high-quick" philosophy.

by Captain D. B. Peat

Reprinted by Permission
FLIGHT OPERATIONS (January 1977)

It is difficult to believe some of the concessions the aviation industry has made in recent years to satisfy complaints about airport noise from aircraft fight operations — that is if you view some of the flight procedures now in effect in comparison with those used prior to the advent of jet aircraft. Back then sound principles of safety were employed in all procedures and safety considerations came first. It was still considered prudent to take off and land into the wind, for example. Who would have dreamed that pilots today would be forced to make downwind landings in bad weather — *with opposite direction takeoffs being conducted on parallel runways at the same time* as is the recent situation at Los Angeles? It would have been inconceivable way back then that we pilots, as a group, would go along with such a scheme.

Likewise it would have been difficult to conceive that a noise abatement takeoff procedure would be adopted for the industry in 1972 that specified a climb to 3000 ft above ground (AG) at V2 + 10 (which is 1.3 Vs and also certified minimum maneuvering speed). Compared with the previous takeoff procedure, which specified acceleration and flap retraction at 1000 ft AG, the climb to 3000 AG at V2 + 10 tacked on an additional 2000 ft of climb while on the ragged edge of a stall — it increased by a factor of 2/3 our exposure to the elements of wind shear, turbulence and engine failure. In some cases certain carriers went beyond the industry's procedure and specified that the noise abatement climb be flown at V-2 (which is 1.2 Vs or only 20 percent above stalling speed and which is defined by FAR as "takeoff safety speed" intended to be used in the event of an engine failure to set up out of the trees, certainly not an everyday standard operating procedure). There were even some cases where the 15-degree body angle (pitch) limit, which was established with the advent of the jet airplane, was eliminated. All of this is an effort to get up quicker or, in the downwind landing case, to fly over unpopulated areas, to reduce noise.

One may ask, why has safety been compromised in the interest of noise abatement? No one wants to have an accident. I suggest that there are several reasons. From industry's point of view, the following are a few examples: (1) the ever-increasing pressures by the environmentalists have resulted in lawsuits against airport operators from which have come unfavorable court decisions; (2) airport operators' threats of curfews and other restrictions to flight operations that would have an adverse economic impact; (3) fear of future adverse governmental regulations and (4) aviation's desire to establish a "good neighbor image" for more favorable public relations.

Some believe there are operational reasons why we pilots have allowed ourselves to be forced into this situation, for example: (1) the jet engine is more reliable and the probability and actual frequency of engine failure are less than we expected with piston engines and (2) the jet airplane has a more favorable thrust/weight ratio which results in fantastic performance when compared to piston aircraft. This performance tends to lull us into a state of blind obedience to flying "by the numbers" simply because they are "in the book."

FACTS FOR PILOTS — NOT PRESSURE

However, here are certan basic aerodynamic facts that all pilots should keep in mind, regardless of the favorable thrust/weight ratios or airspeed computations. This is especially true in view of the aforementioned factors that have affected the development of current procedures. For example, consider a takeoff in our jet aircraft at maximum takeoff gross weight with an initial climb airspeed V2 + 10 kts which would require a body angle (pitch attitude) of let's say 10 degrees. Now, compare a takeoff at a very light gross weight, with a V2 + 10 kts climb which would compute to be a slower airspeed and would require a pitch attitude of let's say 24 degrees.

If, in either of the above configurations, a turbulence encounter was experienced, whether it be mechanical, heavy jet wake or wind shear, we would find ourselves in a much more precarious predicament at the very light weight. This is due to the increased time it would take to decrease the pitch angle while simultaneously experiencing a *rapid* loss of airspeed. It should be obvious that our exposure to a low-altitude stall is greater when flying at the higher pitch attitude. It should also be kept in mind that with a tapered wing, as a stall progresses, the tip stalls first, and when that tapered wing is swept back and a stall is approached, the aircraft tends to pitch up further.

The foregoing seems to suggest that the original body angle limit of 15 degrees was and still remains a sound policy. Further, if our aircraft is light enough to require a body angle of more than 15 degrees to maintain an airspeed of V2 + 10 kts, we will probably be climbing like a hot air balloon at a body angle of 15 degrees, while accepting whatever increased airspeed that is obtained, and we would still be making less noise than at the heavy gross weights.

The argument that a steep climb to 3,000 ft AG produces less noise (than 1,000 ft), therefore justifying whatever increased exposure to a turbulence-induced stall, is not significantly valid. Many noise tests have been conducted, and while agreement on the interpretation of the results does not exist among all the parties involved, it is clear that there is a better way. The alternative to the "get-high-quick" philosophy is to "get-clean-quick" (flaps up at 1,000 ft and then reduce thrust). The advantages to the latter are many and they include, in addition to the greatly improved stall margin: (1) less overall noise and (2) more economy — due to less time at high engine thrust settings, there is a significant reduction in fuel consumption and high engine stresses, therefore less maintenance cost.

The suggested procedure: (1) Climb to 1,000 at V2 + 10 kts, 15-degree body angle max (complies with obstruction clearance criteria); (2) accelerate airspeed and retract flaps to 0 degrees (on speed schedule); (3) climb at 0-degree flap, minimum maneuvering speed; (4) reduce thrust to "quiet EPR" (i.e. computed to produce a minimum climb gradient, as prescribed in FAR certification criteria, in the event of an engine failure); (5) at some altitude between 3,000 ft AG (if over non-noise sensitive area) or 5,000 ft AG (if over noise sensitive area), reset climb thrust and accelerate to 250 kts.

The advantages of the above-suggested procedure far outweigh the one small disadvantage. Test results have shown that there is a *slight* increase in noise under the flight path *close in* to the runway (1/2 to 3/4 mile) due to the lower altitude obtained while accelerating airspeed and retracting flap. However, when comparing the significant reduction in noise over the remainder of the takeoff climb noise footprint area (approximately 10 PNDB less noise), it is clear that his procedure is the most desirable from the environmental point of view.

Implementation of this procedure would greatly reduce our exposure to a low-altitude stall, such as one that did occur involving a transport category aircraft which resulted in a near accident. It would also contribute to a more economical operation for our employers while creating the least possible amount of noise above our airport neighbors.

Hopefully we will soon see an industry-wide adoption of this philosophy for a standardized procedure. This would be for the best interest of everybody concerned.

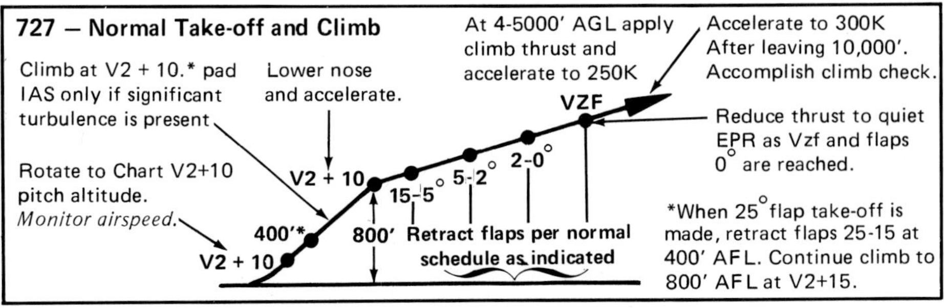

Air Line Pilots Association noise abatement climb procedure is charted here for B-727. In circumstances requiring higher thrust levels for a particular SID or high terrain, the first thrust reduction after takeoff shall be to climb EPR.

CONCORDE'S FIRST YEAR

by Bill Sweetman
Edited for Space

Reprinted by Permission
FLIGHT INTERNATIONAL
(February 12, 1977)

First years in service for any airliner are always untypical. The more so when the fleet is still in single figures and the route for which the vehicle was designed and the fleet ordered remains blocked by political and environmental troubles. Add to that an industrial dispute which effectively prevents deployment of half the total fleet on substitute services, and a heavy financial loss becomes inevitable.

Concorde has now become a pawn in a dispute involving local environmental groups (to which Concorde is a symbol as much as a source of noise nuisance), New York politicians, the Port Authority and the perennial antipathy between the State administration and Big Government in Washington.

In the absence of rights into New York — expected to absorb two round-trips per airline per day, which works out as the normal output of two aeroplanes for each operator — utilisation has, not surprisingly, been low. Air France flew 2,670 revenue hours in the first year with a fleet increasing from one to three aircraft. Once in service each aircraft averaged a rate around 1,190hr/-year. The British fleet, rising from one to four, achieved barely one hour per day. Target for both airlines on the full network is 2,750hr. Utilisation at less than half the expected annual rate is hardly the most realistic test of an airliner's operational, technical and commercial performance and potential.

Almost paradoxically, Concorde's operators and manufacturers are highly encouraged by the first year's operations in airline service. It has tended to confirm the predictions of Air France and British Airways that Concorde will make money if the load factor can be kept above 60 per cent (and experience on Paris-Rio, the only complete route to have been operating for the full 12 months, suggests that it can) and if the aircraft can achieve target utilisation. Technically and operationally Concorde has presented no extraordinary problems. The manufacturers are "92.5 per cent happy" with the performance of the aircraft.

Dispatch reliability is showing signs of settling at around 92.5-93 per cent (percentage of flights departing within 15min of schedule, excluding non-technical delays) across the fleet. "We're not satisfied with the present figure," comments BAC sales manager Concorde Ron Bailey. "It should be up to 97-98 per cent."

One effect of low utilisation has been a stretching out of the learning curves for last-minute trouble-shooting and repairs. Neither does low utilisation mean that the aircraft are "coddled," and always kept in peak condition. "We haven't actually had a flat battery," comments British Airways Concorde director Gordon Davidson, but he compares a British Airways Concorde to "a car that's been sitting in the garage too long." John Masters of the BAC sales team amplifies the point: "The worst thing you can do with most complex electronic systems is switch them off, leave them for 24 hr and switch them on again."

Air France has kept its figures slightly better than British Airways by having a non-flying flight engineer start the checklists three hours ahead of scheduled departure time instead of the usual one hour. But in the past month British Airways has copied Air France — the *entent Concordiale* is more than a public-relations front and this procedure is one of the many things which the airlines have learned from each other's experiences — and Gordon Davidson claims that delays have been dramatically cut down. The extra two hours between detection of a fault or a hypochondriac warning light is often enough to allow an LRU (line-replaceable unit) change, and a great many gremlins are exorcised simply by removing an LRU and putting it back again.

Concorde has entered service more smoothly than the last long-hauler introduced by British Airways and Air France, and compares reasonably well with the more recent wide-bodies. The 1975 endurance-flying program and the total 5,500hr of flight-testing played their part in identifying potential problems, although not all the modifications found necessary had been incorporated when service started.

Encouragingly, no major reliability problem has been identified. The engines have performed well, although their controls, like the other electronic systems, have been among the more common culprits when services have been held up. But Bailey of BAC stresses that there have been no major problems, and no one system stands out as being unacceptably failure-prone.

For Air France, M Jean Cremet confirms that Concorde maintainability has been "better than expected." No problem area has shown up in the Air France fleet, and the airline has reduced its estimates of maintenance cost for 1977 on the basis of this experience.

Even after a short first year in terms of hours flown, maintenance-check intervals are beginning to stretch out as they would for a subsonic aircraft. The first routine check, set initially at 75hr, will soon be moved out to 100hr.

Concorde operations probably benefited more than the engineering side from the 1975 endurance flights. The airline operating regime explored then differed slightly from that encountered in pure flight-testing, with more emphasis on mixed subsonic/supersonic operations. Airline-type operations also pointed out the need for accurate inertial navigation to set descent points, and this has been achieved in service. The airlines have also contributed significantly to the development of noise-abatement techniques.

What did emerge clearly from the first year's operations is that the Concorde seat is a highly marketable commodity even at first-class plus 20 per cent, rather more than what the two airlines regarded as the optimum towards the end of 1975. Paris-Rio is perhaps the most mature route; it has lasted the full year and, unlike London-Bahrain, it is complete. Cremet describes the present 62 per cent load factor as a "cruise level" and expects that frequency will be increased to match an anticipated traffic growth. Loads on Paris-Rio westbound flights are as high as could be expected, at 72 per cent, while Rio-Paris is running at 53 per cent. The strong preference for the westbound flight runs through all the Concorde operations, and is likely to put a ceiling on route load factors. On Paris-Rio-Paris 37 per cent of the traffic originates in South America and 20 per cent in France. Fourteen per cent is British, and Germany and Italy account for a total ten per cent.

London-Bahrain, the other year-old route, is still doing better than British Airways expected. Some 47 per cent of available seats have been sold, and a higher level would be an embarrassment on what is after all the first of three sectors of the London-Melbourne sector.

Air France launched the third route, Paris-Caracas, because there was enough traffic to pay the direct costs of the service; that is to say, Air France would lose less money serving Caracas than it would if its aircraft and crews were idle. Load factor has run at 36 per cent, but Cremet is hopeful that traffic will improve, basing his optimism on recent signs of pick-up in the subsonic market.

But the big success story has been the Washington service. "The Washington market itself is really quite small," says Gordon Davidson, "and a lot of the traffic is US Government and has to fly the flag." Despite that, British Airways has sold 90 per cent of its available seats. But a breakdown of the sales reveals another indicator of Concorde's pulling power, according to British Airways marketing manager Concorde, Alan Beaves: "Two-thirds of our passengers are transferring either at Washington or London, and one-third are transferring at both ends of the journey." This, he says is a very remarkable figure; only on exceptional subsonic routes do 20 per cent of travellers change aircraft at both ends. Even more significant are the 15 passengers per flight find it worth their while to use Concorde to fly from London to New York itself.

Washington, like Rio, is beginning to reach a "cruise level" for Air France, according to Cremet. Air France's share of the Washington Concorde traffic has been 45 per cent, despite the fact that permissible payloads are higher than British Airways'. Cremet points out that Air France has no subsonic services and is less well established in the market.

Despite its commercial success on the present routes, Concorde needs New York if it is to demonstrate its profit-making potential. Cremet speaks for both operators: "We have no spare plan if we don't go to New York. We don't have any other line that can absorb two flights a day." Like British Airways, Air France's purchase agreement includes the provison that its government will negotiate necessary landing rights.

British Airways hopes that the Braniff International London-Dallas interchange will support loads on the Washington route after New York is opened. The two airlines now have a firm understanding and fully intend to seek Civil Aeronautics Board approval for the service, Air France is discussing a similar service with Braniff, but negotiations are not as far advanced.

Under the interchange agreement Braniff would dry-lease Concordes from British Airways for the Washington-Dallas-Washington service, flown by Braniff flight-deck and cabin crews. The aircraft would revert to British Airways on arrival at Washington from Dallas, and would be flown to London by a British Airways crew. Transit time at Washington would be 50min and total Dallas-London block time would be 6-1/2hr. The existing Washington schedule would be unchanged.

The Braniff operation, Davidson points out, "would be the first interchange with a US carrier. It would also be the first time that a 'non-captive' airline operated Concorde." But the agreement, he stresses, is subject to various approvals.

The past year's Concorde experiment has resolved some of the doubts which afflicted its operators before service entry. Although the utilisation achieved so far could hardly be described as "normal service conditions," all the indications are that Concorde can fit into airline service with a minimum of special treatment; that maintenance presents no extraordinary new difficulties to airline staff; and that dispatch reliability is up to subsonic standards. Above all, the first year has proved that Concorde seats sell.

BOURGET '77: UN BON -- MAIS NON PARFAIT -- SALON de l'AIR

by Ed Mack Miller
Photographs by Katherine Duffy

Reprinted by Permission AIR LINE PILOT (October 1977)

"Science, freedom, beauty, adventure: What more could you ask of life? Aviation combined all the elements I loved."

Thus wrote Charles A. Lindbergh in "Spirit of St. Louis."

This year's Paris Air Show combined all these elements, as well as lovely airplanes and excellent flying. For the professional pilot the show was the best ever.

Lindy's spirit blew through the salons and chalets of Bourget in 1977 in a wave of nostalgia that left the Franco-Yank feeling better than it has been in years.

Anne Morrow Lindbergh came to the airshow which commemorated the 50th anniversary of her late husband's feat. She displayed the quiet beauty and dignity of a true queen.

The display salons were chock-a-block with all the new gadgetry of military, commercial and general markets.

As a retired airline pilot, I was, of course, most interested in things in my own line: What were the new Russian airliners like? How would the A-300B Airbus look like in "Eastern" colors? What kind of a stretch was Fokker looking to put on its F-28? What was new in cockpit bric-a-brac for the busy big-plane pilot? What did the Dash 7 look like in flight? How loud was it? How short could the new Boeing and Douglas cargo planes take off and land?

One airline captain in the U.S. asked me: "What is the show like? What are the chalets they talk about — tents? What is the flying like?"

First of all, Le Bourget does not completely shut down during the airshow, of which this was the 32nd (It is held every two years, alternating with England's Farnborough). The airport is used to serve Paris about the way La Guardia does New York, supplementing near-by Charles de Gaulle (north of Paris), and Orly (south of Paris), like La Guardia does JFK and Newark.

Large, permanent buildings, at airshow time, are used to house the displays of everything from huge radars, missiles, and transport concepts to the smaller electronic aviation gadgets.

On the apron, of course, are the aircraft (civil and military) from participating nations.

The lines of wooden chalets nearest the grassy south end of the airfield are leased by aerospace manufacturers as places to wine and dine potential buyers. From the chalets the participating corporations and their guests have as good as possible a viewing area of the air events. Some even reinforce the ceiling, add a ladder, and sport

АН-32

Як-50

EASTERN

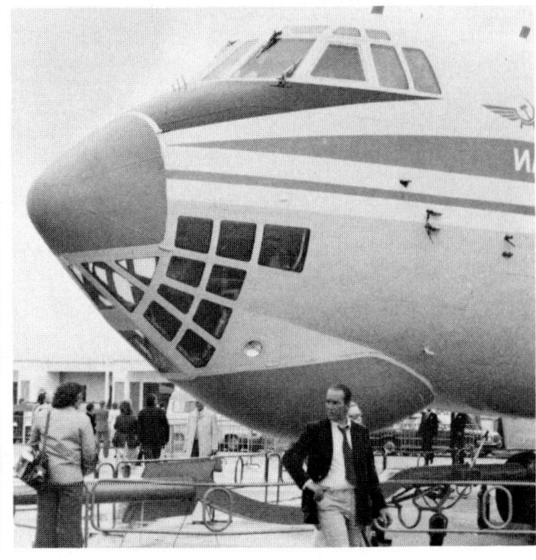

an upper deck for airshow viewing.

For any pilot, the flying, of course, is the main attraction. Everything from helicopters to the latest in fighters is "*air*ticulated," breathtakingly and brilliantly, every day.

The tragedy of this year's show was the crash of the USAF/Fairchild Republic A-10 attack plane on June 3rd just after the official opening of the show. Killed in the accident was Howard W. (Sam) Nelson, Fairchild's chief test pilot and director of flight operations.

I talked about the accident with Bob Hoover, the dean of aerobatic pilots, who said: "I don't think there's any way you can say you can have an event like this and rule out the possibility of an accident . . .You could brief all day long and caution about good judgment, but there's no way in the world that all of your talking is going to change anything . . .simply because it's an individual judgment factor, and no pilot goes into a maneuver with the thought in mind that he's taking a chance — that he is going to kill himself."

The Russians, showed up with lots of hardware, but did not fly it and pulled the most interesting exhibits out early.

The Soviet surprise was the new jumbo IL-86, a kind of combination DC-10 and 747 that looked, as one observer said, like "a fattened-up 707."

The TU-144 was back, four years after its Le Bourget tragedy, with questions of range, fuel, and design still to be answered.

The Russians also had the TU-154 trijet at Paris for six days. The IL-86, TU-144, and TU-154 suddenly returned to Russian on June 8, leaving on static display the YAK-42 trijet, the IL-76 four-engine heavy freighter, the single-place YAK-50 aerobatic plane, which took the top two slots at the Kiev world aerobatic meet a year ago; and the An-32 rear-loading STOL turboprop freighter.

Paris Airshow visitors were pleasantly surprised by the interior of the "Fat Albertski" IL-86 widebody jet. Passengers ascend to the cabin via three stairwells, after depositing their luggage below, and pass down either of two aisles which divide the three rows of 2-4-2 seats (with 3-3-3 planned). The plane accommodates 350 passengers. The plane was designed by G. Novozhilov, who was at the airshow.

The IL-86 is 196 feet long, has a 159-foot span, is 52 feet high and has a 37-foot track. The four NK-86 engines rate at 28,600 pounds of thrust. Maximum takeoff weight is approximately 450,100 lbs., and optimum operating attitude is 33,000 feet. The original design range was about 1500 miles but plans are to extend this to about 2200 miles (with a 40-ton payload). Maximum cruise is specified as 555 miles per hour. Approach speed is about 135 miles per hour. The plane has 14 wheels, an odd troika of main (4-wheel) bogies, plus a dual nose-wheel. The extra main gear was designed to obtain "a lighter foot-print" for fields where runway im-paction problems exist.

Another surprise was the "front office" of the IL-86. To those who remember the Russ opposite numbers to the original 707s and DC-8s, with their almost laughable cockpits, heavy control forces and crude instrumentation, the IL-86 comes as a pleasant surprise — modern in every respect. How well it flies, and how well the hardware works is, of course, still a question. The plane is designed for a four-man crew — pilot, copilot/navigator, flight engineer and spare "long-ranger" fill-in.

The IL-86 at Paris was the prototype. By the time of the next Paris show, it is hoped certification will have been accomplished so that the plane can be flown in exhibitions and sales orders taken.

The YAK-42, which will begin service in 1978 with Aeroflot, is based on the smaller YAK-40, of which 900 have been delivered. It seats 120, with three seats on either side of a center aisle, like the 727. Designed to cruise at .7 Mach (the -40 is .15 M slower), the 42 is 119 feet long, 32 feet high, with a span of 112 feet and an 18.5 foot track. It will gross 115,000 lbs. at takeoff, fly at about 25,000 feet at about 450 knots for a range, with full payload of 6,600 pounds, of nearly 600 miles. It has a fairly slow approach speed of 120 kts. The powerplant is the Lotarev D-36 turbofan, rated at 14,300 lbs. of thrust.

For the air carrier pilot, Le Bourget had enough to keep him busy just boning up on his own industry: a static A-300 Airbus that he could look through; a turbo-powered Fairey Britten-Norman "Islander," being readied for production; NASA's Convair 990 from the Ames Research Center at Sunnybrook, Calif., with a "walk through" exhibit that drew big crowds.

British Aircraft Company was pushing its proposed 156-seat BAC X-11 concept, based on BAC 111 components, but considered a new, not a "stretch" concept.

Also in the "taffy-pull business" are the Fokker-VFW people. They had models at Paris of a Super F-28, which would stretch to fill the gap caused by McDonnell-Douglas and Boeing 7N7 variant programs. Fokker's proposal would have five-abreast seating and grow to about 130 seats. There is a chance, spokesmen told us, that the Fokker could go to a three-engine configuration.

There was an excellent seminar on advances in the art of head-up display (HUD) instrumentation and application.

Featured were presentations of papers by Marconi Elliott Avionics; T.G. Foxworth of IFALPA; R.E. Hillman of the British Aircraft Corp.; Charles Legrand of Thompson-CSF; and Dr. J.P. Tymczyszyn of the U.S. FAA.

According to Hillman (BAC), the interest shown by airlines in HUD seems to be increasing. Some uses being investigated are: visual approach monitoring; windshear protection; takeoff and go-around directors; lower weather minima operation; and runway guidance. He noted that a number of customers for the BAC 1-11 have requested heads up display installation.

Legrand noted that in wind shear such as the JFK accident of June 24, 1975, "the HUD would have secured a safe landing." He described the equipment structure, possible architecture and proposed symbology, and covered Thompson-CSF experience so far.

Foxworth noted that HUD will be used for all landings so the pilot will have confidence in it from repeated usage. It will thus avoid the dangerous transition between instrument and visual control and also protect against "killer" windshears. It will also allow the pilots to have both instrument and visual cues simultaneously. "Thus today's split crew duties — and the loss of monitoring and redundancy that it implies — will be avoided."

He noted that W.R. Casey of McDonnell-Douglas described YC-15 STOL HUD experience with the Sundstrand VAM as "very desirable . . .even essential . . .for the conduct of safe, accurate, night STOL approaches . . ."

The Marconi-Elliott presentation covered its "multi-combiner" approach, and the FAA representative described that agency's evaluation programs, which includes participation by NASA. "Full crew operational simultation will begin later this year and continue during most of 1978. It will be jointly planned by FAA and NASA and will be conducted in an operationally configured simulator representing a medium or heavy turbojet at a NASA or industry facility." The final two program elements will follow: the engineering flight evaluation and the operational flight evaluation and demonstration. The joint FAA/NASA HUD program final report is tentatively scheduled for publication in 1979.

Also on the program at Paris was the Brazilian Embraer EMB110, Bandeirante, which was displayed and flown. It seats 15, and claims commuter airlines in Colorado and Texas as prospective buyers. It is powered by two Canadian P & W PT6A-27 engines.

The Canadian DHC-7 Dash 7 STOL transport, which was flown almost aerobatically, is a handsome and quiet aircraft, employing four P & W PT6A-50 turboshaft engines; it seats 50. Its first potential user in the U.S. is Rocky Mountain Airways of Denver, which has two on order.

At 43,500 lbs., it will take off in under 2500 feet, and at 41,500 lbs. land in little over 2000 feet. It will cruise at over 200 knots at 15,000 feet and, with a full load of passengers and baggage, fly some 700 nautical miles, according to de Havilland spokesmen.

Le Bourget '77, Salon XXXII, was a lot of things. It was, first of all, a great airshow. It was also Paris, beautiful, expensive and rewarding. It was fantastic food, great wines.

What more could you ask of life?

It was a good Bourget. Not perfect, but good.

AEROFLOT'S STOL TRIJET

by Bill Sweetman

Reprinted by Permission FLIGHT INTERNATIONAL (March 19, 1977) Edited for Space

Yakovlev was a latecomer to the design of commercial jet aircraft. The bureau did not develop a successor to the Yak-25/28 military aircraft, and instead branched out into Vtol research and airliner design. In the mid-1960s the bureau designed its first commercial aeroplane since the ten-seater Yak-16 of the late 1940s. Once the Yak-40 was firmly established in production — with a planned run of no fewer than 2,000 aircraft — The Yakovlev bureau could start thinking about a follow-on airliner.

The result of those deliberations, the Yak-42, should be in Aeroflot service by the beginning of next year at the latest. By 1979 it should begin to be available for export. Although Aeroflot is traditionally jealous of its priority, the Soviet Union is beginning to recognise the balance-of-payments benefits of aviation exports, and efforts to put Soviet aircraft into foreign markets are accelerating.

The Yak-42 is a highly specialised replacement for the An-24 and those short-range types such as the Tu-134s and Il-18s which operate on shorter routes. It was described as a "Siberian route transport" when it was first announced, and it is designed with a typical lack of compromise for operations from short fields in extreme climatic conditions. In the design of the Yak-40 very high value was placed on mechanical simplicity, and the Yak-42 follows this trend.

In power and wing area the Yak-42 is closer to the Boeing 727-100 than anything else, but its gross weight is slightly less than that of the 737. The result is that the Yak-42's power/weight ratio and wing loading are far outside the Western spectrum.

Not surprisingly, other aspects of the design are very reminiscent of the Yak-40, even though the new aircraft is more than three times as heavy. The basic layout and the accent on simplicity are shared features, and the control system is similar in philosophy. The engines are from the same bureau, albeit under another name. The first prototype was even fitted with an 11°-swept wing for comparison with the 23°-swept wing of the second prototype. The more swept wing will be used for the production aircraft.

The low gross weight limits fuel capacity (the 727 gets far more fuel into a wing of roughly the same size) and hence range. The Yak-42 payload-range line is not unlike that of the Fokker-VFW F.28 Mk 4000, with a full 120-passenger load being carried some 1,000 n.m. On the credit side, the low design gross weight and accent on simplicity have permitted a commendably light structure, only 8,500lb heavier than that of the more compact Boeing 737.

The wing is big, simple and light, and it is designed for a 30,000hr life with a one-hour average sector. There are no leading-edge devices. The trailing-edge flaps are of a new type, simpler and lighter than Western tracked and multi-slotted designs. The Yak-42 has a simple slotted type with a fixed slat on the leading edge and a plain drooping trailing edge. This layout is claimed to give the same sort of performance as a triple-slotted flap and is far simpler.

Yak-42 controls are manual, and the only powered surface, apart from the flaps and spoilers, is the hydraulically actuated trimming tailplane. The control system is triplicated. Hawker Siddeley proposed manual controls for the HS.146 with the exception of a powered rudder and powered roll spoilers to supplement the ailerons at low speeds. Like the HS.146 and the Yak-40, the Yak-42 probably has electrically actuated trim tabs on the ailerons and rudder. Western certification authorities tend to be a little wary of manual controls with geared tabs; this is one area in which changes have been made to the Yak-40 for Western markets.

The Yak-42 is designed for two-crew operation, with a simple and uncluttered flight deck drawing on Yak-40 experience. Initial operations will be to Category 2 limits, with Category 3a following later.

The D-36 engine, of 14,500lb thrust with a 5.34:1 bypass ratio, is the product of a design team led by Vladimir Lotarev — like the Soloviev bureau the Lotarev team uses the simple D *(dvigatel* = engine) designation rather than the leader's initials — which in effect is the former Ivchenko bureau that produced the AI-25 of the Yak-40. The D-36 is designed for a service life of 18,000hr — a departure from earlier Russian practice, particularly bearing in mind the short cycles — and is claimed to have a cruise s.f.c. of 0.65lb/lb/hr.

The D-36 is the key to the Yak-42's Stol performance, giving the aircraft a very high thrust/weight ratio, even at high temperaures and high elevations, without making fuel consumption worse than that of the types which it will replace. On Moscow-Leningrad, for instance, the Yak-42 will burn 25 per cent less fuel per passenger than the Tu-143A. The outer engines of the production aircraft will be fitted with thrust reversers, which are not installed on the prototype.

But will the Yak-42 appeal to Western operators? The question will not be important if the Soviet Union cannot certificate the Yak-42 to Western standards, and the main standards which the aircraft will have to meet are the British Civil Aviation Requirements (BCARS) and the US Federal Aviation Regulations (FARs).

STOL FROM DOWN UNDER

Reprinted by Permission
PROFESSIONAL PILOT (April 1977)

Australia's Government Aircraft Factories' Nomad, a twin turboprop utility STOL, is moving along in production and receiving acceptance in countries where semi-prepared strips are the basic airports. Two prototype N22 aircraft first flew in mid and late 1971 and an Australian Type Certificate was issued in August '73. The aircraft was designed to FAR Part 23 requirements but three FAA-required safety corrections which had held up U.S. certification have now been cleared and FAA Type Certificate is a matter of completing paper work.

Idea behind the Nomad is to provide high volumetric capacity and interior flexibility associated with local airlines requiring economical short range aircraft for low or high passenger traffic operations as well as using these craft where landing facilities are minimal.

The Nomad N22B is the production configuration and is available as 12 passenger commuter or mixed passenger/cargo model or seven-eight passenger VIP type. A quick change interior allows conversion from passenger to cargo in 10 minutes. There is also a military version known as Mission Master with MIL spec or civil avionics. wing hardpoints and stores pylons, armor-protected 25G crew seats, self-sealing fuel tanks, floor dropping hatch, medivac fittings and parachute static lines.

N24 test craft. Deliveries to start late '77.

GAF has also developed the N24, a stretched version, capable of carrying 14/16 passengers in a high density configuration. Two of the latter craft are currently undergoing certification testing and are actually production aircraft, having been built on production jigs. These particular craft are first of a batch of six ordered by the Australian Northern Territory Medical Service to replace its aging piston-engined deHavilland Doves. Like the N22, the N24 is powered by two 400 shp Allison 250-B17bs. NTMS is also having their craft fitted with long range tanks for an additional 600 lbs of fuel to provide a max VFR range of 1100nm. Avionics will be Collins, including the 107 AP and weather radar will be Bendix RDR 1400. Cabin will be set up with eight passenger seats for both patients and medical personnel, plus two stretchers and provision for a third.

GAF's initial production run calls for 100 N22/N24 of which 49 are now sold, plus one option and one lease. The Australian Army has purchased 11 Mission Masters, the Philippine Air Force an additional

12 and the Indonesian Navy six. Other Mission Masters have also gone to the Papua New Guinea Defense Force. Passenger and freight configured Nomads have been purchased by Nationwide Transport; Austirex and Missionary Aviation Fellowship in Australia; while that country's National Mapping Div. of Dept. of National Resources has a version with four passenger seats, lavatory and camera equipment, plus full inertial nav system. Additional Nomads are in service or going to Nordic Aviation in Denmark; Republic of New Guinea Independent Air Transport: Douglas Airways, Papua, Sabah Air in Malaysia and Air Tasmania.

A float-equipped Nomad is due to fly later this year and another Nomad is presently completing anti-icing tests in Scotland. GAF also has marketing teams in Poland and the Philippines to further expand sales. Current Nomad production rate is three per month.

Aeropelican, Australia's most successful commuter airline has evaluated the N22 on its high density Newcastle to Sydney route, a distance of 62 nm. Their normal aircraft fleet consists of four Cessna 402s and three C402As flown by a chief pilot and six line pilots. In addition, there are 24 administration, traffic and engineering personnel. Aeropelican carries an average of 4200 passengers per month with a break even load factor of 45% in the Cessnas. However their average load factor has been 82% and their ontime factor better than 98%.

In using the Nomad on the 62 nm leg they found their direct operating cost for the craft was higher than other operators due to their short haul, low altitude operation, with frequent takeoffs and higher fuel consumption at 1500 ft. the usual operating altitude.

A price of $410,000 is currently quoted for the N22B and $550,000 for the N24A in basic configuration. The N24 has the same engines but incorporates a 6 ft plug in the fuselage enabling it to carry up to 18 passengers, including crew. Nose compartment has also been lengthened two feet increasing volume 43% and wing span increased to 54 ft. Even with the increase in size the N24 retains the performance and operating costs of the smaller N22B.

EUROPE'S MAJOR CIVIL AIRCRAFT PROGRAMS

Reprinted by Permission
INTERAVIA (June 1977)

Here — in capsule form — is a summary of the principal civil transport aircraft programs currently under development or production in Europe — JSAY

Aerospatiale/BAC Concorde — 120-seat supersonic transport; four Rolls-Royce/SNECMA Olympus 593 engines. First flown in March 1969, the Concorde was certificated late in 1975 and entered service with Air France and British Airways the following January. These carriers, with orders for four and five units respectively, are so far the only customers and production will be halted early in 1978 after the completion of No. 16 series machine. Five aircraft are available for sale or lease. Plans for a moderately-improved "Concorde B", with slightly greater range, are unlikely to be fulfilled.

Aerospatiale Corvette — Six/fourteen-seat executive and light transport; two Pratt & Whitney Canada JT15D-1 engines. Certificated in March 1974, the Corvette has had problems in the marketplace, chiefly due to lack of success in the USA. Construction of 40 production aircraft has been authorized, with 33 completed by the end of 1976. Although about 30 aircraft have been sold or leased, the programme may be terminated when the 40th example is completed this summer.

Airbus Industrie A.300 — 250/320-seat short/medium-haul transport; two General Electric CF6-50 engines. Offered in two similar models, the B2 and the longer-range B4, the A.300 was first certificated in March 1974. At end-April 1977, orders totalled 39 units, with over 30 delivered; output was slowed earlier this year from two to one unit per month because of the slow influx of orders. A convertible/freight version of the A.300 B4 has been offered, but the most promising derivative of the design is a 200/220-seat short/medium-haul model, the B10, with shortened fuselage and redesigned wing. Airlines are currently being canvassed, although a go-ahead decision is not expected before the end of 1977 at the earliest.

British Aircraft Corporation BAC.111 — 89/119-seat short/medium-haul transport; two Rolls-Royce Spey 512 engines. Nearing the end of its production life, barring unforeseen development, the BAC.111 is currently offered in two models, the short-fuselage Series 475 and the Series 500. To date, 222 aircraft have been ordered and nearly 220 completed of 225 authorized. A Series 670, based on the 475, has been proposed for Japanese domestic airlines, but the principal design effort involves the X-Eleven, an aircraft using certain elements of the BAC.111 and powered by two General Electric/SNECMA CFM56 engines. Three versions have been proposed: Series 100, seating about 120 passengers; Series 200, about 135 passengers; and Series 300, about 155 passengers. Discussions are in hand to undertake the X-Eleven as a European programme, but its future depends largely on forthcoming British Government policy decisions.

Dassault-Breguet Falcon 10 — 7/10-passenger executive transport; two Garrett AiResearch TFE 731 turbofans. Certificated in September 1973, the Falcon 10 (known as the Mystere 10 in France) has sold well, with 107 on order (plus 70 options) at the end of 1976, when 90 had been delivered. Production is currently running at about 3.0 per month. The bulk of sales have been made in North America, where marketing is handled by the Falcon Jet Corp, a joint Pan Am/Dassault-Breguet venture.

Dassault-Breguet Falcon 20 — 10/14-seat business jet; two General Electric CF700 engines. First certificated in June 1965, the Falcon 20 series (known as the Mystere 20 in France) has sold well, particularly in North America. As of the end of 1976, a total of 366 were on order (plus a further 106 on option) and 353 had been delivered. The current model is the General Electric-powered Falcon 20F, but a G version, with Garrett AiResearch ATF-3 engines, is now in development. This aircraft won a US Coast Guard competition for 41 units in January, to be delivered from 1979. When work on the Coast Guard order is well advanced, the Falcon 20G will be offered on the civil market; prior to this, older models will be converted.

Dassault-Breguet Falcon 50 — 6/14-passenger executive transport; three Garrett AiResearch TFE 731 engines. The Falcon 50 is derived from the Falcon 20, with the addition of a new wing and rear fuselage, and the first of three prototypes flew on November 7 last year. Certification is scheduled around the end of 1978; orders for 20 aircraft have so far been reported and deliveries will start early in 1979.

Dassault-Breguet Mercure 200/ASMR — Short/medium-haul transport seating about 170 passengers; two General Electric CFM56 engines. Derived from the unsuccessful Mercure 100, the new model, also known as the ASMR, has been proposed as a coproduction exercise involving McDonnell Douglas and Aerospatiale.

Dornier Skyservant — 12-seat light transport; two Lycoming 0-540 piston engines. Certificated in February 1967, the Skyservant is being built at a rate of about 20 a year and orders are reported to total about 250. Sales continue to be made in the Third World (over 20 sold in Africa last year) and Dornier is considering a refined, turboprop-powered derivative.

Fairey Britten Norman Islander — 12-seat light transport; two Lycoming 0-540 piston engines. Certificated in August 1967, the Islander has sold well, with 730 on order by the end of February 1977; at end-1976, some 684 had been delivered and output, in Belgium and Romania, is running at about eight units per month. Product improvement continues, the latest being development of a version powered by two Lycoming LTP 101 turboprops, which was first flown on April 6 this year. A militarized version is the Defender, which is now offered for maritime surveillance, in addition to other roles.

Fairey Britten Norman Trislander — 18-seat light transport; three Lycoming 0-540 engines. Certificated in May 1971, this is a stretched Islander with a third engine added. Manufactured in Belgium, 48 had been delivered by the end of 1976 and output is now being raised from one to two per month. A military version, the Trislander M, has been proposed but not yet sold.

Fokker-VFW F.27 — 40/52-seat transport; two Rolls-Royce Dart 536 engines. Europe's best-selling civil transport, the F.27 was also built in the USA (204 units by Fairchild); sales now total 654 to military and civil operators and output is running at 1.5-2.0 units per month. Fokker-VFW sees production continuing into the 1980s, with the latest version, the F.27 Maritime, intended for sea surveillance.

Fokker-VFW F.28 — 60/85-seat transport; two Rolls-Royce Spey 555-15H engines. Sales of this short-haul

Fairey Britten Norman Islander

airliner total 120 units, with about 110 delivered by March 1977; current production models are the MK.4000 and 6000 and output is 1.0-1.5 per month. Fokker-VFW is now working on a derivative of the F.28, seating 115-125 passengers and powered by two 9-10,000 kp (20-22,000 lbt) engines, to enter service around 1983. Tentatively known as the "Super F.28," development cost is estimated at some $250 million.

Hawker Siddeley HS.125 — 8/14-seat executive transport; two Garrett AiResearch TFE 731 engines. Sales of the earlier Series 600, with Viper turbojets, stagnated but demand has picked up again with the introduction of the Series 700, first flown on June 29, 1976 and scheduled for certification by June 1977. First delivery will be made in July. As of mid-March, 372 HS.125s has been sold, including 15 Series 700s.

Hawker Siddeley HS.146 — 70/100-seat short-haul transport; four Lycoming ALF502 turbofans. Work started on this programme in 1973/74, but full-scale development was abandoned by the maufacturer in October of 1974. However, the Goverment has continued to provide

Airbus Industrie A.300

low-key funding and a decision on the future of the venture may be made later this year.

Hawker Sidley HS.748 — 40/58-passenger short-haul transport; two Rolls-Royce Dart 534/535 turboprops. Certificated in 1962, the HS.748 has enjoyed moderate success, with 316 sold by the end of 1976, when 300 had been delivered. Output is running at about one per month and a special maritime surveillance model has been developed; known as the Coastguarder, it was first flown on February 18 this year, and orders are now being sought.

Hawker Siddeley Trident — 97/152-seat medium-haul transport; three Rolls-Royce Spey 512 engines (Trident 2E), plus a Rolls-Royce RB.162 booster engine (Trident 3B). Certificated in 1964, the Trident was over-shadowed in the marketplace by the Boeing 727. Sales total 117 and production is now running down at about one per month; 106 had been delivered by the end of 1976 and the last should be completed around the end of the year.

Piaggio P.166-DL3 — 10-seat light utility transport; two Lycoming LTP 101 turboprops. First flown on July

3, 1976, the P.166-DL3 is a turbine-powered derivative of a successful piston-engine design. Offered in a variety of rules, both civil and military, the DL3 should be certificated this Summer and production has been authorized of an initial batch of four. Piaggio's other aircraft product, the PD-808 executive transport is out of production, although the company has planned a turbofan-engined derivative, with Garrett AiResearch TFE 731s replacing the original Vipers.

Short Skyvan — 19-passenger light transport; two Garrett AiResearch TPE 331 engines. First flown in 1963, the Skyvan has had a patchy sales record; as of the end of 1976, a total of 117 had been sold to civil and military customers and all but two delivered. Production is continuing on seculation to complete the 135 units authorized and sales efforts continue.

Short SD3-30 — 30-seat short-haul transport; two Pratt & Whitney Canada PT6A-45 turboprops. First certificated in February 1976, this aircraft is designed for commuter-type services and nine had been sold by March 1977, with five delivered. Production is continuing at about two per month and marketing efforts are concentrated in Europe and the USA. A military transport and a coastal surveillance model have been designed.

VFW-Fokker VFW 614 — 44-seat short-haul transport; two Rolls-Royce M.45H engines. Certificated in 1975, sales have moved slowly, with 16 on order at the end of last year, when five had been delivered. Long-running negotiations for licence production in Romania are reported close to completion, with 100 units planned for sale in Romania and other East Bloc nations. A maritime surveillance model was proposed for the US Coast Guard competition but lost out and the manufacturer has also proposed an anti-submarine derivative to the Federal German Government.

FOUR RB.211s

by John Belson

Reprinted by Permission
FLIGHT INTERNATIONAL(May 21,77)

"Speedbird 041 is cleared to Johannesburg as filed; Seaford 26 departure. Squawk 2024 just before takeoff." An air traffic control clearance like this will soon be routine for British Airways' new Boeing 747-236s, the Rolls-Royce-powered version of the 747 which allows the British flag carrier to fly non-stop from London to Johannesburg.

Customer acceptance flying with the first of British Airways' six 747-236s got under way at the beginning of this month, and I was privileged to visit the manufacturer's Everett base and fly on the first acceptance sortie. The aircraft, No 302 off the line, is the second -236 built and the first for delivery. The first, No 292, was used for most of the certification flying and is now being refurbished before delivery.

Boeing head of production flight test Sandy McMurray commanded the flight, with British Airways deputy chief pilot technical Capt Brian Lillyman occupying the left-hand seat. I was later invited into the left-hand seat when the acceptance schedule had been completed.

The Rolls-Royce RB.211-524B is the shortest of the three big fan engines, the major external recognition feature of the new 747 variant. The few flight-deck differences are confined mainly to changes which affect all 200-series aircraft rather than just the Rolls-Royce version. Unique to the RB.211 747 are the N3 gauges (which measure h-p shaft speed.) They displace the N2 gauges, which previously occupied the topmost of five rows of engine instruments. There are two turbine-overheat detector circuits, and a strut-overheat warning system is also provided.

Although comfortable, the 747 flight deck is surprisingly small and in some ways hardly worthy of an aircraft offering such capability. On the -236 the space available on the port side behind the flight deck has been extended aft and partitioned, and could be made into a very useful crew rest area. Up front the captain's and first officer's T-layout instrument panels are duplicated. Alongside the servo altimeter is a conventional circular-scale radio altimeter incorporating an insistent 2,500ft aural terrain warning. This stop gap warning will be removed when a new GPWS computer is fitted.

Starting is straightforward and usually begins with the No 4 engine, followed by Nos 1, 2 and 3. With a minimum duct pressure of 30lb/in², N3 readily spooled up. At 15 per cent N3 the captain selected the start lever to rich. For a normal light-up the exhaust gas temperature (EGT) is always less than percentage N3x10. At 50 per cent N3 the starter cut out and the ignition was switched off. Peak temperature was around 420° EGT, well below the maximum 550° allowed. As the engine stabilized I read N1 = 23 per cent, N2 = 40 per cent and N3 = 60 per cent. EGT dropped briefly to 309°, rising to 340° when an air-conditioning pack was switched on, and fuel flow was 680kg/hr. Starting with cross and tail winds is no problem, and the RB.211-524B's characteristic note during light-up, similar to the sound of an accelerating diesel engine, is unmistakeable.

Our taxi weight was 505,000lb, 300,000lb less than the maximum permitted, and we had 154,000lb of fuel aboard (only fixed furnishings and carpets are fitted as yet; seats will not be put aboard until the aircraft reaches London). The centre of gravity was calculated at 20.1 per cent mean aerodynamic chord, putting us in the middle range of the recently introduced triple-green-band take-off trim display. Before starting engines the appropriate forward, mid or aft rim band is selected with a switch just aft of the manual stabiliser trim selectors, and an aural warning sounds if the trim is not set within the correct limits for take-off. Other items included in the take-off configuration check are trailing-edge flaps between 10° and 20°, leading-edge flaps extended, speed brakes retracted and body-gear steering disconnected.

Following normal British Airways practice, a 10°-flap take-off was planned, with an engine pressure ratio (EPR) of 1.64 and air-conditioning packs switched off. The speeds worked out at V^1 140kt, V_R 140kt, and V^2 154kt. Other significant speeds, indicated with "bugs" on the ASI, are V^2 + 10kt as the target initial climb speed, V^2 + 40kt as the minimum speed for 5° flap, and V^2 + 80kt clean. Wherever possible the airline uses graduated power for take-off. The reduction in take-off EPR setting is calculated from a sliding scale of the difference between the maximum (or regulated) and actual take-off weights. No adjustment is made for weight differences of less than 30,000 lb, and the greatest reduction allowed is 0.10 for a weight difference of 150,000lb. An EPR limit computer fitted to the aircraft automatically indicates the EPR, depending on the selected

mode. The program is not yet finalised for the Rolls-Royce engine, however, and the computer was not working for my flight.

The weather was poor, with rain showers, broken cloud below 1,000ft and much more cloud above, and a gusty surface wind about 30° off the runway heading. We filed an IFR plan for a round-robin taking in Portland, Yakima, Spokane and back to Seattle, and programmed the triple inertial navigation systems accordingly. Probably the only navaid introduced during the last 30 years to have delivered all that it promised, INS is extremely reliable and simple to operate. For the first time we know where we are rather than where we were a while ago, and each set soon pays for itself in fuel savings. But sound operating procedures are required to ensure that correct information only is fed into the system.

A glance at the flight engineer's panel showed a single green "ignition on," but otherwise the panel was clear. British Airways practice is to abort below V^1 only for a major failure. A single warning light on the master warning system unaccompanied by other indications of engine malfunction is not seen as sufficient reason for discontinuing a take-off. For take-off the captain advanced the throttles to about 1.5 EPR and the flight engineer adjusted to the actual value required. We used 1.64 EPR, somewhat higher than the 1.44 which is standard on the airline's Pratt & Whitney-powered 747-136s. The higher figure results from measuring EPR at both ends of the engine and integrating the result instead of using the traditional fan-stage reading. Rate of change of thrust with throttle angle is more linear than on the other big fans, and any given EPR can be quickly and accurately set with a single throttle movement. During flight testing this feature (which was criticised initially simply because it was different) allowed drag measurements to be made far more easily and quickly than was possible on aircraft powered by the other big fans.

During the take-off run the co-pilot maintained wings level and slight forward pressure on the control column up to 80 kt. Directional control with rudder is usually achieved by about 80kt (there is no rudder/nosewheel steering link), but in strong crosswinds a large amount of nosewheel steering is required initially before the rudder is fully effective. The nosewheel is lightly loaded (nearly all the weight is taken by the four main undercarriage legs) and prone to skidding when taxiing on wet surfaces. For this reason British Airways recommends the use of differential take-off thrust at low speeds on some occasions.

Comfortable climb

After rotation, 16° pitch attitude allowed a comfortable climb at V^2 +10kt to 3,000ft. It is possible, by gross over-rotation, to scrape the rear fuselage during take-off or landing, and so an over-rotation sticker-shaker is provided. While awaiting further climb clearance at 3,000ft we accelerated to 250kt and retracted all flaps. Our clearance to FL350 was soon forthcoming and we set EPR at 1.51 initially. Climb EPR increases with Mach number and by FL350 it had reached 1.71. Climb performance was of course very good at our relatively light weight — when approaching FL410 later we still exceeded 600ft/min.

At FL350 we paused for pressurisation checks and a level-speed run. The "barber pole" limit on the ASI was 312kt and we quickly accelerated to this speed; the digital Mach indicator was showing 0.907 when the high-speed warning sounded. Some turbulence was felt, so we reduced to 300kt/MO.84, about 10kt faster than for the lighter -136. Being well ahead of and somewhat above the centre of pressure, the flight deck is prone to large lateral gust movements, especially in clear-air turbulence. Boeing has developed a Gust Response Suppression System which senses lateral disturbances at the nose and applies rudder to minimise movements of the aft fuselage. The system operates only when the flaps are up, and flight-test observers report favourably on the improved ride which it gives.

One hour after take-off we were descending from F410 with engines at flight idle, 300kt/MO.827 and a rate of descent of more than 5,000ft/mn. The 747's downhill ability is phenomenal, with selection of flight idle resulting in a marked deceleration. If speed is to be maintained, a high rate of descent has to be demanded of the autopilot just before the throttles are closed. On an operational descent the gradient is 400ft/n.m. down to 10,000ft, then 10 n.m. horizontally to slow down at that level before continuing at 300ft/n.m. and 250kt. Emergency descents are flown at 320kt/MØ.82 after gear and speed brakes have been extended, and descent rates of 10,000ft/min are easily achieved. Bank is usually applied when initiating an emergency descent to avoid an uncomfortable pushover.

We levelled at F280 and checked the engine-relight performance, choosing the corners of the relight envelope at two speeds and two levels. All went well and we heard that distinctive sound as each engine accelerated back to flight idle. As with all big fan engines, some 80 per cent of the RB.211's thrust is produced by the fan. Fan inertia is so high that the effects of an engine failure are felt only gradually, and even after complete seizure the thrust takes at least three seconds to decay.

After a maximum-weight take-off in ISA conditions the Boeing 747-236 can make FL310, with an absolute maximum of F340; F180 can be reached on three engines at the same weight. These levels are roughly the same as for British Airways' -136s, but the new aircraft can fly about 700 n.m. farther and its RB211-524s offer the lowest cruise fuel consumption of the three big fans.There are very few passenger sectors on which the 747 is structurally limited. Full tanks are rarely required, although of course the aircraft is often operated at its maximum structural take-off weight.

Although the -236 can fly the 5,500 n.m. from Johannesburg to London direct, its payload is limited to an uneconomic 20,000lb. Johannesburg's 5,557ft elevation imposes weight/altitude/temperature and tyre-speed limitations which are hard to beat with any aircraft. British

Airways therefore prefers when flying northbound to call at Nairobi, where it has additional traffic rights.

Operating out of Nairobi, perhaps the best known hot-and-high field on the airline's network, the -236 will lift 752,000lb, 80,000lb more than the -136 when tyre-speed limited on Runway 06 in an ambient temperature of 16 °C. An additional 15,000lb of fuel is required by the heavier -236 for the leg to London, leaving an improved payload of about 65,000lb. The airline will continue to maximise payload by such operating practices as improving climb performance after take-off and en-route refiling.

When a hot-and-high airfield has a long runway available, second-segment performance can often be improved by climb procedures even though the operation remains WAT-limited. A five per cent increase in V^2 (and the associated V^1 and V_R speeds) allows a better engine-failure climb performance and hence a higher WAT limit.

Refiling in the air (nominating a closer airfield as destination and then, if fuel permits, re-planning to the original destination) can be very attractive on the 747. The required sector fuel load is calculated by adding together the route fuel, the fuel required to divert to an alternate, a reserve of 16,500lb (allowing 30min holding at 6,000ft plus 3,300lb remaining) and a contingency reserve based on the route distance. It is this last element that offers greater range for the same fuel load. If the contingency fuel remains unused after, say, three-quarters of the distance has been covered, it can be used to extend the sector by as much as 500 n.m.

Seattle Centre is well accustomed to aircraft with Boeing callsigns which go from Mach 0.9 to little more than 100kt within a few minutes. Clearance was therefore soon forthcoming for a low-speed exercise, and 5,000ft of airspace beneath us was kept clear just in case. At this point our weight was 470,000lb, with 120,000lb of fuel remaining. In a clean configuration we set 1.26 EPR and reduced speed while flying level at FL200. The stick shake came at 176kt (as predicted) and a pitch attitude of 7-1/2 °. With leading-edge flaps extended and 10 ° of trailing-edge flap we recorded 127kt; increasing trailing-edge flap to 30 ° and lowering the undercarriage allowed us to fly at 112kt before the stick-shaker fired, and at 110kt with power at about 1.40 EPR. I understand that the 747's actual stalling manners are above reproach.

Before retracting the flaps we checked the flap load-relief system, which raises the flaps from 30 ° to 25 ° and in turn to 20 ° if and for as long as the limiting airspeed is exceeded. An amber caption illuminated on the master-warning panel and the operation was entirely automatic, the flaps returning to their selected position when the airspeed came down. All flaps and undercarriage legs can be lowered by alternative means, and these functions were also thoroughly checked.

Fuel is used from the fuselage tank first, and it is usual to keep a reasonable weight of fuel in the outboard tanks to provide wing bending relief. A speed restriction is imposed if reserve tanks Nos 2 and 3 are used, another of the new features of the extended-range 200-series aircraft.

Two hours after take-off we were preparing for autolanding checks back at Paine Field. Before the first approach we bugged the go-around EPR at 1.60 (less than the static take-off EPR of 1.64 because of our forward speed), checked the ILS tuning and set the decision height of 100ft on the radio altimeter. V_{REF} for our weight of 460,000lb was 126kt, which we set on the edge of the ASI. We also bugged V_{REF} +10kt (target approach speed about equal to V_2 + 10), V_{REF} + 40kt (minimum speed for 5° flap) and V_{REF} + 80kt (minimum speed clean). We flew the intermediate approach at V_{REF} + 40kt and with 10° flap, giving a pitch attitude of about 7°. With the pre-landing check completed, all three autopilots were selected (only one is used for the cruise); the mode annunciator showed a white TRIPLE, NAV and G/S and a green ALT caption. The mode annunciator is split into two parts, the left-hand side indicating flight director and the right-hand side showing autopilot performance.

The autothrottle is controlled from a glareshield selector and V_{REF} +5kt (131kt) was dialled up. Boeing has developed a second-generation full-time autothrottle which is integrated with the EPR limit computer. The system provides automatic throttle control from take-off to landing using three primary modes: EPR control, Mach hold and speed control. Graduated take-off power, automatic climb EPR reduction and power reduction for landing are available, as well as a minimum-airspeed protection feature designed to prevent a stall if power is not applied when levelling from a descent. The new autothrottle can be fitted in any 747 but British Airways has opted to stay with the original system to preserve fleet commonality.

Once established on the localiser, we lowered the under-carriage with one dot to go to the glideslope and took full flap as we captured the glideslope. By about 1,500ft all three autopilots were switched in, the TRIPLE, NAV and G/S capitions were green, and the white FLARE-armed indication was showing. The aircraft stayed tightly coupled to the ILS; at about 75ft above the selected decision height (radio altimeter) an aural warning began and quickly rose in volume until it was abruptly cut off at decision height, leaving just a warning light on. The flare began at 53ft as the throttles closed, and it only remained for the captain to disengage the autopilot after touchdown. Throughout the approach the co-pilot remained head-down, calling significant events as they occured.

Palm switches on the inboard throttles were used to feed a go-around command to the autopilot as the captain advanced the throttles to go-around EPR. Gear and flaps were handled normally while the autopilot levelled the wings and ensured a positive climbing angle. With the trailing-edge flaps still at 30° the climb rate bettered 300ft/min; this improved to over 1,000ft/min when the flaps were raised to 20°.

British Airways had expected to develop the Boeing 747 automatic

landing system to Category 3 capability, but reluctantly decided two years ago that the high cost of the programme could not be justified in view of the limited number of times that the system would be needed. A tendency towards excessive lateral displacement from the runway centreline after touchdown led in any case to Category 3 approval being limited to Heathrow's 300ft-wide Runway 28L. The airline now removes the third autopilot from its 747s (after acceptance flying) and downgrades the autoland capability to Category 2 with a decision height of 100ft and runway visual range of 400m.

With the acceptance flying almost complete, I took the left-hand seat for the remainder of the sortie. After entering service the 747 quickly earned itself a reputation for easy handling and exceptionally good flying qualities, and I was in no way disappointed. The rudder is heavy but very powerful, and its authority decreases as airspeed rises. Roll control is remarkably light and responsive, and its characteristics are kept constant throughout the flight envelope by varying the amount of spoiler used. The jolt usually produced when the spoilers extend during autopilot-controlled turns can be avoided when flying manually by limiting control-wheel movement to 10°.

Pitch control is also good and the response of the electric trim is first-class. The 747 flies nose-up most of the time, and during manoeuvres at approach speeds the pitch attitude can become quite exaggerated. At first I made insufficient allowance and lost height, but during the next turn I did much better. Extension of undercarriage and flaps called for little retrimming, and during final approach I held the ILS glideslope with the nose about 2-1/2° up. Patches of cloud around 800ft obscured the runway, the surface wind remained gusty and the runway was wet.

To check the automatic braking system we selected "maximum" from a choice of three settings, probably the only time this setting will ever be used. The spoilers were armed, ready to deploy automatically as soon as the mainwheels were firmly on the runway.Whatever the setting, the system allows only medium auto-braking until the nosewheel is on the runway. I aimed for a touchdown spot farther into the runway than usual (no VASIs available) but otherwise flew a normal approach.

"Land with drift on and don't worry much about flaring." said Sandy McMurray. He called out the radio height in the final stages and at 30ft I closed the throttles and flared, finding no difficulty in using only one hand to hold the nose well up. At this point we had to go briefly into conference to decide whether I had in fact touched down! The spoiler lever then automatically moved to the "deploy" position, confirming that the main oleos were compressed.

I armed and then selected reverse thrust while lowering the nosewheel and allowing the automatic wheel-braking system to slow the aircraft very quickly. I just managed to cancel reverse thrust by 80kt, slowed to less than 10kt and, using nosewheel steering, positioned at the edge of the runway for a 180° turn.

The view from the flight deck is good but the ground within 80ft of the nose is invisible. Sitting 30ft above the tarmac, frequent checks of the INS groundspeed readout are recommended while taxiing and especially before sharp turns. About 35 percent N1 is needed to get the aircraft rolling, but once under way idle power is usually sufficient to keep it moving. I found the toe brakes very powerful and I used them far too coarsely at first. Earlier 747 variants suffer from an excess of idle thrust but the Rolls-Royce engine is good in this respect. Ground manoeuvring is also simplified by steering the body-mounted landing gear:the two body legs turn in the opposite direction to the nosewheel, significantly reducing turning circle and tyre scrub.

By any yardstick the Boeing 747 is remarkable aeroplane. Doubts about its size and complexity, voiced before the first example was built in 1968, have long since evaporated. It is a delight to operate, easy to fly and almost invariably provokes lasting admiration among pilots despite the plethora of bells, horns, chimes and warbles on the flight deck. The manufacturer has incorporated various airframe improvements in the 200-series aircraft which are offered for a minimum of 20 years' service. The four Rolls-Royce RB.211-524Bs represent 25 per cent of the value of the aircrft, underlining Britain's ability to compete with the best the US can offer.

NOTAMS

This section of the JEPPESEN SANDERSON AVIATION YEARBOOK was set up for those outstanding stories or photographs that would not easily fit into one of the four established categories. In this section will appear varied subjects: aviation greats, unique events or photographs, significant historical items, those items that touch more than one area of the world of aviation.

NOTAMS this year offers something new — a collection of outstanding aviation photographs in full color. Covering the total range of the aviation world, these photographs were chosen to illustrate the drama, the beauty and the humor of the world of flying.

In addition, this year's section covers a wide spectrum of aviation topics: aviation great Charles Lindbergh, a fascinating collection of almost-ran airplanes, and a photo story on the fabulous Hughes Hercules, the "Spruce Goose."

The NOTAMS section is planned to provide special coverage of many different areas of aviation. We hope that you find it so.

1

2
3
N266GL

5
N-X-211
RYAN
NYP
Spirit of St.Louis

6

8

9

10

11

12
NASA
905
NASA
13
Enterprise
NASA
United States
N905NA

NASA
905

14

15

COLOR PHOTOGRAPHS

1 Sky Dreams. Photographer: Shelby Stover.
Reprinted by permission: Wind & Sea Graphics, Box 728, San Luis Obispo, CA.

2 Photo courtesy of Jeppesen Sanderson Inc., Product Development Division.

3 Learjet Longhorn in Flight
Reprinted by permission: Gates Learjet Corporation.

4 A hang gliding enthusiast alone in the sky. Photographer: Bettina Gray.
Reprinted by permission: Bettina Gray, Box 32, Rancho Santa Fe, CA.

5 Spirit of St. Louis replica in flight. Photographer: Jack Cox.
Reprinted by permission: SPORT AVIATION.

6 The commuter bus approach to skydiving. Photographer: Tony Gonzales.
Reprinted by permission: Tony Gonzales, Box 2407, Phoenix, Arizona © 1977

7 Just before dawn at Stapleton International Airport, Denver, Colorado..
Photographer: Mikel Peterson. Reprinted by permission: AIR LINE PILOT.

8 Gossamer Condor in flight.
Reprinted by permission: POPULAR SCIENCE © 1977 Times Mirror Magazines, Inc.

9 Seagull Ten exit over Tahlequah, Oklahoma. Photographer: Jerry Irwin.
Reprinted by permission: PARACHUTIST.

10 A United Airlines 727 on final approach to O'Hare International.
Reprinted by permission: AIR LINE PILOT.

11 Mirror Image 8-man formation during the 1977 Nationals.
Reprinted by permission: PARACHUTIST.

12 The Space Shuttle Orbiter and the 747 ferry plane in the Mate-Demate Device.
Official NASA photograph.

13 The Space Shuttle Orbiter during a successful ferry flight test. Official NASA photograph.

14 The Space Shuttle Orbiter and the 747 ferry plane in the Mate-Demate Device.
Official NASA photograph.

15 Royal Air Force Jet Provosts in line astern. Ministry of Defense photograph, reprinted by permission: AEROSPACE (Royal Aeronautical Society).

THE LINDBERGH LEGACY

By Robert J. Serling

Reprinted by Permission FRONTIER (The magazine of Frontier Airlines)

The date was May 20, 1927.

On that day in history, at the hour of 7:52 a.m. EST, a small silver airplane waddled down a wet, muddy clay runway at an airport called Roosevelt Field and staggered its way up into a gray overcast.

In that tiny aircraft was a lone pilot and ahead of him lay 3,640 miles of danger — some 33 hours of tedium mixed with terror. And thus was born the Lindbergh legend, a feat which to most people seemed foolhardy but which to the man they called The Lone Eagle was simply calculated risk.

Millions of words have been written about that New York to Paris flight just 50 years ago — including a few that disparaged the feat. Just a crazy stunt, they said, pulled off mainly because Charles A. Lindbergh was not much more than a lucky barnstormer, a gambler defying the odds because the payoff would be world fame and incredible success.

It is only too easy, given the hindsight of a half-century, to agree with that cynical verdict. A brief glance at what Lindbergh had to work with on his flight to glory, compared to the tools of modern trans-ocean air travel, and one has to marvel that he got away with it.

Lindbergh jammed 450 gallons into the fuel tanks of the *Spirit of St. Louis*, as much as they would hold. His decision to take off was based on a single weather report which turned out to be so inaccurate that he later admitted he never would have left if he had known what lay ahead.

By contrast, a jetliner carries up to 30,000 gallons and gets up-to-date weather information from more than 30 different sources.

The Lone Eagle took along one air raft with a pump and repair kit, a four-quart canteen of water, a hunting knife, one hack-saw blade, a bail of cord and string with a large needle, five cans of Army emergency rations, a flashlight and four red flares — and he had to be talked into carrying this much because he thought it added excessive weight.

The emergency equipment aboard a transocean jet includes a radar reflector that unfolds like an umbrella, enough life rafts for up to 400 persons, life jackets, a police whistle, gloves, sea water desalter, first aid kits, signal flares, mirror signaler, fishing kit, sunburn lotion, yellow marking dye, candy, chewing gum and vitamins, flashlights, pliers, repair kits, storage water, a powerful radio transmitter that can be heard 150 miles away, inflation pumps, oars, life raft canopies and Protestant, Catholic and Jewish Bibles.

The *Spirit of St. Louis* (incidentally, its official title was the "M-2") grossed 5,130 pounds fully loaded. On May 20, 1927, it was dangerously close to being overloaded; Lindbergh used nearly 2,500 feet of runway and almost didn't make it.

A 707 or DC-8 has a takeoff weight of some 300,000 pounds and a 747 about 700,000, each burning more fuel in the first three minutes after takeoff than Lindbergh did on his entire flight, and requiring runways of up to 12,000 feet in length.

The little Ryan monoplane that became so much a part of the Lindbergh legend cost less than $13,000. The cramped cockpit, containing only 11 instruments, had no forward vision; Lindbergh either peered out a side window or used a small periscope to see straight ahead. His Wright Whirlwind engine developed 220 HP.

The cost of a large jetliner ranges between $8 million and $35 million at 1977 prices. Its cockpit is equipped with more than 200 instruments and just one engine on a 707 weighs 1,000 pounds more than the Ryan grossed on takeoff. The $1.1 million engine of a 747 develops up to the equivalent of 75,000 HP.

Lindberg had only one navigation instrument — a relatively new

New York, May 20, 1927; en route to Paris, Lindbergh takes off from Roosevelt Field.

device known as an earth induction compass. He hit the coast of Ireland only 25 miles off his intended couse, but remarked later that it was mostly luck. He had no radio, the only voice contact coming near the Irish coast when he swooped down near a fishing boat and yelled at the startled occupants, "which way is Paris?" His top speed for the entire trip was less than 120 miles an hour and his altitude ranged between 50 and 10,000 feet. After six hours — the time of the average eastbound New York-Paris flight — the *Spirit of St. Louis* had just reached Nova Scotia, 600 miles from New York, with 27-1/2 hours of flight still ahead.

The crew of an ocean-flying jetliner has seven different means of determining and maintaining position, including Doppler and Loran — both extremely accurate electronic navigation systems — as well as the new Inertial Navigation System which is the most accurate yet developed. The average trans-atlantic or transpacific jet has four separate radios, makes about 20 air traffic control position reports, a dozen or so contacts with the airline's own communications facilities, and over the Atlantic receives fresh weather reports from a half-dozen European or U.S. cities depending on the flight's direction. Whereas Lindbergh never could climb above the weather, jets cruise routinely at 40,000 feet; in the 33-1/2 hours it took Lindbergh to cross the Atlantic, a jet could be two-thirds of the way around the world.

The contrasts, however, do not tell the whole story. In truth, Lindbergh took every advantage of the state of the art available to him 50 years ago. He had the best, most reliable engine then developed and his plane — although the size of a Piper Cub — was sturdy and well-designed for its time. What is more important to realize is the impact his dramatic flight had on aviation itself, one far transcending the adulation he received personally.

Lindy flew the Atlantic at the time when a rather jaded American public, already disenchanted with the Prohibition experiment and the disillusionment of World War I's failure to bring real peace, needed a legitimate hero. But aviation needed one even more. The nation's airline industry was only one year old when Lindbergh flew nonstop to Paris; on May 20, 1926, President Calvin Coolidge signed the Air Commerce Act that created an Aeronautics Branch within the Commerce Department, and for the first time in U.S. history gave the goverment authority to establish federal airways; install, operate and maintain air navigation aids; issue airworthiness certificates for airplanes, and investigate accidents.

That legislation was handmaiden to the previously passed Kelly Act, which gave private operators authority to fly the mail. At the time the Air Commerce Act was enacted, five companies were carrying mail and passengers; by the end of the year there were 12 airlines operating in the United States. They were struggling, inadequately financed, poorly equipped and badly patronized — there are no exact figures available on 1926 passenger traffic, but the total is believed to be less than 2,000 and

In 1926, Lindbergh became an airmail pilot. One night in September 1926, he departed from St. Louis bound for Chicago. The weather was bad and after trying for hours to find a hole in the clouds he ran out of gas. He was forced to bail out and landed in a corn field. Lindy made two emergency parachute jumps during his airmail days.

some historians estimate it at closer to 1,000. Whatever the modest volume, it has to be compared with the nearly 13,000 who flew the U.S. airlines in 1927, and 53,000 carried in 1928. The latter alone gives some indication of the impact Lindbergh's flight had on the public's attitude toward flying.

That the airlines themselves were in part responsible for this incredible surge is without question — through improved aircraft, more reliable operations and greater comfort. But public confidence was another major factor, and this is where the Lone Eagle's miracle had its greatest effect. Before it took place, the U.S. was not an air-minded nation, in the sense that most Americans regarded the airplane as a kind of toy, usually approaching the status of romantic fiction. Flying was something to be enjoyed vicariously — in pulp magazines relating the make-believe exploits of World War I pursuit pilots, in movies like *Wings*, in barnstorming appearances at county and state fairs. It was not to be personally experienced, but was something to be read about — usually with much head-shaking, for the majority of newspaper stories dealing with aviation dealt with death.

It was different in Europe, where the airplane was a real instrument, not a toy. There its power had been seen and felt, not just read about. And when peace came, the transition from military to civilian use was accepted as a logical development. This wasn't true in the United States, where the airplane was regarded with either fond amusement or outright contempt and fear.

Lindbergh's epic voyage changed all that virtually overnight, wiping out all the previous years of commercial aviation's immaturity and

Photo courtesy of Bell Aerospace Textron, Buffalo, N.Y.

The Bell P-59 Airacomet was America's first jet airplane and made its first flight Oct. 1, 1942. Very few people know that Lindbergh once flew the P-59.

the public's indifference. Lindy's flight was followed by other successful ocean ventures which gave the airplane a reputation for safety it didn't really deserve but badly needed — and this furnished the infant airlines a welcome stimulant. Suddenly, flying became respectable and even acceptable and with this achievement came a growth in air travel that has never ceased to this day.

Charles A. Lindbergh's life was a kaleidoscope of controversy; a genuine hero, to many he became a real villain. His pre-Pearl Harbor isolationism cast him in the false role of an arrogant ingrate who paid back his country for its adoration, fame and countless honors by becoming a neo-Nazi, contaminated by anti-Semitism and totally unconcerned by the menace of Hitler. This was the mantle cast upon him by Franklin Roosevelt who, from all objective accounts, actually feared Lindbergh. Walter Ross, author of a brilliant Lindbergh biography (*The Last Hero*, Harper & Row) is a writer who is frankly critical of certain Lindbergh statements and speeches made before World War II. However, he also points out that not all isolationists, the Lone Eagle included, were Fascist-minded or pro-Hitler; most were patriotic Americans sincerely opposed to American involvement in another European war. Ross does not believe Lindy was anti-Semitic, and largely blames FDR for labeling him as such. As he put it in *The Last Hero*:

> One cannot avoid the conclusion that Roosevelt built Lindbergh up as a pro-Nazi so he could break him. Lindbergh was a strong and dangerous political antagonist, never a potential traitor. But Roosevelt had to get rid of him at any cost. . .

Objective authors like Ross have disclosed elements of Lindy's humanitarianism, intellect and warmth that have long been hidden under the oversimplification of the Lindbergh legend. The Lone Eagle admittedly had his faults, the chief of which was his airman's propensity for viewing Hitler's Germany solely as a military threat while failing to grasp the even greater moral threat.

But Lindbergh also was a vastly underrated aeronautical scientist whose contributions to both military and civil aviation have never been adequately appreciated.

His exploits as a civil adviser in the field of fighter tactics in World War II remained almost unknown until years after the war ended — and he really *did* shoot down two Japanese planes in the course of that assignment. As a technical consultant to both TWA and Pan Am, he performed brilliantly. Pan Am, in fact, thought enough of him to have him serve on its board of directors for a long time.

As a pilot, he had few if any peers — exceptionally skilled is the only apt description. Najeeb E. Halaby, former FAA administrator and later head of Pan Am, tells a story about Lindbergh dating back to the days when Halaby was a young wartime test pilot. He was called before his commanding officer one day and told that a certain civilian had been given permission to fly what was then the nation's first jet aircraft.

"You're the only one who knows much about the plane," the officer said, "so check this guy out before he takes it up."

The civilian, of course, turned out to be Lindbergh. And Halaby remembers to this day how he handled a plane he had never seen before.

"I gave him only the briefest of instructions and a few words of caution," he recounts. "He took off and flew that jet as if he had been born in one."

Later, they became close friends, even though they differed occasionally on aviation policy

George Dade, an airplane buff and president of Long Island Early Flyers Club, located the remains of Lindbergh's first Curtiss Jenny in an Iowa pig barn. This photo shows George reconstructing Lindy's Jenny. On July 4, 1976, the Jenny entered the Long Island aviation museum.

matters. For example, Halaby as head of FAA was a strong supporter of the U.S. supersonic transport program while Lindbergh always opposed the SST as an environmental threat. In his later years, in fact, the man so responsible for much of aviation's progress seemed to fear and even resent that very progress. Yet it was not so much an inconsistency, but rather his deep belief that environment must not be sacrificed simply because of technology. He always retained pride in his airman's background; one of the few times he agreed to attend a public dinner in his honor was the occasion when the Air Line Pilots Association made him an honorary member in 1966.

He died August 26, 1974, on the island of Maui, Hawaii. The burial services were as modest, simple and dignified as the man himself — Lindbergh never understood why so often and for so long he was regarded in extremes, worshipped by some and hated by others. With his later years, however, came a greater awareness of his stature — both as a man and as a hero.

Many veterans in aviation today credit their first interest in aviation to the flight that took place 50 years ago. How many youngsters were inspired to become pilots by the Lone Eagle's defiance of the odds will never be known. How many ordinary Americans whose decision to at least try flying stemmed from Lindy's exploit is impossible to judge. It is certain, however, that commercial aviation derived enormous benefits from an achievement that combined pure luck with magnificent skill, that recognized danger mitigated by careful planning.

Only five decades have elapsed since one man flew from New York to Paris. Today, there are nearly 80,000 flights a year over the North Atlantic, carrying in excess of 10 million people.

And *that* is the real legacy.

(Editor's note: Mr. Serling was aviation editor for United Press International for many years before becoming an authoritative and compelling writer on aviation. He is the author of the novels THE PRESIDENT'S PLANE IS MISSING and SHE'LL NEVER GET OFF THE GROUND, (the story of a woman airline pilot). He has written many highly-acclaimed nonfiction books on air safety. His writings have earned him numerous awards for aviation reporting, and he is an associate member of the Society of Air Safety Investigators. Mr. Serling is editorial adviser for this magazine and was instrumental in its conception.)

GALLERY OF GHOSTS

Reprinted by Permission
AOPA PILOT (APRIL 77)

The road to success is littered with mistakes, both grand and small. In the airplane building business, mistakes often have wings, which is to say they're often expensive.

It is as much to honor those companies who have succeeded despite their missteps as it is to remind them promises are not always kept, that the AOPA Pilot does herewith humbly present,

Their Mistakes.

BELLANCA TRAINER

The biography for this nameless waif is brief but complete. Its birth announcement, a 1973 news release, read, "Flight testing is under way on Bellanca Aircraft Corporation's new two-place trainer. The engineering prototype of the high-wing, tricycle-geared trainer made its first flight on Friday, Oct. 26, with Bellanca Executive Vice President Don O'Mara at the controls. The new aircraft, which the company plans to start producing by late 1974, is powered by a 115-hp Lycoming engine and will feature a full professional panel, dual control wheels and 360-degree visibility through large wraparound windows."

But before the trainer, likened to a "fabric 150," was put into production, a recession hit, Arabian gas became scarce and Bellanca's financial footing got slippery. The trainer's death came, ignominously, last fall when it was cannibalized at the place where it was born.

"I think our new little airplane will be able to offer some stiff competition," one Bellanca executive boasted back in '73. It might have; we'll never know.

CHAMPION LANCER

The goal was admirable — a super-low-cost twin trainer — but the reality was something less. It was called the Lancer 402 and was built by Champion Aircraft of Osceola, Wis.

Its merit was obvious. When the Lancer went into production in 1963, it bore a price tag of $12,500. Its closest price competitor was Piper's Apache which then was selling for $37,990.

But its demerits were just as plain and, unfortunately, more numerous. To begin, it was ugly, a regular pelican with boils. It had long main gear legs which hung permanently from its high strut-supported wing. Its small snout, tall windshield and high perched nacelles merely accentuated its compromise design. Add to this the fact that the Lancer was fabric-skinned and its narrow cabin required tandem seating, and you've got an idea of the saleman's problems.

The Lancer's most serious failing, however, was not cosmetic. It was muscle, or lack thereof. The Lancer was powered by two 100-hp Continentals which turned fixed-pitch props. Needless to say, its performance was neither exciting nor reassuring. Sales followed suit. About two dozen were sold before production was permanently halted.

BEECH TWIN QUAD

The "Twin Quad" was one part Bonanza, one part Twin Beech and five parts ambition. The year was 1947 and everything seemed possible to aviation's soothsayers The Beechcraft Model 34 created to handle the anticipated growth of the feeder airlines, was as advanced in design as in hope. This V-tailed, all-metal, 20-passenger transport could be quickly converted for cargo hauling and its two props were spun by *four* 400-hp Lycomings (thus the name "Twin Quad") which were completely recessed in the wings.

"An extensive flight testing program proved the "Twin-Quad" to be the plane the air transport industry was then frantically demanding," reported Beechcraft's official biography. "But financial problems overtook the airlines; the potential market shrank drastically and the Model 34 was reluctantly shelved (in 1949)."

The single prototype 34 was destroyed and its pilot killed during a flight test in 1949.

PIPER APACHE

Designing airplanes is much like any acquired art — you do it again and again until you get it right. This aircraft is a case in point. In 1951, William Piper, Sr., agreed to enter his little company into the twin-engine airplane business. His plane, he insisted, would cost less than $17,000. To meet that figure, the aircraft would be constructed of metal and fabric, be powered by two, four-cylinder engines, have fixed-pitch props and, possibly, fixed gear.

The finished product was a twin-finned, four-place machine powered

by 125-hp Lycomings. Flight tests, which began in March, 1952, revealed the boxy craft suffered from serious vibration and engine cooling problems.

Back to the drawing boards. The double tail was replaced with a conventional single fin, retractable gear was agreed to, 150-hp engines with constant speed props were added and the fuselage and wings were re-skinnned with aluminum. The price for this airplane was $30,000, a staggering sum to Piper whose company had survived — unlike so many others — thanks to inexpensive airplanes like the Cub. But they crossed their fingers, named their twin "Apache" and went to market. This time, they'd got it right.

CESSNA CH-I

In the summer of 1952, Cessna aircraft purchased the Seibel Helicopter Co. and the following year that marriage produced an unfortunate offspring, the CH-l, a whirlybird trapped in a Skylane's body. Cessna hoped to sell the little choppers to the Army and the Pentagon did purchase 10 improved models in 1957, but never came back for more. So, Cessna dressed up its 270-hp orphan with a four-place "executive interior," named it "Skyhook" and tried to sell it to civilians at $80,000 per copy. The drums rolled, the trumpets blared, the press releases flew, but only 23 Skyhooks got sold. So finally in December 1962, after having spent 10 years and several million dollars. Cessna abandoned the helicopter business. It recalled and then scrapped the Skyhooks it had sold and went back to building airplanes with fixed wings and wheels.

CESSNA MODEL 620

All Cessna's come in two-flavors, single-engine and twin-engine. Right? Wrong. There was one brief but magnificent moment when Cessna conducted a four-engined symphony; the movement was entitled the Model 620. It had four, supercharged, 320-hp Continentals, a stand-up cabin, seats for 10, pressurization, airconditioning and a wash room. Cruising at 18,000 feet, the 620 could fly 1,450 sm at 247 mph and arrive with 45 minutes reserve in the tanks.

The 620 first flew in August 1956, and its flight trials proved it a good airplane, equal to its design goals. The 620's flaw wasn't one of engineering, but rather one of timing. As the big Cessna neared production, the airlines; began

buying jets and dumping their prop-driven Convairs and Martins on the civilian market. A cost analysis by Cessna revealed that corporations could purchase large, surplus airliners for less than a new 620 would cost. That finding was fatal; the 620 program was cancelled in 1957.

MOONEY MARK 22

The engineering that precedes prototype construction is often so detailed and so complex it can take years to complete. But the thousands of dollars invested in these "paper" airplanes can prevent a badly designed prototype from being built, a far more expensive proposition.

With that in mind, the original Mooney Mark 22 comes into focus. The Kerrville, Tex., planemaker's single-engine models had earned an enviable reputation for their low cost and high performance. The next logical step for Mooney was to build a twin, which it did, lickety split.

Armed with a minimum of engineering data and a maximum of energy, the Mooney makers pulled the nose-mounted engine off an M20A, hung two 180-hp Lycomings on the wooden wings, installed a dorsal fin and sent the insta-twin skyward. That was in 1959. In 1960, the twin Mooney was disassembled, never to be seen again.

Recalled one Mooney engineer, "that thing would never have been certified. . .it had no directional control if you lost an engine. It was just an assemblage of parts that made it look like a twin."

MOONEY MUSTANG

Why doesn't someone build a pressurized single? You've not heard tell of the Mooney Mustang. The Mark 22 — the same designation as that of the short-lived Mooney twin — first took flight on Sept. 24, 1964, and it was one screaming single. Powered by a turbocharged 310-hp Lycoming, the Mustang could buzz along with five souls on board at a cool cruise of 229 mph. Maximum speed was 256 mph, range was over 1,000 miles and its service ceiling was a storm-topping 24,000 feet.

Performance was no problem for the Mustang, but money was. The Mark 22 was an expensive bird to build and buy. When sales began in 1967, the Mustang went for $34,000 but the price climbed as fast as the plane. By 1970, the year Mooney aircraft shut down completely, the Mustang was selling for $47,000.

During its four-year production run, only 28 Mustangs were built and when Republic Steel restarted the Mooney line in 1973, it passed over the Mark 22. The Mustang was just too much pony, even in Kerrville, Tex.

PIPER POCONO

The Big Airplane has often become the Big Headache. For Beech it was the Twin Quad. For Cessna it was the Model 620. For Piper, the migraine was called Pocono.

Design work on the mini-airliner began in 1966 and two years later an impressive piece of metal rolled into the sunshine at Piper's Vero Beach, Fla., plant. The craft's long, nearly stand-up cabin could seat 18, its 51-foot-wide wings could lift 9,500 pounds and its twin, 500-hp supercharged Lycomings could pull it along at 210-230 mph. A lot of airplane for $200,000.

The plane flew but was never placed in production. It has a weakness — power. William T. Piper, Sr., had commanded his flagship to piston-powered, reasoning that turbines were too expensive and Jet-A was not universally available. But the piston engines then available simply weren't equal to the Pocono's needs and so the big Piper was tied down in limbo. Waiting. And so it waits today, sun-bleached and weathering on flat tires and without engines in an obscure corner of the company's Lakeland, Fla., ramp.

McDONNELL MODEL 220

James S. McDonnell, head of the giant aerospace firm that bears his name, was the recipient of a meritorious service award presented during the National Business Aircraft Assn. convention in 1966. When he stood to accept the honor, "Old Mac" said, unabashedly, he'd paid for it. Dearly.

You see, 10 years earlier the Air Force had invited companies to submit flying entries to compete for an executive jet contract. McDonnell sank about $11 million into developing its entry, a four-engine beast called the Model 220. Unfortunately for McDonnell, Lockheed entered the competition. It called its entry the JetStar and the Air Force called it the winner.

McDonnell considered marketing his .79 Mach jet as a corporate carrier, but the project fizzled, the prototype was donated to the Flight Safety Foundation and McDonnell went back to building F4 Phantoms. The 220, by then grounded, changed hands several times until 1974 when it was purchased by Richard Archer, a wheat farmer, for $160,000. Archer says he's not quite sure what he'll do with his mini DC-8, but then neither was James S. McDonnell.

THE SPRUCE GOOSE IS ALIVE AND WELL!

by Ed Schnepf

Reprinted by Permission
AIR PROGRESS (August 77)

Stupendous — Collossal — Magnificent — Gargantuan! Pushy superlatives that Howard Hughes himself may have chosen to publicize one of his films? Perhaps so, but when used to describe one's reaction at actually seeing the enigmatic Spruce Goose, even these supersell lines fail to recite its awesome visual impact.

Yes, the Hughes Hercules, as it is properly known, is alive and well in Long Beach. In fact it's better than ever and according to Summa Corp. spokesmen, could indeed fly in a few weeks time if it had to. A statement of this kind may come as a bit of a shock to sceptics who, for years, have argued that in all probability Howard Hughes' biggest pet toy didn't exist anymore. After all, who in his right mind would spend a million dollars a year carefully maintaining a mammoth aircraft that flew less than two minutes over thirty years ago? But this is exactly what Hughes did, and to this day the world's largest seaplane — with a wingspan greater than the length of a football field — is kept beautifully maintained in an air conditioned hangar at Long Beach, California.

The enigmatic story of the Spruce Goose will end if a California museum group accomplishes their goal of putting the Hercules on permanent display beside the Queen Mary. Formed as a non-profit organization called "Air Museum of the West," the group has tentatively been assured that the Hughes Summa Corp. will donate the Hercules to them once negotiations are completed with the City of Long Beach. As soon as that happens, the Hercules will be taxi/towed to the Queen Mary area and placed on display for the world to see what, up until now, has been often called aviation's best kept secret.

"You can't believe what beautiful shape that boat is in," said Clay Lacy, one of the museum's founders. "Every instrument is operable, every system is go. The only things missing were the eight props removed a few years ago to stem corrosion from the salt air. But they are like brand new — just like the rest of this wooden wonder."

Lacy and several other members of the museum group were accorded the rare privilege of actually seeing the seaplane in its giant aluminum hangar recently. The visit culminated two years of work by the museum members, who Summa officials felt should at least see that what they have so diligently worked for actually exists. At the end of the tour, one member, William Ade, commented that he felt he had just visited a national shrine. "This aircraft is more than a testimony to Howard Hughes," Bill Ade commented. "It reflects all of the achievement of the aircraft industry itself. It deserves to be memorialized."

This kind of dedication typifies the group's attitude toward saving the seaplane from an uncertain fate. Up until now, several offers to take over the project have been turned down by Summa Corp. They prefer to see the aircraft remain in California on public display rather than be cut up and pieces sent to various other museums, as the Smithsonian suggested.

The Air Museum of the West has plans to raise from $1.5 to $2 million to build an air museum to house the giant flying boat. Other exhibits pertaining to western aviation history will also be included. Several colorful aircraft are already pledged for display and industry-wide cooperation is anticipated, according to Municipal Judge Gilbert Alsten, one of the organizers. "We hope to make this the finest air museum on the West Coast. We have everything going for us. Plenty of easy parking, the attraction of the Queen Mary opposite us, and city officials in Long Beach who are eager to see it all happen. Nothing like the Hughes Hercules has ever been built before and nothing like it probably will ever be build again."

Indeed this is so, for the era of the giant prop-driven seaplane is long gone, likely never to return. So, too, is Howard Hughes, a man best remembered for his pioneering drive in making air travel the commonplace transportation mode it is today. It cannot be said that the "Spruce Goose" contributed much to the history of aviation, since it never fulfilled the role for which it was

built. But it is an engineering wonder, a silent Goliath that contains in its cavernous birch hide the soul of everything that is and was the aviation industry's greatest hopes and finest hours. It is a rare example of aviation's primordial past, and as such, millions will undoubtedly come to pay homage when it goes on display.

MILITARY AEROSPACE

AVIATION YEARBOOK

INTRODUCTION

Developments in military aviation and aerospace in 1977 can either scare you to death or offer promise of solutions to many of the problems confronting life on planet Earth, depending on whose views you choose to peruse.

If you're the fearsome kind, you may want to skip a couple of the entries in this chapter — the *Aviation Week* report by Clarence A. Robinson, Jr., on USSR development of a laser beam weapon to destroy missiles and satellites in orbit; and the comments of Maj. Gen. George J. Keegan, former chief of U.S. Air Force intelligence, who describes disturbing Soviet progress on the beam weapon and sundry other projects.

On the brighter side, the space shuttle made its first free glide landings at Edwards Air Force Base, Calif. this year, and Dr. Krafft Ehricke discusses shuttle applications that will, he's confident, free us from earth's material limitations by colonizing the moon and drawing upon its resources and those of more distant planets to fill our ever-growing needs.

The Air Force's B-1 bomber is the subject of several pieces in this chapter leading off with President Carter's decision to cancel its production. Was that a good decision? The President may want to cancel it, but Congress *does* have a say in the ingredients of our defense posture, and as this volume went to press there was some question whether Congress would consent to see the B-1 die, or whether, if Congress *did* vote funds to continue it, the President would allow the Defense Department to spend the money.

And that the B-1, or some other new bomber, is an invaluable military tool is argued with foresight by Dr. G. K. Burke in "The Case for the Manned Penetrating Bomber," and with hindsight in a dialogue of rare historic value between Lt. Gen. Ira Eaker, USAF's Eighth Air Force commander in World War II, and Albert Speer who, as Hitler's production boss, saw his factories pulverized by Eaker's bomber crews.

Meanwhile, the alternative to the B-1 favored by President Carter is the cruise missile, whose characteristics, along with those of an imposing variety of other "Cost Effective and Fearless" unmanned vehicles, are described by Stefan Geysenheyner.

The Army and Navy suffered no major setbacks comparable to the B-1 during the year. The Navy went along with the Army in choosing Sikorsky to build its Light Airborne Multi-Purpose System (LAMPS) for duty aboard its surface combat vessels and particularly on a new class of mini-aircraft carriers under development, called sea control ships.

The LAMPS helicopter will be a variant of the Army's UH-60 Black Hawk assault helicopter which will begin replacing the Bell UH-1 Huey this summer. According to current programs, Sikorsky will build some 1,100 UH-60s for the Army and 200 LAMPS for the Navy.

Aircraft requirements of the nation's fifth military service, the Coast Guard, seldom receive much notice because Coast Guard planes have traditionally been hand-me-downs from the other services. You'll still find Grumman HU-16

Albatross amphibians in the Coast Guard, for instance, even though they've long been retired from the Air Force and Navy.

But now the Coast Guard is acquiring a new patrol craft not in the inventory of the other services. It has chosen the Dassault-Breguet Falcon 20G to replace the Albatross, to be fitted with Collins search and surveillance radar. We were unfortunately unable to include an article on this plane.

For an insight into what's going on in the rest of the world, and an outsider's look at some U.S. aviation programs, don't miss John W. R. Taylor's "Aerospace Review," by the editor of *Jane's All the World's Aircraft.* As all Jane's readers know, Taylor seems privy to all the world's aviation secrets, and he lets us in on many of them in this all too brief roundup.

In his article, Taylor comments on the state of Soviet technology as uncovered in the MiG-25 flown to Japan by Soviet Air Force Lieutenant Viktor Belenko late in 1976. But because the MiG-25's makeup is of more than passing interest to several groups of readers — aeronautical engineers, aviation buffs in general, and military pilots who may have to confront the MiG-25 or a descendent in the future — we have added a more detailed evaluation by Georgiy Panyalev.

Britain is also the source of "Nimrod Is Go," in which we depart from our normal content to report on a British plane built for the Royal Air Force. The reason is that the Nimrod referred to in the title of this piece was selected by the British as their airborne command post in preference to the Boeing E-3A AWACS favored by other NATO allies, thus introducing a complicating factor in NATO's integrated air defense fighter support operations.

The "aerospace" portion of this chapter is somewhat slighted, but in defense we note that, with the notable exception of the space shuttle, not much was going on in U.S. space activities this year. Across the Atlantic, however, the European Space Agency, now in its third year, was planning and carrying out several projects, including the Spacelab payload for the Space Shuttle. The U.S. managed to mar that year when a Delta booster carrying ESA's Orbital Test Satellite exploded shortly after launch, destroying the $42 million payload.

These are what we see as the military/aerospace highlights of the year. Undoubtedly readers will find subjects we've overlooked or underemphasized. But before readers write to protest our omissions, we suggest they take a look at last year's volume, for one year is a relatively brief span in the evolution of aviation and space programs, and what we covered last year we hesitate to use our limited space to report on again. On the other hand, what we overlooked this year may just be coming into focus for inclusion in next year's volume. Maybe the best way to find that out is to check in with us again next year.

Allan R. Scholin
December 1977

Edited for Space

Washington — Soviet Union is developing a charged-particle beam device designed to destroy U.S. intercontinental and submarine-launched ballistic missile nuclear warheads. Development tests are being conducted at a facility in Soviet Central Asia.

The Soviets also are exploring another facet of beam weapons technology and preparing to test a spaceborne hydrogen fluoride high-energy laser designed for a satellite killer role. U.S. officials have coined the term directed-energy weapons in referring to both beam weapons and high-energy lasers.

A charged-particle beam weapon focuses and projects atomic particles at the speed of light which could be directed from ground-based sites into space to intercept and neutralize reentry vehicles, according to U.S. officials. Both the USSR and the U.S. also are investigating the concept of placing charged-particle beam devices on spacecraft to intercept missile warheads in space. This method would avoid problems with propagating the beam through the earth's atmosphere.

Because of a controversy within the U.S. intelligence community, the details of Soviet directed-energy weapons have not been made available to the President or to the National Security Council.

Recent events have persuaded a number of U.S. analysts that directed-energy weapons are nearing prototype testing in the Soviet Union. They include:

- Detection of large amounts of gaseous hydrogen with traces of tritium in the upper atmosphere. The USAF/TRW Block 647 defense support system early warning satellite with scanning radiation detectors and infrared sensors has been used to determine that on seven occasions since November, 1975, tests that may be related to development of a charged-particle beam device have been carried out in a facility at Semipalatinsk.
- Ground testing of a small hydrogen fluoride high-energy laser and detection of preparations to launch the device on board a spacecraft.
- Test of a new, far more powerful fusion-pulsed magnetohydrodynamic generator to provide power for a charged-particle beam system at Azgir in Kazakhstan near the Caspian Sea. The experiment took

place late last year and was monitored by the TRW early warning satellite stationed over the Indian Ocean.

- New test site at Azgir under the direct control of the Soviet national air defense force (PVO Strany). Since the PVO Strany would be responsible for deploying a beam weapon to counter U.S. ICBM warheads, its involvement indicates a near-term weapons application for these experiments, U.S. officials believe.
- Shifts in position by a number of experienced high-energy physicists, who earlier discounted the Soviet capability to develop the technology for a charged-particle beam device. There is now grudging admission that the USSR is involved in a program that could produce such a weapon.
- Recent revelations by Soviet physicist Leonid I. Rudakov during a tour last summer of U.S. fusion laboratories that the USSR can convert electron beam energy to compress fusionable material to release maximum fusion energy. Much of the data outlined by Rudakov during his visit to the Lawrence Livermore Laboratory has since been labeled top secret by the Defense Dept. and the Energy Research and Development Administration, but it gave a clue to U.S. scientists that the USSR is far ahead of the U.S. in controlled fusion by inertial confinement (compression of small pellets of thermal nuclear fuel) and weapons based on that technology.
- Pattern of activity in the USSR, including deployment of large over-the-horizon radars in northern Russia to detect and track U.S. ICBM reentry vehicles, development and deployment of precision mechanical/phased-array antiballistic missile radars and massive efforts aimed at civil defense.
- Point-by-point verification by a team of U.S. physicists and engineers working under USAF sponsorship that the Soviets had achieved a level of success in each of seven areas of high-energy physics necessary to develop a beam weapon.

U.S. officials, scientists and engineers queried said that the technologies that can be applied to produce a beam weapon include:

- Explosive or pulsed power generation through either fission or fusion to achieve peak pulses of power.
- Giant capacitors capable of storing extremely high levels of power for fractions of a second.
- Electron injectors capable of generating high-energy pulse streams of electrons at high velocities. This is critical to producing some types of beam weapons.
- Collective accelerator to generate electron pulse streams or hot gas plasma necessary to accelerate other subatomic particles at high velocities.
- Flux compression to convert energy from explosive generators to energy to produce the electron beam.
- Switching necessary to store the energy from the generators in large capacitors.
- Development of pressurized lines needed to transfer the pulses from the generators to power stores.

The lines must be cryogenically cooled because of the extreme power levels involved.

There is little doubt within the U.S. scientific or intelligence communities that the Soviets are involved in developing high-energy technology components that could be used to produce a charged-particle beam weapon, but there is a great difference of opinion among officials over whether such a device is now being constructed or tested in the USSR.

In increasing numbers, U.S. officials are coming to a conclusion that a decisive turn in the balance of strategic power is in the making, which could tip that balance heavily in the Soviets' favor through charged-particle beam development, and the development of energetic strategic laser weapons.

The Semipalatinsk facility where beam weapons tests are taking place has been under observation by the U.S. for about 10 years.

The total amount invested by the USSR in the test project for the 10 years' work there is estimated at $3 billion by U.S. analysts.

For several years, Air Force Maj. Gen. George J. Keegan, who until his recent retirement headed USAF's intelligence activities, has been trying to convince the Central Intelligence Agency and a number of top U.S. high-energy physicists that the Soviets are developing a charged-particle beam weapon for use in an antiballistic missile role.

Evidence was gathered by Air Force intelligence from a variety of sources, including early warning and high-resolution reconnaissance satellites, published USSR papers on high-energy physics and visits between Soviet and Free World physicists. In contacts with scientists deeply involved in developing components necessary for beam weapon application in both the USSR and the U.S., data was gleaned that clearly showed the Russians to be years ahead of the U.S. in most areas of technology, one U.S. physicist said. He added that it became increasingly clear that the Soviets were making a concerted effort to develop the technology in each area so that, if it was pulled together, a beam weapon and possibly related laser weapons could result.

All of the evidence that Gen. Keegan and his small team gathered about Soviet designs on charged-particle beams was presented to the CIA and its Nuclear Intelligence Board, which has so far rejected their conclusion that beam weapons development is evident.

"Keegan refused to accept CIA's evaluation of the USAF intelligence data," one U.S. official said, "So, he systematically set about acquiring talented young physicists to analyze the information and to probe the basic physics of the problem — an area in which U.S. scientists were notably deficient."

"These young physicists gathered to his cause by George (Gen. Keegan) were a very sharp group of young turks, and some have since gone on to gain stature within the high-energy physics crowd," one official said.

It was anticipated by Gen. Keegan and his advisers that the USSR would be forced to vent gaseous hydrogen from the experiments at Semipalatinsk and that early warning satellites could detect it.

"Explosions of such gaseous hydrogen discharges are now being detected with regularity from Soviet experiments," a U.S. official said, "and scientific studies of the gas releases and explosions have confirmed their source as being near the Semipalatinsk facility."

"There is now no doubt that there is dumping of energy taking place at the site with burning of large hydrogen flames," one official said. "What bothered the Nuclear Intelligence Board at first was that it was hard to imagine that some seven technologies critical to the weapons concept could be perfected there within the time frame presented and not be detected by us.

"In each case, the Air Force was able to disprove the theories advanced, at least to USAF satisfaction," one U.S. official said. "But along the way Keegan became an outcast within CIA and the Defense Intelligence Agency. This was despite the fact that many times in the past it turned out that his intelligence information proved correct when it was not accepted at first. He (Keegan) made some great intelligence breakthroughs," another official said.

A number of influential U.S. physicists sought to discredit Gen. Keegan's evidence about Soviet beam development. The general attitude within the scientific community was that, if U.S. could not successfully produce the technology to have a beam weapon, the Russians certainly could not. "It was the original not-invented-here attitude," one U.S. physicist said.

As evidence of Soviet intent mounted, the Air Force convened a munitions panel of its Scientific Advisory Board to examine the problem. The panel met at Livermore Laboratory for three days to study the data of Gen. Keegan and his technologists.

"The panel of experts rejected virtually all of the Air Force's hypotheses. In an emotional meeting, they denigrated all suggestions of nuclear explosion generation, power storage, power transmission and collective acceleration," an official explained. "The net result is that evidence about possible beam weapons development was rejected."

In an effort to prove that USAF intelligence estimates were correct, Gen. Keegan and his young physicists set about trying to prove Soviet technology exists in areas necessary for beam weapons.

After isolating the theoretical roadblocks, identified by the Scientific Advisory Board's munitions panel, the physicists, along with several new groups recruited by Gen. Keegan, went to work exploring possible USSR technologies.

Within a few months the team, under the direction of a young Air Force physicist, found that all the munitions panel's objections could be overcome "and had already been solved in the Soviet Union. Several breakthoughs in the high-energy physics were involved," an official said.

In 1975, Gen. Keegan disclosed his findings on Soviet technology related to beam weapons development to William Colby, then head of the CIA, and to a number of its nuclear scientific advisers.

"On the strength of Keegan's information that the Soviets were on the verge of developing a weapon to neutralize our ICBMs and SLBMs, Colby directed the formal convening of the CIA's Nuclear Intelligence Panel to consider the disclosures," according to a U.S. official.

In a final meeting last year with the panel, Gen. Keegan and his associates presented evidence over a three-day period to the panel. The panel went into executive session to study the data and then wrote its report.

"What the report said was that there were no technological errors in USAF's analytical work. It was agreed by the board that there is a massive effort in the USSR involving hundreds of laboratories and thousands of top scientists to develop the technology necessary for production of a beam or other energy weapon for use against U.S. ICBMs and SLBMs," an official said. The report also said the board was unable to accept USAF's detailed conclusions regarding the experimental site at Semipalatinsk. It reasoned, according to several sources, that since none of the key subtechnologies involved had been perfected in the U.S., it was implausible that the Soviets could be so far ahead. In any event, the U.S. scientific advisers to CIA were unwilling to concede that the Soviets could harness such advanced technology into a working weapon or demonstration system.

"After 10 years of work at the site and after developmental testing of the beam for over a year, the only thing required is to scale the device for weapons application," an official said. That could be accomplished by as early as 1978 with a prototype beam weapon, and it could be in an operational form by 1980, some officials believe.

Another big objection offered by some U.S. physicists and other scientists is that the beam from such a weapon will have to be propagated and bent to intercept incoming warheads in reentry vehicles, an extremely difficult task.

One possible solution is that a "magnetic mirror" can be used for beam bending to intercept reentry vehicles.

Despite strenuous objections from U.S. scientists over the feasibility of beam bending, USAF intelligence established a Soviet solution to the problem for the Soviet beam concept, an official said.

Precise pointing and tracking may not be required. "All that is needed is for the Soviet long-range precision radars now deployed in violation of the ABM agreement to detect avenues or windows for reentry vehicle trajectories against targets in the USSR. By aiming rapidly pulsed proton beams into these windows, ICBMs and SLBMs could be quickly saturated and destroyed," he explained.

The windows would be located from 1,000 to 2,000 naut. mi. out in space. "With this method, many acquisition and tracking problems could be overcome. By using the window concept to scatter the beam over a wide area through which warheads must transit, it is believed that not many beam weapon devices would be required to protect the USSR from a U.S. retaliatory strike," the official said.

Many deployment schemes of great simplicity are open to the Russians. One such scheme would be to place the collective accelerators vertically inside silos that the USSR now claims are for command, control and communication.

There are at least 150 of these silos that the U.S. is now overlooking by accepting the Soviet definition as command and control centers for their use. Using nearby silos linked to those with the accelerator for containment of the explosive generator, the Soviets could deploy such a system within a few years, an official said.

"Since the necessary radars are nearing operational readiness, all of the needed system components could be emplaced," he added.

"The one thing that George (Gen. Keegan) finds so pernicious about this whole thing is that CIA and other top U.S. officials scoff at the idea that the backward Russians can develop technology that we have been unable to develop in the U.S.," one official said. "He (Keegan) admits that he could be wrong, but he is not wrong about the Soviets' will to produce such a weapon and about the national assets they are devoting to it."

"From all of this evidence we have a good idea of where the Soviets are in development and where they are headed with beam weapons and high-energy lasers. Not much has been done in this country since Seesaw," a U.S. physicist said. "But there is certainly a lot of new interest now within the scientific community."

There is an effort under way to establish an agency in the U.S. to coordinate the development of directed-energy weapons. Some congressional staff members as well as officials within the Administration are pressing for this to be accomplished.

John L. Allen, deputy director of Defense research and engineering for research and advanced technology, said: "the Defense Dept.'s Advanced Research Projects Agency and the services are investigating the application of high-energy lasers. Both the Army and Navy are pursuing terrestrial applications. The Air Force is pursuing airborne applications, and the Defense Advance Research Projects Agency is looking at the possible application of lasers in space defense with emphasis on chemical lasers. It is still too early to determine the potential cost effectiveness of high-energy lasers as weapons, but the next two or three years will yield a great deal of insight."

"Many possibilities are open for the U.S. but remain unexplored," a senior U.S. official said. "Whether this results from lack of interest, lack of funds for research, lack of national focus for efforts in this field, or a belief that the possibility that such weapons may adversely effect détente is unclear. It does seem that the Soviets have taken a very different course which may eventually prove most U.S. planners and analysts to be wrong."

STRATEGIC BALANCE: TRENDS & PERCEPTIONS

by Maj. Gen. George J. Keegan, Jr., USAF (Ret.) Reprinted by Permission
Edited for Space

General Keegan, former chief of U.S. Air Force Intelligence, challenged U.S. intelligence estimates at a press luncheon sponsored by the American Security Council on March 11, 1977. Significant portions of that speech are excerpted below.

—JSAY

Today, I speak as a private citizen — expressing my own personal views regarding the Soviet threat and the evolving world power balance.

This is the thrust of my concerns: Persistent underestimation; diminution of our retaliatory punch through an unperceived war survival effort of unprecedented scope; anti-satellite weapons to deny our use of space for warning and command; pioneering research in directed energy weapons to kill our retaliatory missiles; an omnipotence of land; an ability to deny our use of the seas; and a continuing projection of power into the Third World representing the greatest imperialism in history. Deny these realities if you will. I cannot.

The Question of Superiority

When I retired, I expressed some judgments that are in need of clarification. First, I would like to set the record straight regarding the strategic judgments which have been attributed to me in recent months. And I would like to explain further why I have considered it necessary to speak out not only about the poor quality of our estimative judgments but about the serious nature of the evolving threat which faces the free world.

First, let me deal with the question of "superiority." Frankly, the word is neither a very meaningful nor a very helpful one, in that it has been subjected to such loose interpretation and definition. As a pre-condition to further discussion of superiority I must make it absolutely clear that I believe the free world has a formidable array of physical power, of psychological strength, and of economic viability at its command. No opposing consortium of nations can afford to challenge us beyond the more dangerous thresholds without entailing some very serious risks indeed. We are a very powerful nation.

Now, as to the difficult question of who is number one and who is not. It is my considered belief that the United States is superior in only one major combat area — and that is in its ability to respond quickly and efficiently to a nuclear initiative by the Soviet Union. And that's where our superiority stops! Otherwise we are almost totally unprepared to wage and sustain strategic warfare — and the Soviets *are* prepared. The Soviets have more staying power. They are far better able to protect their society, their forces, their economy, and their war-oriented industrial base. Their hot production base is capable of generating, and is now generating, a vastly superior quantity of high quality armaments for all levels of military application.

Soviet forces today are premised exclusively on the principle that they must be prepared to wage war at all conflict levels and emerge successfully. Such a strategic philosophy is totally different from our own.

U.S. strategic forces and policy, in contrast, have been premised on the view that nuclear war was so horrible that it could not be contemplated in any rational environment. Therefore, we have based our military preparations, at the strategic level at least, on the view that we must be prepared to deter such aggression. The Soviet view, in contrast, has been that they must be prepared to wage nuclear war if they are to satisfy their own security requirements.

While I have no quarrel with the view that nuclear war must be avoided at all costs, the question for me is which strategy is more likely to assure avoidance of war? If it was our intent to deter by a process in which we would inflict unacceptable punishment upon the Soviets for their initiation of global war — then I am unaware of a single definitive effort ever conducted by the United States to determine precisely and in great detail what it would take to deter. The American view of what is necessary to deter was, in fact, based largely on loose generalizations regarding the logic of city busting. Such logic was created in large part by academicians and theoreticians for whom the sound of battle was but an abstraction. The Soviet "mindset" on the other hand — documented, and massively ignored by this country until recently — is based on an entirely different set of considerations.

A Nation Can Survive a Nuclear War

After World War II Soviet military professionals undertook the most extensive examination of the lessons of that war, along with analysis of the impact of modern technology on any future war ever undertaken. A decade or so after the Soviet studies were undertaken, the Soviets concluded that a nation could survive a nuclear war and emerge as a viable society. Virtually all Soviet strategic planning and investments since that period have been premised on that one fundamental conclusion.

The Soviets published their conclusions and made them available in some of the most learned and sophisticated strategic literature in modern history. Few in this country would pay attention.

Thus having decided that they could wage nuclear war and emerge with some margin of advantage — however pyrrhic — the Soviets enunciated a strategic policy premised on the view that war at any level of conflict could be waged, and won — and that they could emerge as a viable, controlled, surviving military entity. Fundamentally, that is what the famous Penkovsky documents were all about.

Oleg Penkovsky was the Soviet colonel who in the early 1960s, supplied U.S. intelligence with a remarkable array of top secret documents regarding Soviet strategic planning — until he was arrested and shot. Regrettably, those remarkable documents — clearly reflecting Soviet long-term plans — have not yet been published and made available to the American people — for reasons which I am at a total loss to comprehend.

However, you don't have to have the Penkovsky documents in order to understand Soviet strategic planning. All you have to read is Soviet Marshal V.D. Sokolovsky's book *Military Strategy*, now in its third edition. What it provides is a detailed, unmistakable blueprint of how one prepares an economy and a society for the acceptance of total war — premised of course, on the bedrock principle that such capabilities have one fundamental end in mind — namely, to help the Soviet political leadership impose its way of life over the rest of the world.

The Sokolovsky book is the single most comprehensive blueprint on the requirements and preparations incident to waging total war ever published.

As a result of the Soviets having taken a largely differing view of strategic balance and war-fighting capability in the nuclear age, the Soviets have emerged with an entirely different strategy, an entirely different conception of war and peace. When we talk of superiority in the United States, we are talking of two things: One is a superior ability to respond very quickly to a warning that an attack is underway and, if we choose, to launch our retaliatory forces before they are crippled by such an attack. "Launch under attack" before we are in fact struck is the only meaningful choice remaining open to us in the event of war — yet, this is a doctrine which many in high places allege has been renounced as a form of strategic response. It is hardly a matter which should be left in doubt. The other is that we are a technologically superior society capable of building and producing anything that we wish. It is a promissory situation quite removed from the reality of existing military capability.

USSR On War Footing

The Soviet Union today is on a virtual war footing. Soviet industrial, scientific, and economic investment remains largely subordinated to the conflict purposes of the State. Until the United States is willing to understand this simple and fundamental truth, we will continue to deceive ourselves about the Soviet Union and its capabilities — not to speak of its intentions.

Now, what do I believe about relevant Soviet fighting capabilities? In my considered judgment, the Soviet Union today has a capability to initiate, wage, survive and emerge from a global conflict with far greater effectiveness than the United States and its Allies. That is not to say that if we retaliated in a timely fashion to a Soviet attack, Soviet cities would not be burned to the ground. However, when you hear the phrase, "They would cease functioning as a viable society," I believe that is an unstudied recalling of the language utilized by some of the Whiz Kids during the McNamara era. It is not based on an in-depth examination of the extraordinary changes which have taken place in the Soviet Union during the past decade. The truth of the matter is that after a nuclear exchange today the Soviets would continue to function as a viable, controlled military entity.

Regrettably, it remained largely for me — as Chief of Air Force Intelligence — to bring to the attention of the intelligence community and our national leaders significant new data on Soviet strategy, weapons developments, forces, research and development, civil defense preparations, advances in chemical warfare, Backfire bomber capabilities, and possible violations of the SALT accords.

In many instances, our Air Force data was ignored, dismissed or taken under advisement, because it did not coincide with the mindset of an intelligence process that, in my opinion, has become highly politicized in the last twenty years. That judgmental process has tended to reflect the hopes and aspirations of those in diplomatic leadership.

Strategic Warning Deficient

Our greatest single deficiency today is in the strategic warning area. Since the Pearl Harbor attack, this country has spent tens of billions of dollars in order to assure that a surprise attack would never happen again. Yet today I submit that we are not much better off than we were on the eve of Pearl Harbor.

Let me explain. As successive political administrations have come under ever greater compulsion to "control" the intelligence process, increasing civilianization and control of that process has taken place. Preparations for war are best understood by those whose profession it is to understand and to be prepared to cope with war in all of its dimensions. As our military intelligence capabilities in the Services have been centralized, and allowed to wither away in the name of more efficient centralization — our military assessment capabilities have declined precipitously. Over the years warning analysis has been confined increasingly to an ever smaller number of highly placed civilians at the top of the structure. These are people who never go to war. These are people who understand little about the horrors of war. And these are people who understand even less about military doctrine, strategy and weapons.

Finally, for good measure, we have imposed severe compartmentalization upon the entire process of judgment-making and the passing of warning information and intelligence.

There is an unstated rule today which requires the various agencies of government to use "agreed national intelligence" in making its various judgements regarding the strategic balance. The result is that our judgements tend to lag the reality by anywhere from five to fifteen years. "Agreed national intelligence" as derived by committee, contains little that is controversial, except in some footnote or dissent. And it rarely contains meaningful references to the dynamics of strategic competition in terms of the new weapons, new forces, and new capabilities being evolved by the Soviet Union.

The result is that when the Joint Chiefs of Staff must reply to an inquiry from Senator Proxmire regarding allegations attributed to me by the *New York Times,* they are restricted in their response by that intelligence which is published in agreed national estimates. Those estimates bear no more relationship to reality then did our estimates of evolving Japanese capability in 1938 and 1939.

During the decade of the Fifties when the military services retained most of their capabilities and had a powerful and viable attaché program throughout the world, the President of the United States was seldom if ever surprised by an evolving conflict crisis or situation. During the period of the 1960s and 1970s just the reverse has been true. We have been met by one surprise after another, diplomatically, militarily, and economically. The 1973 war in the Middle East was typical. Most of the lieutenants in the Bar Lev Line in the Sinai knew that war was coming, as did most of the company and battalion G-2s in World War II in the days preceding the Battle of the Bulge. However, if the facts do not conform to the particular mindset of those at the top, then the situation tends to be dismissed.

The Problem of Mirror Image

Another problem that afflicts the intelligence community is that of the mirror image, in which we persistently try to view what the Soviets are doing in the light of our own logic, experience and strategic preconceptions. In my judgment, there is no way in which we can understand what is transpiring within the Soviet Union unless we do it from the Soviet point of view. And it was for that reason that Air Force Intelligence moved to obtain original Soviet writings, to translate them and to make them available to the American public.

Not-Invented-Here Bias

Some years ago, I was the first not to discover but to suggest that certain extraordinary weapon development projects in the Soviet Union, far beyond anything ever undertaken in the free world, might seriously inhibit or neutralize our strategic potential in the next decade. Each one of my findings and allegations was the result of the most intensive analysis and research. Yet the response in almost every instance was a reflection of America's scientific egocentricity: if it has not been invented in the United States, then it is not likely to be invented in the Soviet Union.

It has been my unbroken experience since World War II that most scientific consultants brought into the intelligence community to render advice about the potential of Soviet technology have been wrong. Yet the practice of inviting scientists in to advise the intelligence community continues unabated.

Despite broadened travel opportunities within the Soviet Union today, many of our senior scientists remain functionally incapable of recognizing that the "peasants" behind the so-called Potemkin facade are capable of some of the most original and scientifically creative work in the world — a great deal of it beyond the scope of our own technical and scientific capability or understanding.

I ask you, who is superior in the quality of its chemical warfare weapons and in their ability to defend against same? I ask you, who produces the largest quantity of high quality battlefield weapons in the world? Where are the ground jamming equipments in the West to equal those of the Soviet Union and the Warsaw Pact? Where are the mobile SAMs which approximate those of the Soviet Union in quality, effectiveness, and supportability?

When people in the United States talk about our technological superiority — I suggest to you that they do not know what they are talking about. Our superiority is something which exists only in the minds of men and in the productive and laboratory potential of this great country. It remains to be bought and paid for, to be manufactured, distributed, and deployed to our forces.

Now what are some of the intelligence judgments that have occasioned my views? The ultimate one has to do with whether the Soviets are seeking parity or superiority. The evidence on this subject has been conclusive for better than fifteen years. Soviet documents in the thousands have been readily available and even more readily confirmable through observable hard intelligence. Yet a vast mythology about Soviet strategic objectives has been imposed upon the intelligence community, mostly by members of the NSC staff, CIA, the State Department, and on occasion — the Rand Corporation.

The failure to assess Soviet superiority aims when they were thoroughly documented remains one of the most disgraceful chapters in the annals of American intelligence.

Soviet Civil Defense

Another important intelligence judgment which has influenced my views has to do with civil defense. There are some who think that I violated security in calling the nation's attention to information which, in my opinion, has altered the strategic balance in significant, if not decisive, ways. Little of what I divulged could be called sensitive. All, or virtually all of it, was available in the public sector to any enterprising researcher and analyst.

There is a vast body of readily available literature in the Soviet Union regarding their civil defense effort. It remained for my small staff to acquire this literature with its thousands of photographs, sketches, periodicals and pamphlets. We then proceeded to avail ourselves of the Soviet regulations and directives on the subject of civil defense — all readily available for anyone who would make even a cursory effort.

And then we talked to a few of the highly knowledgeable human sources who now reside in the free world — people who have had an extensive experience with and involvement in virtually every aspect of Soviet hardening, war survivability and civil defense preparations. Why did not the CIA undertake such initiatives and analysis when it was important for them to do so?

Soviet Union Has Gone Underground

During the past twenty years the Soviets have proceeded to place their most essential war support and fighting capabilities underground. Those strategic, economic, population, industrial, and command assets essential for prevailing in war have been hardened beyond our ability to seriously damage or cripple.

The United States today lacks the firepower, accuracy and yield of weapons to overcome the extensive military and civil hardening and sheltering which has now taken place in the Soviet Union.

When people tell me that we have greater accuracy, my response is, "Yes, we have more accurate weapons. But not much greater in accuracy than do the Soviets." But what have we done with our accuracy? We have reduced the yield of our weapons! And so while those weapons are somewhat more efficient, they can really do very little more than destroy cities. We are assured that we have a superior MIRV technology. I take very little comfort from the fact. The Soviet target base has been doubling almost every ten years — the most explosive military growth in history. We have not been keeping pace and we have long since run out of nuclear firepower to neutralize Soviet strategic and general purpose combat forces. All that we have done with our MIRVs is to go to smaller yields. The end result is that what we can do with our MIRVs is to burn cities. Gone is our ability to destroy command posts. Gone is our ability to destroy essential Soviet civilian and military command and control. Gone is our capability to destroy Soviet strategic communications systems. The ultimate tragedy is that while we have been accused of having an "overkill" capability, the fact of the matter is that we have long since run out of nuclear weapons with which to destroy the bulk of the Soviet Union's fighting capabilities — and especially their ground divisions, reserves and stockpiles.

From the evidence which we have examined to date, it would seem that the entire industrial population of the Soviet Union is now fully protected. Every daytime working industrial shift in the Soviet Union has nearby underground bunkers hardened to 145 psi. Now unless you get a direct nuclear hit against one of these, the occupants are going to survive.

Despite the extensive evidence which my small staff uncovered regarding the measures taken by the Soviets to protect their civil population — in place — against all but a direct nuclear hit, the State Department and foreign service officers serving in the Soviet Union have failed to report any signs of civil defense shelters over the years. This despite the fact that many of our foreign service officers were in fact living in Soviet apartment houses in which civil defense shelters had long since been erected, and which they as tenants were required to inhabit during frequently scheduled exercises!

ABM Policy Based On Politicized Study

In 1972, based on a totally politicized civil defense study of 1970 (which found no serious evidence of Soviet civil defense activities) — we concluded an ABM treaty premised on explicit assumptions that neither side would defend its civil populations — thus holding each hostage to the nuclear threat. Yet even a cursory examination of the "hard" intelligence would have confirmed the folly of this premise.

Could we conceivably have pursued our aims in detente and SALT, had we assumed from the beginning that the Soviets were bent upon the attainment of strategic superiority? Yet I am assured that our negotiators entered the 1969 arms control negotiatons upon the express assumption that the Soviets would seek no more than parity. I consider that to have been one of the most fundamental failures of U.S. intelligence and diplomacy since World War II.

U.S. Superior Bomber Payload — Another Myth

We are told that we have a vastly superior bomber payload capability. Regretfully, that is another myth. For fifteen years as a member of this intelligence community, I watched the accountants and econometricians make some 800 Soviet medium bombers disappear from the strategic equation. Most of these have the same one-way intercontinental capabilities that our own medium bombers once had. I know. I was required to fly such missions and to practice them day after day while I was a member of the Strategic Air Command.

It has been labeled "worst case" to consider that in anything as horrible as global conflict, the Soviets would not do what is realistic by using all of their forces. The bomber payload question is entirely one of how you do your bookkeeping and what you exclude from it. We have no advantage in bomber payload.

Now what else is ignored? Probably the most important item is the Soviet capability to reload and refire offensive ballistic missiles from their soft and hardened silo sites. Every year for the last five years I have dissented to the community's persistent refusal to accord the Soviet Union the intercontinental ballistic missile refire capability which the evidence clearly supports.

The Soviets have designed each of their ICBMs from the very beginning with a refire capability in mind. We discovered in 1961 or 1962, if not the first certainly the second generation ICBM, the SS-7, was designed for refire.

Some three or four years later a few of us acquired some hard, incontrovertible evidence that each SS-7 missile on a launch pad had four missiles in nearby hardened storage for refire purposes. Despite the evidence — which would have forced the community to raise its estimates of the numbers of operationally deployed ICBMs — the reaction was what it has been for better than ten years. "The Soviets may have a capability to refire an additional ICBM or two from some of their launch pads." The fact of the matter was that each ICBM had at least four additional ICBMs for refire capability, which under their nuclear war-winning doctrines make a great deal of sense. Every intermediate, every medium range and every battlefield ballistic missile system deployed by the Soviet Union has been designed with a refire capability in mind.

At the present time the Soviets are deploying three, and possibly four, new land based missile systems. Two of these utilize a "cold launch" technique, which I was the first to report in this country. The community has steadfastly, on purely political grounds in my opinion, refused to acknowledge this refire capability. By popping their missiles out of the silo and igniting the engines in the atmosphere, little or no damage occurs to the silo. After the initial firing the Soviets can reload another missile in a canister into a silo and refire within minutes or a few hours. Such a technique has permitted hardening the silos beyond our capability to damage or cripple them seriously. The result is that a great many silos are going to survive for re-use.

My suggestion is that today the Soviets have somewhere between 500 and 3,000 additional ICBMs that can be refired, which are totally ignored in discussions of the strategic balance in this country's national estimates. The reason they are not discussed or estimated is a very simple one. It would perturb SALT and detente, by according the Soviets greater strength than our diplomats care to recognize for fear, presumably, that the Pentagon will be forced to ask for more weapons — thus exacerbating the arms race.

Some four years ago I produced one of the most extensive studies of Soviet missile development and testing on record. The results of our analysis showed clearly that the Soviets would probably introduce into test, before 1982, at least ten and possibly fifteen major new land and sea based offensive ballistic missiles! The matter was not reported to the national leadership. Rather, it was taken under advisement for additional study. Two years later we had acquired hard confirming evidence that at least seven or eight of these systems would be placed into test sometime in 1977 or 1978. Still, the national leadership was not informed. Why? As of the time of my retirement on 1 January — the President of the

United States, the Secretary of State and the Secretary of Defense still did not know that by 1978 seven or eight major new or modified offensive ballistic missile systems would be placed into test and that within the ensuing few years an additional seven systems would likely be introduced. Using the standard rule of thumb for calculating the cost of developing such offensive ballistic systems, it would be my conservative guess that the Soviets have already invested somewhere between fifty and one hundred billion dollars in these new systems. Yet where is the accounting in our estimates of these Soviet defense expenditures?

Behind The Controversy Over the Soviet "Backfire" Bomber

One of the more serious crises encountered during my incumbency as Chief of Air Force Intelligence related to the role of the new Soviet supersonic Backfire swing-wing bomber.

We first became aware of the Backfire's existence back in 1970. Well over a hundred have been produced to date and introduced into the Soviet long-range air armies and naval air forces. In its profile the bomber resembles and is about four-fifths the size of the American B-1. The central issue all along has been whether this bomber posed a threat to the United States. The Soviets argued that it did not and the CIA and State Department rather conveniently accepted the Soviet point of view — fearing, I presume, that acceptance of the American point of view would perturb another SALT agreement.

In anticipation of such a problem I wanted to be assured that the most competent analysts in the free world were involved in assessing the Backfire bomber's capabilities.

Accordingly, I directed my staff to go to Boeing; North American Rockwell — designers and builders of the B-1; General Dynamics — designers and builders of the supersonic B-58 Hustler; the Royal Air Force; and the Royal Aircraft establishment in England. We asked each, separately, to examine all of the available intelligence and to assess for us the capabilities of the Backfire bomber. Every single one of these organizations independently agreed that the Backfire had an intercontinental capability.

Finally, overwhelmed by the massive weight of analysis, various elements of the intelligence com-

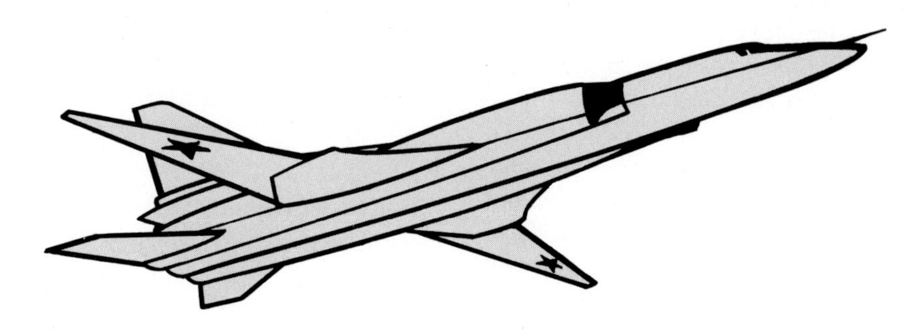

munity reluctantly came to agree with the Air Force definition of the Backfire's intercontinental range capability. However, CIA and State now judge that the Soviet Union had no *intention* of using the Backfire as an intercontinental bomber. Ironically, the Soviets went to great extremes in the second model of their Backfire bomber, the B model, to improve significantly its range capabilities by reducing its drag and extending its wing tips. I suspect that a third Backfire model will soon appear with even greater range. The fact that the Soviets have already doubled plant capacity for producing the Backfire bomber has yet, to my knowledge, to be brought to the attention of our national leadership.

Seemingly undaunted by the most extensive analysis of a foreign bomber ever performed in the United States — the CIA in one final supersecret, uncoordinated effort, proceeded over a period of eighteen months to undertake an analysis designed to prove that the Backfire bomber could not reach the United States. This effort, in which small bits and pieces of controlled information were provided to McDonnell Douglas Aircraft, designer and builder of fighters, represents one of the most artful contrivances I have ever observed. It is one, which I suspect, may have been designed to salvage a SALT accord.

Finally, when the CIA surprised everyone by surfacing its new analysis, months of painful and extensive analysis were required to show that the books had been rigged. Artifically high G loadings had been assigned to the design of the Backfire along with excessive engine drag and other factors which were designed to reduce range. It is for such reasons that a massive audit of CIA's estimative record is absolutely essential if this country is to preclude more of the same kind of chicanery in high places.

Our intelligence assumptions regarding Soviet and Warsaw Pact fighting capabilities are equally disturbing. For years we have deluded ourselves into thinking that we could defend NATO with forces at hand. My own extensive studies of evolving Soviet and Warsaw Pact capabilities and preparations made it rather clear, to me at least, that such assumptions are built on quicksand. One has but to study the doctrines, tactics, and equipments employed in the Middle East War of 1973, then to observe the evolving Soviet maneuver and exercise scenarios, and finally to read the vast wealth of evolving Soviet doctrinal military literature. The end result is to dispel rather quickly our illusions about NATO.

It is apparent that a very distinctive change in Soviet strategy occurred prior to 1970 in the NATO-Warsaw Pact area — as evidenced by the remarkable changes in equipment trends of Warsaw Pact air, ground, and naval forces.

Rather than boring you with the evidential details, let me simply express my concluding judgment:

Soviet-Warsaw Pact Preparation For A Blitz

I think a Soviet-Warsaw Pact planner today — given the forces, capabilities, the combined arms doctrines, the offensive and defensive chemical warfare preparations, the communication jamming capability, the hardness and survivability of East European forces, the great masses of fast armor, the development of new armored personnel carrier regiments — supported by the world's most advanced self-propelled guns, and a nuclear arsenal visible to anyone who will pry around — would have every reason for believing that he could take Europe by force of arms with a minimum of fighting, in 24-36 hours with or without the use of nuclear weapons. The Soviets have been preparing themselves assiduously for blitz warfare on an unprecedented scale. Those are the facts of life facing NATO, and there is no amount of sophistry that can persuade me otherwise.

But we have a mindset. We have morale to worry about. We have an alliance to keep together. We have a diplomacy, and we have had a policy supported by estimates which have tended to drift in the same direction. We have deluded ourselves into believing we could hold NATO with conventional forces, while pressing for removal of the only weapons that have any deterrent value in NATO, which is not to say we could not do the conventional job if we had the adequate resources to do so. We do not, and I see no hope that we are about to make the commitment to provide them.

A similar situation prevails with regard to our estimates of the improving Soviet capability to project their power abroad. In recent years there has emerged a small and powerful Soviet navy unlike that of any other modern naval power. It is a navy that cannot be compared to our own because its functions are quite different. The Soviet navy cannot be compared numerically with those of the Western powers. I would not trade a U.S. Navy fighter, a U.S. Navy fighter pilot, or a U.S. ship's captain for half a dozen of their Soviet counterparts. However, those, it seems to me, are no longer relevant measures of comparision or contrast.

What the Soviets have done is to deploy a technically capable naval force which, within a very short period of time, could cripple or neutralize the bulk of the free world's major naval surface combatants deployed within 3,000 miles of the Eurasian periphery.

Soviet Cruise Missile Rules The Waves

Many of my Navy colleagues have shared privately for the past ten years a view that is seldom talked about in respectable society — namely, that the Soviet cruise missile has dominated the world's oceans and waterways for better than ten years. The Western navies have as yet developed no effective demonstrable means of coping with the cruise missile threat. The sinking of the "Eilat" in but a matter of a few moments during the 1967 Seven Days war sent ripples of shock throughout the United States' and free world navies. Although the exercise was repeated in the war between India and Pakistan — little was said publicly about the matter.

It is my belief that the Soviet cruise missiles fired from submarines, surface ships and delivered by aircraft are capable of crippling very quickly the major concentrations of naval power in the Mediterranean and along virtually all of the approaches to Western and Eastern Eurasia. Unless it be United States aircraft carriers — I know of no free world naval weapons, forces or systems capable of coping with a professionally mounted Soviet surprise cruise missile attack upon our main forces. Once some 200-300 major fighting ships are disposed of — free world surface fleets can be quickly placed at risk. In "OKEAN-75" — the most extensive multi-ocean naval exercise in history — the Soviets with the aid of their ocean surveillance satellites were able to conduct hundreds of attacks against naval targets within seconds of each other on a global basis. As to firepower — which free world warships or aircraft carriers would dare to risk exposure to the *Kiev's* (Soviet ASW carrier) 300-400 mile range cruise missiles, their nerve gas, nuclear or conventional warheads?

During the past ten years I have known but one senior naval officer who has had the courage to express similar judgments on a semi-private basis to his naval leadership. He was rewarded for his pains by early retirement.

What the Soviets have done is to assure that they can neutralize our ability to project power ashore from the sea. They have developed a second-to-none ability to interfere with the free world's access to fuel energy, raw materials and to be able to move and transport goods, services and logistics on the world's high seas. For this country to have allowed the British to demobilize much of their fleet — a fleet on which we have been quite dependent for the naval balance — and to allow our own navy to be reduced by half in

the face of what was evolving, is one of the most blind acts of a government in modern history. I could make similar speeches about what has happened in the field of land warfare, air technology, and in penetration of the Third World through ever-improving methods of subversive penetration and economic warfare. But that must remain for another day.

It seems to me that the problem of perceiving what is transpiring in the Soviet Union is analogous to what existed between 1935 and 1939 in the United Kingdom. Those of you who will take the trouble to read Norman Gilbert's latest volume on Winston Churchill's pre-war years will see extensive documentary proof of the intelligence which was available to the British Government on events transpiring in Hitlerite Germany, events that were almost totally discounted by the leaders of the free world and which were believed but by a very few such as Winston Churchill. The intelligence was not heeded. In fact, it was ignored. The free world's leaders allowed themselves to be victimized by their hopes. Although conditions have changed, I submit to you that intellectually we are about where the English were in the 1930's.

Anti-Satellite Efforts Critical

We are well aware of the Soviet anti-satellite efforts which are designed to deny us the use of space for our warning, our surveillance, our communications, our navigation, our precise positioning, and for our command and control functions.

The Soviets have prepared to deny the U.S. its use of space — without which it cannot be warned, nor deploy and employ its forces.

Soviet High Energy Beam Research

Finally, you are aware of recent accounts about Soviet high energy beam research. The Soviet Union, irrespective of what any scientist in this country tells you — and I say this because I have done more work on this subject than any other official in this country — is 20 years ahead of the United States in its development of a technology which they believe will soon neutralize the ballistic missile weapon as a prime element in our deterrent forces. It is my firm belief that they are now testing this technology. For five years the intelligence community has said: "No, Keegan, you're wrong. Our scientists say it is not possible." Our scientists never really tried to find out. Little basic research was undertaken which might have allowed them to judge what the Soviets were doing. It was left to my little organization to sponsor the most advanced basic research, since the development of the A-bomb, to prove to the intelligence community that what the Soviets have been writing about for 15 years is, in fact, feasible. And I submit that the Soviets, on the basis of what I have examined, have every expectation that well before 1980, if they don't blow themselves up — and they may — will perceive that they have technically and scientifically solved the problem of the ballistic missile threat.

The intelligence community was wrong about parity — just as it has been wrong about virtually every

great Soviet scientific and military advance since World War II. The intelligence community was wrong about the A-bomb. The intelligence community was wrong about the Soviet thermonuclear weapon and whether the Soviets would have an ICBM. I say they were wrong because predictions usually occurred — if at all — only on the eve of the event. We cannot operate with lead times like that in today's environment.

The intelligence community was consistently wrong in its assessment of evolving Soviet scientific capability. Today, look at the quality of Soviet weapons being deployed.

Crude, by our standards, maybe.

Not polished, by our standards, maybe.

Difficult to operate, by our standards, maybe.

But in terms of net lethal killing effectiveness, within the combined arms doctrines on night-fighting, deception, nerve gas, jamming and blitz warfare on a scale never before contemplated by anyone in the free world, the Soviets are unmatched. They do not have to match the F-15 or the F-16 fighters in order to succeed. If fight they must, you may rest assured that it will be on their terms, not ours.

War Survival Effort of Unprecedented Scope

Once again, I would like to caution that we have great strengths in this country and that we are not on the edge of the abyss. But because of the failure in our perceptions, we are inviting a global conflict — a conflict which I believe is now in gestation. Some time in the future such a conflict is more likely than not to occur — principally because of what the Soviets are doing and what we are not.

Now we do not have to stand this country on its head to avert another mindless and needless holocaust. We are dedicated to preventing that, but we are not doing what we should.

I disagree totally with those who hold that the Soviets are destined to be the world's leading superpower, that they are destined to far outpace the United States as a strategic power, and that there is nothing we can do about it, and that the sooner we resign ourselves to getting the American people to accept that fact of life and adjust to same through a more enlightened and rational diplomacy, the better off we are going to be. I think that is an odious and alien doctrine ignorant of the traditions under which this country was built. This country, with its creative genius, wealth and unmatched industrial know-how, for the cost of a few gallons of gasoline per person per year can assure that the Soviets will never be tempted (as these new weapons give them a heightened sense of security) to challenge the basic institutions and worth of the Free World.

Major General George J. Keegan, Jr. (USAF, ret.) was the assistant chief of staff, Intelligence, Headquarters USAF, from 1972 to 1977. He is now the executive vice president of the United States Strategic Institute in Washington D.C.

General Keegan, born July 4, 1921, entered active military service in February, 1943. In World War II, he flew 56 combat missions in B-25 aircraft in six campaigns from New Guinea to Okinawa.

Released from active military duty in 1945, he graduated from Harvard in 1947, and returned to active duty that year, serving as a B-29 pilot and combat intelligence staff officer in Guam, Okinawa, and Japan. On his return to the United States, he served as chief of combat intelligence for the Tactical Air Command.

From 1953 to 1957, General Keegan served as chief of special studies for the intelligence office of Headquarters USAF. During that time, he was the author of several major studies on Soviet science, education, and strategy, served as a member of Secretary Stassen's Disarmament staff in the White House, and was Air Force liason officer for Senator Symington's disarmament hearings.

Returning to the field in 1957, General Keegan served with the Strategic Air Command in a variety of flying, staff, and command positions. For three years, beginning in 1961, he served as chief of the air estimates division of the intelligence office of SAC Headquarters.

After graduating from the National War College in 1965, General Keegan was assigned to Headquarters USAF as a deputy assistant dealing with National Security Council matters He was next assigned to a similar position with the Joint Chiefs of Staff.

In 1967, General Keegan became deputy chief of staff, Intelligence, Seventh Air Force. He assumed the compariable position at Headquarters, Pacific Command, in 1969.

Deputy chief of staff, Plans and Operations, Air Force Logistics Command, was General Keegan's assignment in 1970. In 1972, he became assistant chief of staff, Intelligence, at Headquarters USAF, remaining in that position until his retirement.

SOVIET CIVIL DEFENSE & U. S. STRATEGY

By Thomas H. Etzold

Reprinted by Permission
AIR FORCE MAGAZINE (October 1977)
Edited for Space

The author raises some often-ignored questions about the purpose of the USSR's civil defense program; examines its possible implications for US strategy; and suggests a range of practical, doctrinal, and political responses.

Recently, Soviet civil defense and war survival programs have seemed fundamentally to threaten the strategies intended to ensure the security of the United States. Mutual assured destruction and associated ideas about the "sufficiency" of strategic nuclear forces in an era of parity have depended on the idea that, without terminal defenses against ballistic trajectory weapons, the citizens of the United States and the Soviet Union would be hostages, a situation that would enhance mutual deterrence. Yet, Russian developments in civil defense have raised the disturbing possibility that soon only Western populations may be sufficiently vulnerable to deter their governments from effective political-military pursuit of national interests.

Indeed, Russian war survival measures have assumed impressive dimensions. The Soviet government has begun civil defense training for much of the population, and it has continued to train and equip troops for nuclear, biological, and chemical warfare. There are special "civil defense troops" and a civil defense academy in the Soviet military. The Russians are dispersing industry and hardening industrial and military sites including command, communications, and missile installations; they are storing grain; and they are endeavoring to protect high government officials and significant numbers of workers through a program of shelter building and city evacuation planning.

In the context of the Russian civil defense effort, three questions require attention. There is first the question of what problems Russian civil defense may raise for American strategy. Second, there is the deceptively difficult question of just what these programs may mean. And, finally, there is the immediate question of how the US should respond to Soviet activities in this field.

THE PROBLEMS FOR AMERICAN STRATEGY

Most commentators on Soviet civil defense have concentrated on the problems it may pose for Western strategy. Three types of difficulties are evident. One relates to general nuclear war, a second to limited strategic options, and a third to ordinary political intercourse.

The implications of Soviet civil defense have been most alarming to observers who consider the possibility of full-scale nuclear war. Analysts concluded that, as a result of civil defense measures only about four percent of the Soviet population would perish from blast, fire, and initial radiation, vs. forty percent or more in the West. Similarly, these analysts reasoned that the United States is losing the ability to destroy the percentages of Soviet population and industry long thought necessary to deter Soviet leaders from initiating nuclear war or other major aggression.

However, the reasons for anxiety about Soviet civil defense in relation to a strategic nuclear exchange should be offset to some extent by several factors. One little-known fact bearing on the problem is that in recent years plans for employing American strategic nuclear forces have not envisioned the kind of one-time strike usually used as the basis for calculating casualties and damage. As Gen. Maxwell Taylor has noted in his book, *Precarious Security* (W.W. Norton, New York, N.Y., 1976), in recent US strategy, assured destruction capability has meant the ability to kill X percent of the Soviet population and destroy X percent of Soviet industry *X times at intervals*.

In succeeding strikes, due to reduced warning and political direction, depletion of emergency stocks, and damage to transportation and other facilities, the consequences of follow-on strikes would be severe. The more the Soviets concentrated population to begin reconstruction in the aftermath of a first or second phase of attack, the more effective further attacks would be. The more they dispersed to avoid such consequences, the slower recovery would go forward. In addition to the effects of concussion, firestorm, and radiation, there would be the incalculable tolls of disease, infirmity, and disruption of complex communal life. There might also be unexpected consequences from the selfishness and violence that the initial survivors of holocaust could be expected to display.

The second strategic problem facing the US in connection with Soviet civil defense programs grows out of contemporary scenarios concerning crisis bargaining, coercion, and the attractiveness of possessing, perhaps using, limited strategic options.

Those concerned over the effects of Soviet civil defense have suggested that because the Soviets have dispersed and hardened industrial and military targets as well as increased the numbers of launchers and associated facilities, the effects of a limited strike would be trivial, and, therefore, acceptable to the Soviet Union. But, because of the collocation of American military installations and cities, and due to the lack of hardening and population protection measures, similar limited attacks on the United States would produce results by no means trivial or acceptable.

Further, in this view, the growing Soviet capability to evacuate and/or shelter the populations of major cities strengthens Russian immunity to the threat of a limited strike, say, against one or two cities. To trade

New York for Moscow, or St. Louis for Leningrad, never seemed a happy prospect. Now, so the argument goes, in addition to being a catastrophe it may be a bad trade.

In the case of limited option strategy, as in that of general nuclear war, there are countervailing considerations. One is that the effects of using nuclear weapons have always been presumed to be both psychological and physical.

This point is important because the leadership of the Soviet Union is uniquely sensitive — even vulnerable — to internal disruptions. It fears challenges to authority and potential losses of control even in the most minor contexts, as the interesting and pronounced reaction to recent explosions in the Moscow subway demonstrated. The detonation of a nuclear device on or over Soviet territory would pose an enormous hazard to the political stability of Soviet leadership, and is, therefore, something they would want to avoid.

Even more important in keeping limited options open, however, are such easily available technological alternatives as dedicating a specific, small portion of US strategic forces to limited strike operations and fitting them with dirty warheads, or perhaps employing groundburst weapons in limited strike situations.

The third category of strategic problem, that of ordinary political intercourse, may seem both out of place in this discussion of strategy and relatively minor by comparison with the foregoing two topics. However, the difficulty of pursuing national interests in more or less peaceful competition with the Soviet Union is a daily problem, and it is a problem of strategy as well as of diplomacy. For the weight of a state's views has always depended in large measure on its ability to compel agreement. In George Kennan's words, "You have no idea how much it contributes to the general politeness and pleasantness of diplomacy when you have a little quiet armed force in the background."

The argument with regard to ongoing political relations is that if the Soviets believe they possess genuine capability to survive nuclear war, and if correspondingly they feel less than deterred, they may become politically more assertive, more willing to run risks.

WHAT THE RUSSIAN PROGRAMS MAY MEAN

The purpose of introducing this category of questions is to raise doubts, not to provide answers. As noted above, most American comment on Soviet civil defense programs has — perhaps rightly — focused on the problems of strategy. However, the unfortunate side effect of this focus has been the emergence of an unproven but widening conviction that Russian civil defense immediately threatens the West.

As a result of the conviction mentioned above, too few questions about the Soviet effort are coming up for discussion. Many of the undiscussed issues are of pressing relevance to strategic circumstances. Is, for instance, the Soviet civil defense effort in fact an indication of Russian intention to "go to the brink" from time to time, and so to intimidate the West into concession? Or are war survival programs evidence of Russian pessimism regarding the ability of the powers to avoid nuclear war sometime in the future, no matter how hard they may try to do so? Is it a sign of concern over Western intentions, or over Chinese? To what extent is it related to Russian perception of the hazard of nuclear proliferation?

There is another extremely important set of questions. Is it possible that the Russian civil defense programs are more internal than external in their origins and implications? Could they be the result of bureaucratic politicking, as are many of our most costly and visible programs? Could they be designed to give the leadership enhanced control over the population, or the population increased dependence on and confidence in the leadership? Is the dispersal of industry a sign of Russian determination to complicate Western targeting, or is it merely a normal accompaniment to the development of Russia's still-primitive internal transportation system? Are grain stockpiles accumulated to anticipate holocaust, or are they a way of explaining perennial agricultural shortages and, possibly, hedging against price fluctuations in world commodities markets?

So far, the foregoing questions have received inadequate public attention. All of them, however, deserve careful analysis before one reaches conclusions on the meaning, implications, and requirements of Soviet civil defense for the West.

HOW THE UNITED STATES SHOULD RESPOND

With the issues of strategy and the possible meaning or meanings — of Russian civil defense efforts clearly in view, what should the United States do in response to Soviet programs? An answer here must comprise three elements: one practical, one doctrinal, and one political, and in that order of importance.

First the practical. The US should immediately augment its present meager efforts in civil defense. It is evident that the United States is not going to devote resources to such programs in amounts anything like those the Soviet Union has spent in recent years, and that the US cannot really expect to attain equivalency in this area soon, if ever. Indeed, there is no reason to believe that equivalency in such measures is necessary either to stable deterrence or to adequate freedom of decision in political matters. There is, however, reason to think that both friends and enemies would consider increased attention to civil defense an indication that this country was determined to hold its own in working out inevitable conflicts of interest with the Soviet Union.

It is possible that, with further study, the US could determine how to derive the most immediate benefits from moderate increases in civil defense spending. In practice, this would probably mean that protection of high government leaders, military communications and command facilities, and some additional strike forces or other military installations would take precedence over civilian shelter plans. In the short run, to be sure, this would be impressive to enemies, and, if adequately explained, tolerable to the American people.

Second, the doctrinal element. It is essential here to keep in mind that definitions of strategic sufficiency have always been arbitrary. There is no magic attached to such figures as the traditional requirement of killing twenty-five percent of Soviet population and destroying fifty percent of Soviet industry to deter the Russians. Such figures were originally the figures at which force planners of the early 1960's ran out of targets substantial enough to result in significant additional increments of damage.

There is no evidence that the leaders of the Soviet Union are willing to risk casualties and damage in a nuclear exchange, even at much lower levels than the twenty-five to fifty percent formula. On the contrary, they give every evidence of fearing nuclear war and desiring to avoid it.

Deterrence does not depend on equal security for each side, but on unacceptable insecurity for both. Increasingly, recent studies have pointed to the low level of nuclear damage either side may be willing to tolerate rather than to the high level of risk and sacrifice each may accept.

Finally, the political element, the most important of all. Here I return to the difficulties of the third category of strategic problem, that occurring in the context of ordinary political intercourse and relating to the Soviets' potential as aggressive risk-takers.

It is essential to recognize that the more the US makes of Soviet civil defense, the more political advantage it forfeits. By exaggerating the real threats and strategic challenges posed by these developments, Americans run the risk of talking themselves into weakness of will, and in that sense of doing the Soviets' work for them, if the United States scares itself and its friends to the point that it shows political timidity or weakness in bargaining.

The immediate response required of the US by Russian civil defense and war survival measures is not to scurry about in frantic or futile attempts to redress the balance of capabilities, even if some gestures toward improved civil defense should be made for psychological reasons. Instead, the immediate requirement is to affirm that deterrence works. This country possesses adequate strategic systems today. The United States government and the American people ought, therefore, to behave with the confidence that, for the present, the Russians are fully as deterred as we are. They ought also to keep in mind that the continuation of mutual deterrence will depend on the coherence of the relationship between strategic systems and strategic doctrine. Both will require improvement in the coming years.

Once again, AIR FORCE Magazine opens a new year with a review of aerospace developments by the leading authority on the world's aircraft. In this article, he assesses the technological balance between East and West, reports on the status and prospects of the aerospace industry with emphasis on Europe, and examines some novel potential solutions to persistent technical and operational problems.

JANE'S AEROSPACE REVIEW 1976/77

by John W. R. Taylor
Editor, JANE'S ALL THE WORLD'S AIRCRAFT

Reprinted by Permission
AIR FORCE MAGAZINE
(January 77)

Viewed from this side of the Atlantic, 1976 seems to have been a somewhat ominous year.

If the portents now to be seen are a warning, rather than a sentence of imminent doom, that is no reason to ignore them. The threat is real enough, underlined by the MiG-25 *Foxbat* fighter that appeared suddenly at the end of the runway at Hakodate Airport, Japan, on September 6; by Tupolev *Backfire* supersonic bombers that venture out over the Atlantic as far as the Azores; by so strong a desire for a US-Soviet strategic arms agreement that US negotiators appeared willing to close their eyes to differing interpretations of vital issues; and by unrealistic estimates throughout the West of what it costs to stay alive in an age when a thousand strategic thermonuclear missiles can be launched by a single telephone call and the depression of a thousand firing buttons.

This may seem a gloomy way in which to begin this new year, but consider a few facts:

Whatever assessment the Pentagon may have made concerning the qualities of that MiG-25 in Japan, details presented by normally responsible journals have been utterly misleading. Some have exaggerated and some depreciated the aircraft's capabilities and the stature of Soviet technology. One typical example commented that "some sources continue to report a thrust of about 24,500 pounds per engine, rather lower than 31,000 pounds previously estimated." In fact, the now-confirmed correct thrust of 24,250 pounds per engine has appeared annually in *Jane's* since 1968, when the Soviet authorities released the rating in documents relating to a new record set by the MiG.

In November, another journal published the remark that "*Foxbat* demonstrated its superior intercept capabilities by setting a series of official Fédération Aéronautique Internationale world time-to-climb records under the Soviet designation E-266. These records stood for two years and were only recently reclaimed by the USAF McDonnell Douglas F-15." In fact, only sixteen weeks after the F-15 *Streak Eagle* set its spectacular climb records, a *Foxbat* not only recaptured the two top records but climbed in four minutes eleven seconds to 35,000 m (114,830 feet), a height the F-15 pilots had not attempted to reach. This particular Soviet record-breaker was described as an E-266M. Assuming that the suffix stands for *modifikatsirovanny*, as in other known Soviet designations, it is interesting to conjecture what modifications were made to the basic MiG-25 to boost its rate of climb to such a degree.

The dangers inherent in comparing the MiG-25 that landed in Japan with US fighters of the mid-1970s should be obvious. The Soviet aircraft set its earliest international speed record in the spring of 1965, and must have been designed in the early 1960s. Is it surprising that its radar employs vacuum tubes? Are there none in the early F-4 Phantoms that were contemporary with the MiG?

Much has also been made of the fact that the MiG-25 is constructed largely of nickel steel, with no more than about three percent titanium in areas suh as the wing leading-edges and engine nozzles. Yet there is little wrong with steel, provided the designer is prepared to compensate for the weight penalty. The recent out-cry that followed the revelation that key parts of the UK/German/ Italian *Tornado* combat aircraft are made from titanium "sponge" purchased from the Soviet Union hardly suggests any shortage of titanium in that country. And if anyone doubts Soviet competence to fashion titanium a decade and a half after *Foxbat* was designed, they should visit the metallurgical displays that highlight Soviet static participation in the Paris air shows.

Nor should it have caused surprise that the MiG's Machmeter was redlined at 2.8. What other combat aircraft exceeds such a speed with four large missiles on pylons under its wings? The air forces of Israel and Iran have ample evidence that the reconnaissance *Foxbat-B* — with the same basic airframe and engines, but no missiles — can and does routinely exceed Mach 3, making it almost impossible to intercept.

TECHNOLOGY AND SEMANTICS

In any case, it is potentially suicidal to base one's estimate of the latest Soviet combat aircraft on an early-1960s design. Up to 1967, periodical Aviation Day flypasts over Moscow gave Western observers fleeting oportunities to keep tabs on Soviet progress. Much, no doubt, is learned today from satellite reconnaissance and other intelligence sources. However, the only post-1967 military designs of which the public can have even the scantiest knowledge are the Tupolev *Backfire* bomber, Sukhoi Su-19 *Fencer* fighter-bomber, and Yakovlev Yak-36 *Forger* VTOL carrier-based combat aircraft that put in its first appearance during the cruise of the *Kiev* through the Mediterranean and North Atlantic, en route to Murmansk last summer.

If the Soviet Navy was prepared to show off the Yak-36 so blatantly, one must assume that it is regarded as merely a first step toward something better. *Backfire* and *Fencer* reflect a totally different attitude. Both have been in squadron service with the Soviet Air Force for about two years, yet few photographs of *Backfire* have been published in any unclassified magazine; and no photograph of *Fencer* has ever appeared in the press.

This is normal, for the Soviet intelligence machine operates with immense competence in matters of publicity. To their credit, Soviet official information sources have never lied to *Jane's*, and the correctness of the thrust figure quoted for *Foxbat* underlines the reliability of the engine ratings — and all other data — given in their claims for international aviation records. Whether the extreme secrecy usually adopted can be justified is a matter of opinion. It can be argued that the most effective deterrent is one that can be seen and judged to be capable of doing its job. On the other hand, the unknown may sometimes inspire fear or apprehension in the "other side's" military leadership, while permitting politicians to underrate it publicly as an excuse for defense economies — the perfect "heads I win, tails you lose" situation.

The Kremlin may be sincere in suggesting that the *Kiev* is an antisubmarine cruiser and not an aircraft carrier. How, then, does one define an aircraft carrier if not as a carrier of aircraft? Equally, Soviet delegates to the SALT talks are correct in stating that *Backfire* has clear tactical roles. But is a strategic bomber to be defined solely as an aircraft that can attack the US from the USSR, and vice versa? Does it cease to be strategic if potential targets for its bombs are in less-distant allied nations, such as the UK or West Germany? Or is the test whether or not it will cover the distance between the US and USSR without being flight refueled on either the outward or return flight?

Such play with words and definitions is ludicrous. Last July 20, Air Force Secretary Thomas Reed stated that there is "absolutely no question" as to whether or not *Backfire* is an intercontinental strategic weapon. "With no refueling," he said, "*Backfire* could be launched from Soviet soil against targets in the US and then fly on to Cuba for recovery. With only one refueling, the Soviet bomber could be launched from Russia against all areas of the US, except for some parts of Florida, and return to the Soviet Union." For the record, the projection forward of the nose of the *Backfire-B* is a refueling probe.

On such evidence, any SALT agreement that was signed at the cost of accepting *Backfire* as merely a tactical aircraft could only lessen the hope of lasting peace. Whatever the scale of the opposing forces, from extravagant overkill to commonsense basic defense, peace can only be a product of precisely balanced strength. The three immediate essential requirements for the US are to recognize the *Backfire* is a strategic bomber, build the B-1 as its uniquely flexible counterpart, and order as a matter of urgency replacements for ADCOM's time-expired F-106s.

INTERCEPTOR INNOVATIONS

For an Englishman to make such a comment, knowing he will not have to contribute one cent toward the cost of the B-1s and follow-on interceptors (FOIs) may seem one hulluva nerve. But if NATO is to have any significance, the security of each individual member must be the concern of all. Surely there is something wrong when the US considers an interceptor force of twelve squadrons of 1956-model F-106s (six active, six ANG), one F-4 squadron, and three active Army Nike-Hercules batteries in Alaska to be adequate for home defense *and* to support tactical fighters in overseas air defense missions, while Soviet home defense forces deploy 2,600 piloted interceptors and 10,000 surface-to-air missile launchers.

An attractive replacement for interceptors and SAMs was foreshadowed during 1976, when a high-energy laser mounted on an armored vehicle destroyed two target drones during US Army trials at Redstone Arsenal. Operational capability of such "death-ray" weapons must, however, be years away, putting them in the same category as reconnaissance satellites carrying advanced sensors that will end the viability of deep-diving SLBM submarines by keeping their precise positions pinpointed around the clock.

Until that day comes, it will be dangerous to neglect any aspect of conventional defense. As its new interceptor, the USAF is expected to order a variant of the F-14 *Tomcat*, F-15 *Eagle*, or F-16 Air Combat Fighter. This makes sense. If a cure to recently reported engine problems has been found, the F-14 armed with *Phoenix* missiles must be rated the West's most effective interceptor, with the others not far behind. The high cost of the *Tomcat* is offset by the fact that fewer would be needed, as its AN/AWG-9 weapons control system has the ability to guide six *Phoenix* missiles simultaneously against six targets — a feat no other current interceptor can match.

Records show that by last September *Tomcats* had fired 108 live *Phoenix* missiles, achieving a kill rate of eighty-five percent. Twelve of the targets destroyed were flying above Mach 1.3 and above 45,000 feet, some of them simulating *Foxbats* near Mach 3 and 80,000 feet. Twenty-one kills involved targets that were between thirty-five and 125 miles from the *Tomcat* at the moment of launch. Twenty-five were made against simulated cruise missiles. Six of the actions achieved two or more simultaneous kills. There was, however, another series of air combat trials that produced an even more thought-provoking result.

To evaluate the *Tomcat's* full potential, the US Navy arranged a number of simulated air combats against French Air Force *Mirage* F1s over the Mediterranean. F-14 crews were adjudged winners in seven of the eight encounters involving missile "firings," and in six of the eight close-range dogfights. Similar successes were achieved against Northrop F-5E "aggressor" aircraft flown by US pilots using typical Soviet tactics.

Against US Marine Corps *Harriers* the results were startlingly different. Using to the full the V/STOL aircraft's low-speed maneuverability, and rapid acceleration and deceleration, the Marine pilots outfought F-14s in six of the sixteen engagements, losing only three, with the others indecisive. There could be no better incentive for ensuring successful development of the McDonnell Douglas AV-8B advanced version of the *Harrier*; and the US Navy must be relieved to know that the *Kiev's* Yak-36s do not appear to share the *Harrier's* VIFF (thrust vectoring in forward flight) and STOL takeoff capability.

EUROPEAN AEROSPACE INDUSTRY

There is an important lesson to be learned from this. At a period when most NATO and friendly air forces are flying US fighters, with the F-16 soon to follow in huge quantities, there is growing pressure for Europe's aerospace industries to become primarily subcontractors to US manufacturers. By doing so, it is suggested, they could maintain high levels of employment, technological know-how, and capability without incurring the expense of developing their own designs. The cost of US aircraft to the US services would also be reduced by increased overall production, so benefiting everyone.

A flaw in such proposals is indicated by the mere existence of aircraft like the *Harrier* — still the only operational V/STOL fixed-wing combat aircraft in the world — and the *Concorde* — still the only supersonic airliner in scheduled passenger service. While Europe can pioneer such concepts with brilliant success, it is nonsensical to reject the skill of its designers and engineers.

The same is true of engines and equipment. Without the reliable power of Rolls-Royce turbojets for its MiG-15 and Il-28 bombers in the late 1940s, the Soviet industry could hardly have progressed so rapidly from its primitive first-generation jets, such as the Yak-15 and MiG-9, to the formidable first-line types of today. China is currently evolving a new generation of combat aircraft fitted with Rolls-Royce Spey turbofans, already the well-proven powerplants of RAF *Phantoms*, and A-7 *Corsairs* of the US Navy and USAF. RB.211 turbofans, also from Rolls-Royce, power the Lockhead *TriStar* transport and the latest version of the Boeing 747, which set a new world record a few weeks ago by climbing to 2,000 m (6,562 feet) in six minutes thirty-three seconds at a gross weight of 840,500 pounds. Equally familiar, and highly valued in the US, are the Marconi-Elliott head-up displays installed in the A-7 and F-16, and Martin-Baker ejection seats in the F-14 and the Navy's forthcoming air combat fighter development of the Northrop F-17, designated the F-18.

France, too, continues to make an impact in the US. At the time this review was written, it seemed likely that the contract to supply forty-one new medium-range surveillance aircraft to the US Coast Guard would go to Dassault, which entered a specially equipped version of its Falcon 20G business jet with Garrett AiResearch ATF 3-6 turbofans. This is good news for Garrett, too, being the first production application for the engine.

Of far greater potential importance is Dassault's other joint program with a US manufacturer, McDonnell Douglas, on the Advanced Short/-Medium-Range (ASMR) airliner. This began as the "stretched" *Mercure* 200, and is intended to carry about 160 passengers 2,000 miles on the power of two 22,000 lb st General Electric/SNECMA CFM56 turbofans. If the program moves ahead, Dassault will have design leadership and the French government would probably subsidize each of the first 300 production ASMRs by $2 million, to ensure a competitive price.

Britain, Germany, Italy, and Japan are all being wooed by the three major US airliner manufacturers as useful partners in new commercial programs. As a result, there has never been such a profusion of paper aeroplanes, claiming to be quieter, more economical successors to the DC-9, DC-10, Boeing 727 and 737, BAC One-Eleven, and other current transports, whether or not they really need replacing. The average airline executive must be thoroughly confused by designations like 7X7, 7N7, DC-X-200, and X-Eleven, especially as each one might cover a range of "take-your-pick" sizes, configurations, and powerplants. In November 1976, the only one with any immediate prospect of advancing from project to prototype construction seemed to be the Dassault/McDonnell Douglas ASMR. It is, therefore, the only one of which details can be found in the 1976/77 *Jane's*, which has always done its best to avoid becoming *All the World's Paper Aircraft*.

GROWING EXPORT COMPETITION

Surprisingly, perhaps, the French and British governments have already discussed the prospects for a second-generation supersonic transport, with McDonnell Douglas as a potential US partner. Until *Concorde* achieves economic success, this advanced supersonic transport (AST) will amount to little more than three more letters to add to all the "X" projects mentioned earlier. It is, nonetheless, interesting to note how little the basic parameters have changed since 1972, when this writer was told by a Douglas vice president in California that "America's first supersonic airliner will look like *Concorde*, be made of the same materials as *Concorde*, fly at the same speed as *Concorde*, but will be twice as big and will therefore make money."

If any US citizen feels alarmed by the prospect of McDonnell Douglas working in partnership with people in factories thousands of miles away, it may be worthwhile pointing out that fuselage panels of the DC-9 and DC-10 are already being manufactured by Aeritalia in Italy, main fuselage sections of the Lockheed C-130 by Scottish Aviation, and rudders and elevators for the Boeing 727 in far-off Australia. The full list of contracts resulting from shopping around for competitive prices and surplus workshop capacity would fill an issue of AIR FORCE Magazine.

There is no reason why planned nationalization of the major UK aerospace companies should lead to any changes in the overall picture. Success for any organization stems from imaginative and efficient management allied with skilled and conscientious work forces. Whether the money comes from private or public funds is immaterial. What *does* matter is whether public ownership will generate more enthusiastic and consistent government support than the UK industry has received in the past twenty years, and a more responsible attitude from certain trade unions. Until there are improvements in these areas, the industry will continue to fall short of its immense capability.

Britain's aerospace industry desperately needs a new, major commercial transport program to back up a healthy order book for other products. During the first nine months of 1976, its exports totaled more than £ 672 million ($1,122 million), representing an increase of £ 96 million ($160 million) compared with the same period of 1975 and confirming its position as the busiest and much the most profitable industry of its kind in Europe. The first foreign order for the new Hawker Siddeley *Hawk* was announced in November — involving up to fifty aircraft for Finland — following intense competition from Germany, France, Sweden, Italy, and Czechoslovakia. With a potential market for 6,000 *Hawk*-type aircraft in the coming decade, this highly advanced ground attack/trainer promises to become a profitable partner for the *Harrier* and its derivatives.

Despite losing this particular contract to Britain, disappointing foreign reaction to the *Concorde*, and initially slow sales of the multination European *Airbus*, France, too, can report growing aerospace exports by its nationalized Aérospatiale group and a handful of private companies headed by Dassault-Breguet. After a brief excursion into sweptwings with the *Mirage* F1, and too-expensive essays at variable geometry, Dassault is reverting to its classic and widely accepted delta formula with the single-engine *Delta Mirage* 2000 fighter for the French Air Force of the 1980s, and twin-engine *Delta Super Mirage* for export.

Few other countries can afford any longer to undertake such programs on their own. Sweden is one of the exceptions, with plans for a new light attack/trainer, known at present as *Attack Aircraft System* 85, to replace its current Saab 105s (SK 60s) in the 1980s. With production of the tandem-delta *Viggen* scheduled to continue through that decade, Saab-Scania will keep busy. Already, however, it is beginning to encounter unexpected competition from Israel Aircraft Industries in Austria, long a customer for Sweden's combat aircraft. Israel's *Kfir*-C2 began as a somewhat refined *Mirage* delta with a General Electric J79 afterburning turbojet of the kind fitted to the F-4 *Phantom*. The addition of delta canards has now reduced takeoff and landing distances, and improved the aircraft's dogfighting maneuverability to such an extent that the *Kfir*-C2 has become one of the most attractive, available, modestly priced, multirole fighters of its time.

TECHNOLOGY TRANSFER

China's F-9, known to NATO as *Fantan*, also owes much to foreign design, in this case the Mikoyan MiG-19, which has itself been in large-scale production at Shenyang (formerly Mukden) since about 1960.

All that may be published about the F-9 is that its airframe resembles that of the MiG-19 but is scaled up, with semi-circular lateral air intakes and a pointed nose radome. The design is probably being modified at the moment, so that production *Fantans* will be able to take the Rolls-Royce Spey turbofan engines for which China has acquired license rights. Added to experience gained already in producing aircraft like the Tupolev Tu-16 *Badger* twin-jet strategic bomber and MiG-21 fighter, it is clear that China's industry is making rapid progress toward technological self-sufficiency. MiGs exported to Pakistan and Tanzania have long reflected high manufacturing standards.

The Nationalists on Taiwan are equally progressive. Having cut their teeth on the little Pazmany PL-1 trainer, devised originally for US amateur constructors, they advanced via license manufacture of the Bell UH-1H helicopter to production of Northrop F-5E supersonic tactical fighters. The T-CH-1 turboprop-powered trainer and light attack aircraft represented the first attempt at local design and was strongly influenced by the North American T-28. The T-CH-1 went into production in early 1976. The Aero Industry Development Center at Taichung is now building the prototype of a thirty-eight-passenger twin-turbo-prop transport designated XC-2.

Despite recessions, political inhibitions, and other hindrances, the number of aircraft-producing nations continues to grow. Turkey, Greece, and Iran have all begun to establish highly professional aviation industries with the help of foreign aid. Romania and Yugoslavia have demonstrated that even states with comparatively small national companies can work together to produce an effective modern combat aeroplane, by developing the IAR-93/*Orao*, a *Jaguar*-like attack fighter intended to make them independent of foreign types that might arrive with strings attached.

NEW VIEWS OF OLD PROBLEMS

It would be unrealistic to expect such efforts to pioneer important new concepts. There is, however, increasing evidence that the aerospace industries of the major industrial nations are beginning to look beyond the technology plateau on which they have worked in recent years. The new world absolute speed record of 2,189 mph over a 15/25 km course was set by a USAF Lockheed SR-71A reconnaissance aircraft dating back to the mid-1960s; but rollout of the Space Shuttle Orbiter *Enterprise*, at Palmdale, Calif., on September 17 gave the world a first glimpse of a true aerospace craft of tomorrow. Completely independent of fossil fuels, it will use solid boosters and onboard liquid-propellant rocket engines to thrust it to orbital speed, enabling it to supersede expendable launch vehicles as a means of putting satellites in space, and to transport nonastronaut scientists into orbit inside a space laboratory designed and built in Europe.

One of the most disturbing features of the present time is that so little progress is being made toward adapting conventional aircraft and land vehicles to run on fuels like liquid hydrogen, which must eventually take the place of hydrocarbons if the world is to remain brightly lit, warm, mobile, and at work throughout the twenty-first century. It is risky to assume that, having demonstrated its unrivaled expertise by soft-landing two Viking spacecraft on Mars last summer, and using them to photograph and analyze the planet's surface, NASA will be able to solve almost overnight a problem as mundane as a future energy shortage when it becomes urgent.

Reverting to January 1977, recent NASA research aimed at improving the aerodynamic efficiency, and fuel economy, of aircraft is already being embodied in new designs. During the past year, Dr. Richard Whitcomb's supercritical wing has lifted into the air the Boeing and McDonnell Douglas Advanced Medium STOL Transport (AMST) prototypes that will haul 27,000-pound payloads into and out of 2,000-foot strips, pointing the way to an eventual replacement for USAF's C-130s. These strangely configured aircraft, with their wide bodies, uniquely positioned engines, and huge blown flaps, are as revolutionary in their way as anything to be seen in the air at the present time.

Grumman's (now cancelled) *Gulfstream* III, goes an interesting stage further, by adding NASA-developed "winglets" to the tips of its supercritical wing, to ensure an even better cruising fuel consumption. (The Learjet Longhorn also adds winglets to its design.)

Nor are rotating-wing aircraft being over-looked by research engineers. Twenty years of experiment with every conceivable kind of VTOL technique have failed to better the helicopter in terms of payload/range after vertical takeoff. So, while competing types of utility tactical transport and advanced attack helicopters battle their way toward large US Army production orders, much effort continues to be put into improving the overall capability of such aircraft.

Bell is about to resume its tilt-rotor research with the NASA/Army XV-15; and Sikorsky expects to attain entirely new standards of speed and agility with the Advancing Blade Concept (ABC) contrarotating rotor system fitted to its XH-59A.

Meanwhile, in Europe, a new technique known by the acronym STOVL could produce an even more versatile follow-on battlefield support aircraft. RAF thought on the subject led to the formulation of Air

Staff Target (AST) 403, defining the basic parameters of the kind of aircraft that might replace both the *Jaguar* and the *Harrier* before the end of the 1980s. France needs something similar as a partner to its *Delta Mirage* 2000 high-performance interceptors. Belgium, the Netherlands, and Germany have parallel requirements. So the five nations formed a sub-group of the European Programme Group to develop their project, under the chairmanship of the UK.

AST 403 was aimed at a Mach 1.6 close support/air combat type able to destroy any battlefield target in a single pass and to match the agility of anything encountered in the air. Inevitably, it was greeted with the comment: "What you want is the F-16, and think of the money you'll save with more than 1,000 already ordered for the USAF and the air forces of Belgium, Denmark, Iran, the Netherlands, and Norway." Experience in operating the *Harrier*, and in working with European partners on the *Tornado*, prompted other thoughts.

One of the greatest worries confronting any modern air force is how it could stay in business if a preemptive strike by the opposition took out all its runways in the opening minutes of a confrontation. *Harrier* squadrons have no such problems, as they do not need runways or even dirt strips from which to fly. However, the cost of taking off vertically is so high in terms of reduced payload/range that they normally operate in a short takeoff (STO) mode in order to lift a greater weight of fuel and weapons.

On the other hand, all operational experience by the RAF and USMC points to the importance of being able to land vertically. It would be feasible to touch down at around 100 knots into some form of mobile arrester gear; but the risk would be high if circumstances compelled the use of narrow, cambered roads, surrounded by natural or structural obstacles, subject to crosswinds, cluttered with ground equipment, vehicles, and other aircraft. Hence STOVL — short takeoff/vertical landing — making the best of both worlds.

STOVL combined with Mach 1.6 speed, equipment for all-weather operation, and the ability to carry a wide range of air-to-ground and air-to-weapons would seem to meet most anticipated needs for the rest of the present century. Add thrust vectoring in forward flight, and the resulting aircraft begins to sound expensive; but is anything else practicable to preserve balanced forces in a period when the Soviet Union is producing 1,000 advanced tactical combat aircraft every year?

It is much too early to guess whether or not the Future Tactical Combat Aircraft (FTCA) being discussed by the European five-nation group could ever be reconciled with AST 403, and whether the result would be a STOVL configuration. It might encourage the right answer if the members of the group take a close look at the kind of tactical combat aircraft that is in the mind of USAF technical staff who are also looking ahead to an FTCA for the mid-1980s.

USAF
ALCM
USAF
4
USAF
ALCM
USAF
5

THE MIG-25 WEAPON SYSTEM:

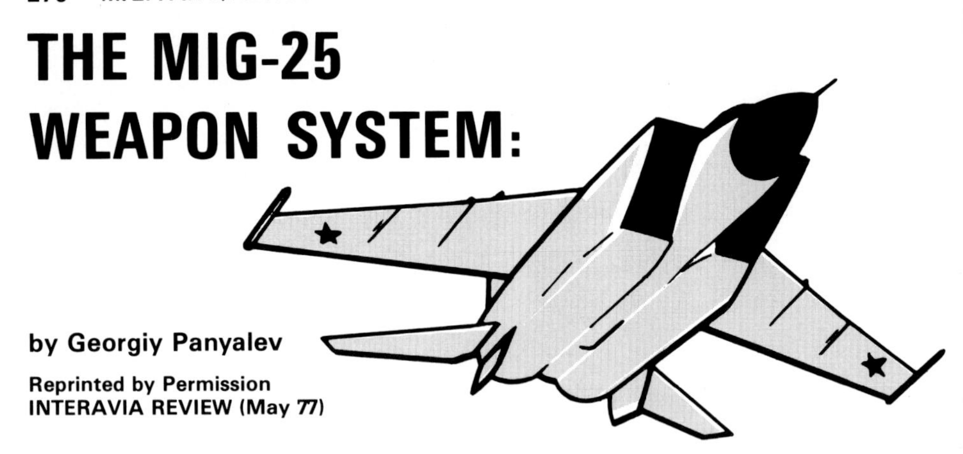

by Georgiy Panyalev

Reprinted by Permission
INTERAVIA REVIEW (May 77)

WESTERN UNDERESTIMATION

In a spectacular action which attracted worldwide attention, a Soviet MiG-25 *Foxbat* interceptor was delivered into Western hands for the first time on September 6, 1976, by its pilot, Lieutenant Viktor Belenko, who landed the aircraft on the airfield of Hakodate on the northern tip of the Japanese island of Hokkaido. A first, superficial look over the MiG-25 was a particular surprise to those observers who had expected a wonder aircraft. During later, more detailed investigations, those who by their specialist knowledge had expected to discover a Soviet combat aircraft of the 1960s were proved correct. But even they were astonished by certain details and equipment which make the MiG-25 a weapon system particularly well suited to its specific mission.

This article is intended to avoid controversial issues and is restricted to technical facts. It reveals the complete flight performance envelope, gives a rating of the MiG-25 as a weapon system, takes a close look at its technological state-of-the-art and discusses the different operational versions and the development potential of the aircraft.

MILITARY REQUIREMENTS

In October 1954 the US Air Force issued its requirements for a bomber for the late 1960s and 1970s. This aircraft had to be capable of reaching Mach 3 at high altitude and was to have a range of 3,800 miles (6,200 km) with a conventional or nuclear bomb load. In December 1957, the North American company was named as main contractor to build the bomber, then designated B-70, and General Electric won the contract to develop the J-93 engines. However, in December 1959, the number of prototypes on order was drastically reduced and in March 1961, President Kennedy stopped the programme completely. Finally, it was agreed that at least three prototypes should be built so that the high cost of development would at least be justified by the acquisition of technological know-how. The first flight of the XB-70 took place on September 21, 1964 and Mach 3 was attained about one year later, on October 14, 1965.

The B-70 bomber was undoubtedly the target which the Soviet military had in mind when it issued its own requirements. Since the Soviets had first received performance data for the B-70 at the end of the 1950s, they had to push development of the MiG-25 with maximum urgency, as the 1964 first flight data of the B-70 indicates. In all probability, even the usual request for proposals phase was dispensed with and selection of the Mikoyan design bureau — designated to develop high-speed aircraft — was made immediately. The MiG-25 high altitude interceptor was then developed by Mikoyan in close collaboration with the Central Institute for Hydromechanics and Aerodynamics (TSAGI) in Moscow. This compression of the timescale was certainly the reason why proven design principles and available technology were utilized, and why even certain components and systems already in service were incorporated. The objective of having an air defence system ready for series production by the end of the 1960s, when the B-70 was expected to enter service, was fully attained with the MiG-25.

TECHNOLOGICAL STATUS

The shortage of time forced the Soviet designers to resort to their proven constructional techniques, which specified the use of steel for all highly-stressed structural members. The airframe was certainly heavier as a result, but it was designed relatively quickly, and only the zones subject to high thermal stresses, such as the leading edges of the wings and tailplane, were made in titanium. It should be noted here that at the beginning of the 1960s even the aircraft industries of the West — with the exception of the US industry — were hardly capable of machining titanium.

The technological state-of-the-art of the electronic equipment on the MiG-25 does not in general correspond to Western standards; this is in any case well known and was not expected. It should be stated, in this connection, however, that tubes and capacitors can have quite superior characteristics; the lack of printed circuits should not, consequently, be assumed to imply inferior performance capability. Nor is the fact that the Russian aircraft has only half the number of instruments of the US F-4 Phantom valid support for an assumption that the MiG-25 is of lower technological standard. It does, however, have a very adverse effect on pilot workload, though the overall cockpit layout, which bears a resemblance to that of the MiG-21, gives the pilot a high level of familiarity from the outset.

Contrary to the points mentioned above, which are often incorrectly interpreted, there are a series of other characteristics which make the MiG-25 appear to be equal, if not superior, to its Western counterparts. The overall aerodynamic design of the *Foxbat* is comparable to that of the American F-15; this is true of the external configuration and also of the electronically controlled double-wedge engine air intakes. The Tomansky RD-31 engine itself is a first-class product of the Soviet aerospace industry. Of prime interest is the five-stage trans-sonic compressor, which has no Western equivalent. Also a revelation was the quality of the airborne computer, which as part of the ground-based flight control system, ensures the automatic navigation and vectoring of the aircraft on to targets over the maximum range.

Viewed as a whole, the MiG-25 does not make out as badly as many have been eager to proclaim. It is representative of Soviet combat aircraft construction of the mid-1960s, with all its weaknesses and strengths. It is unrealistic, if not dangerous, to compare the *Foxbat* with the latest US designs such as the F-14 and F-15, which have only just entered service. From their design year to their mission requirements, these aircraft are more comparable with the MiG-25M, a development of the *Foxbat* described later in this article.

THE MIG-25 FOXBAT-A

The technical construction of the *Foxbat-A* high altitude interceptor will be described in detail, as being representative of the whole Mig-25 family, while only the differentiating characteristics of the other variants will be mentioned.

The high-set wing of double delta shape has a leading edge sweep of 40° on the inner section and 38° from the outer pylon. The anhedral is 4°. The wing has no leading edge flaps, only simple trailing edge flaps which extend over 37 per cent of the free span. Structurally, the main spar is designed as an integral fuel tank and is fabricated in steel; the wing leading edge is of titanium, whilst dural is used for the landing flaps and ailerons. Flutter damper bodies are fitted at the wingtips, which carry a continuous-wave target illuminating radar in their forward ends, with the receiver antenna for the *Sirena 3* 360° radar warning receiver somewhere in the centre section.

The *Fox Fire* X-Band air target tracking radar, with a dish diameter of 0.85 m, (2.7 ft) and a reported peak power of an astonishing 600 KW is housed in the voluminous fuselage nose, with a large electronics bay behind it. The single-seat cockpit is of similar layout to that of the MiG-21, with its somewhat basic instrumentation. In the event of radar failure, a simple optical sight is used as emergency back-up, though a target could then only be engaged with IR missiles. The ejector seat fitted is an older model from the MiG-21 which can still be used at zero altitude but requires a minimum forward speed of 150 km/h (93 mi/h). The bay for the forward-retracting, twin-wheel nose gear is located beneath the cockpit. The large area engine intakes, fitted with two electronically controlled variable ramps, provide optimum conditions for the compressor over the whole high Mach No. envelope. In the low speed bracket, ie, at high angles of attack, the lower air inlet lip is moved downwards to ensure better air flow. At higher speeds the excess air is bled via doors on the upper surface of the intakes. The sectional area of the engine intakes indicates an air mass flow of about 170 kg/sec (374 lb/sec), which according to Japanese reports is divided into cooling air for the engine compartments and combustion air. The actual air throughput of each engine should therefore be about 130 kg/-sec = 286 lb/sec.

The fuselage centre section is of rectangular shape and is mainly constructed of steel; it takes the loads from the wings, the main landing gear and the front engine mountings. The fuel is stored in tanks in the fuselage centre section between the air intakes and in saddle tanks around the air intake ducts. The main gear, linked to the fuselage, has single wheels of 1.20 m (3.9 ft) diameter and its high-pressure tyres only permit operation from concrete runways. The gear retracts forwards into the fuselage, in such a way that the wheel when fully retracted lies in a vertical position between the air intake duct and the fuselage outer skin.

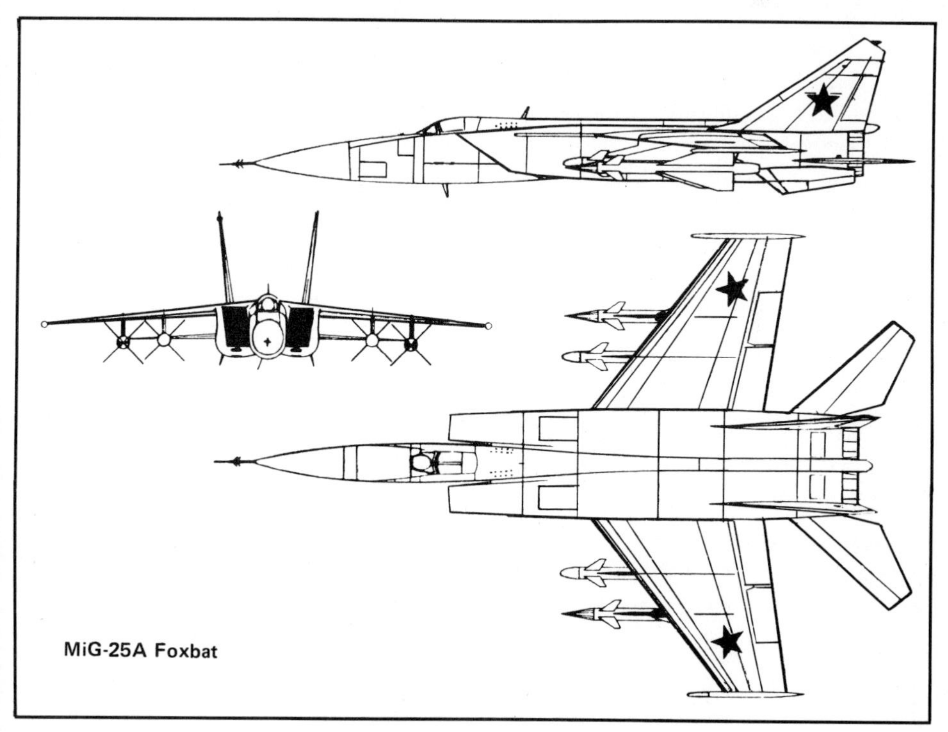
MiG-25A Foxbat

The rear fuselage holds the afterburners, which has an outside diameter of no less than 1.50 m (4.9 ft). On the top of the fuselage, between the two exhaust nozzle sections, is the container for the two braking parachutes. The highly-swept elevator surfaces are designed to be all-moving, and here again the main spars, including the bearings, are of steel, the leading edges are of titanium, whilst the rear sections are of dural. The twin fin and rudder assemblies are of similar construction but, as with the fins on the underside of the fuselage, large sections are designed as surface antennas.

The exclusive armament of the *Foxbat-A* is four long range AA-6 *Acrid* air-to-air missiles, the pylons for which are not suitable for carrying external fuel tanks. It cannot be excluded, however, that suitable pylons will be fitted subsequently and other missile launch rails mounted on the fuselage.

Judging from its major design parameters, the thrust-to-weight ratio, wing loading, electronic equipment and armament, the *Foxbat-A* is clearly designed for the high-speed, high-altitude intercept mission and about 400 examples have already entered service with the Soviet "Air Defence of the Homeland" Command. A look-down, shoot-down capability for the *Foxbat-A* is highly probable, but it must have considerable limitations. In addition to its principal role of interception, a secondary role for the MiG-25 could well be as a weapon carrier for air-to-ground missiles with around 100-200 km (62-125 mi) range. In this role the MiG-25 could be primarily used to attack radar sites and shipping targets in coastal zones.

UNHINDERED RECONNASISSANCE WITH THE MiG-25R FOXBAT-B

Although the B-70 bomber, primary target of the MiG-25, no longer existed, the MiG was put into series production as an interceptor though in more limited numbers than the standard interceptor of the Air Defence of the Homeland, the Su-15 *Flagon*. The extraordinary possibilities of the MiG-25 as a reconnaissance aircraft were very soon recognized, however. Today, the MiG-25R is in service with the tactical air forces, such as the Frontal Aviation units of the Group of Soviet Forces in East Germany. In contrast to the *Foxbat-A*, the *Foxbat-B* has the *Jay Bird* radar, which probably is used for anti-collision purposes at medium and low altitudes. It carries absolutely no armament, so it can be expected that the underwing hardpoints will be used in the future for expendable external fuel tanks.

The reconnaissance equipment of the *Foxbat-B* consists of five cameras mounted in the nose section, of which two are inclined to the left and two to the right at angles of 15° and 45° respectively, to the vertical. From 24 km (14 mi) altitude, useful photographs can be obtained with the 45° slant cameras over a 70 km (43 mi) wide corridor. The fifth camera takes vertical photographs, and probably has a long focus lens to provide sufficiently sharp images of point targets to be attacked. In addition to the cameras, the *Foxbat-B* also carries passive IR linescan equipment, the sensor doors for which can be clearly seen on both sides of the fuselage nose. Nothing is known concerning the performance capability of these reconnaissance equipments, but it can be presumed that optical and IR photos cover the same area out to 70 km (43 mi) depth and their results are complementary.

MIG-25U FOXBAT-C

As with other Soviet heavy combat aircraft and bombers, a trainer variant of the MiG-25 has also been developed by adding a second cockpit. As a result of the very limited electronic equipment — even the *Jay Bird* radar is not fitted — it can be clearly established that the *Foxbat-C* is being used, at least for the time being, exclusively for training purposes. There is still a possibility, however, that a two-seater variant with a more suitably faired-in second cockpit will be developed. Performance capability, size and take-off weight make the MiG-25 appear predestined for other operational missions, and the fitting of external fuel tanks and the resulting range increase would in any case necessitate a second crewman as navigator on air-to-ground missions. With suitable electronic equipment and an armament of free-fall or guided bombs, or air-to-ground missiles, this variant of the MiG-25 could be used as a bomber to attack strategic targets in Central and Western Europe. Attacks on shipping at greater ranges from the coast could also be carried out. This verson of the MiG-25 will certainly be taken seriously into consideration by NATO, since its development represents a logical step which could be taken quickly and without technical problems.

MIG-25RE FOXBAT-D

This variant of MiG-25 differs from the *Foxbat-B* in that, in place of the cameras and IR linescan equipment, a side-looking radar of considerable dimensions is incorporated. Little is known regarding the performance of this radar equipment, but it can be deduced from the operating altitude and the available antenna dimensions that a strip of 100 km (62 mi) depth on either side of the aircraft's flight path could be reconnoitred. Further variants of this type, primarly for electronic warfare, may exist.

MIG-25M FOXBAT E(?)

According to statements made by Belenko, the pilot who defected to Japan, a developed version of the *Foxbat-A* inceptor already exists, this being confirmed by the record flights of the E-266M in May, 1975. He said that in this aircraft, the airframe, engine, avionics and armament had been considerably improved. The structural strength of this version, designated MiG-25M, had been improved so much that near ground level speeds of over Mach 1 were attainable (as against Mach 0.85 for the MiG-25). An engine of greatly improved performance (RD-F or RD-F31) delivers a thrust with reheat of 14,000 kp (30,000 lbt) instead of the 11,200 kp (24,600 lbt) of the earlier model. In addition, the *Foxbat-E*, as a result of its new or considerably modified radar and improved missile seeker heads, now has a real capability to attack low-flying targets. The number of missiles has been increased from four to six by the addition of two fuselage hardpoints, so that for the intercept mission it can be assumed that a mix of AA-6 *Acrid* and AA-7 *Apex* missiles will be carried. The MiG-25M is probably also fitted with a cannon, which is either the GSh-23 twin barrel 23 mm gun or the six-barrel Gatling type gun of the MiG-27. It should also be possible to carry additional fuel, in two external underwing fuel tanks or in one large-volume tank under the fuselage.

The increase in structural integrity, thrust-to-weight ratio, electronic equipment and armament has permitted the operational spectrum of the MiG-25 to be widened, so that from a pure interceptor it has now evolved into an air superiority fighter. In the latter mission it would probably be armed exclusively with the AA-7 *Apex* missile.

CONCLUSION

Using the proven technology in the MiG-25, the USSR has succeeded over a relatively short period of time in developing a high-performance combat aircraft which can adequately fulfil its role as a weapon system. The fact that, in a few points of design, the technology used is not as advanced as modern Western developments, cannot alter the conclusion that to engage it successfully in combat will be extremely difficult. The alleged appearance of the MiG-25M also confirms that the *Foxbat* has entered a modernization phase, which makes the argument that it is a product of inferior technology appear somewhat fragile.

For the immediate future it can be assumed that, based on the individual variants of the MiG-25 in existence today, further versions will be developed and a family of combat aircraft in line with previous Soviet practice will be built up. For the present, the MiG-25M warrants particular attention, since as an air superiority fighter it is at least suitable for the medium altitude bracket and therefore could represent a danger to the F-14 and F-15. Its primary threat in the European theatre will, however, result from the use of the *Foxbat* as a bomber or guided missile carrier. What initially appeared to be a costly example of faulty planning after the cancellation of the B-70 programme may, therefore, soon become an important component of the Soviet air attack potential.

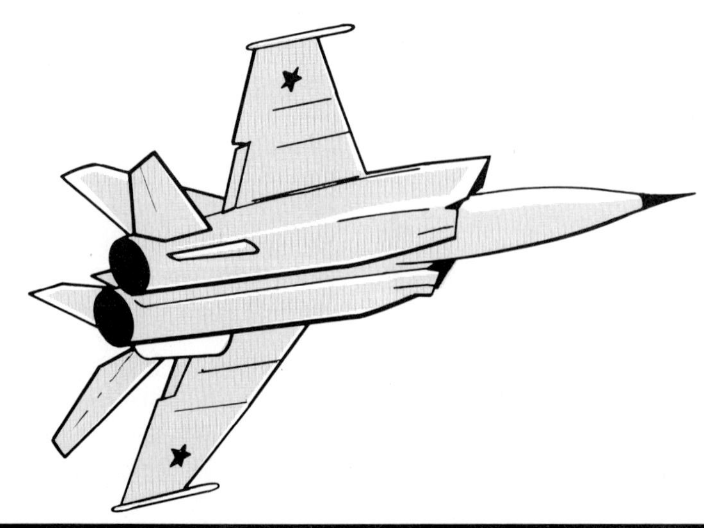

A CASE FOR THE MANNED PENETRATING BOMBER

Edited for Space
Reprinted by Permission
AIR UNIVERSITY REVIEW (July-August 77)

by Dr. G. K. Burke

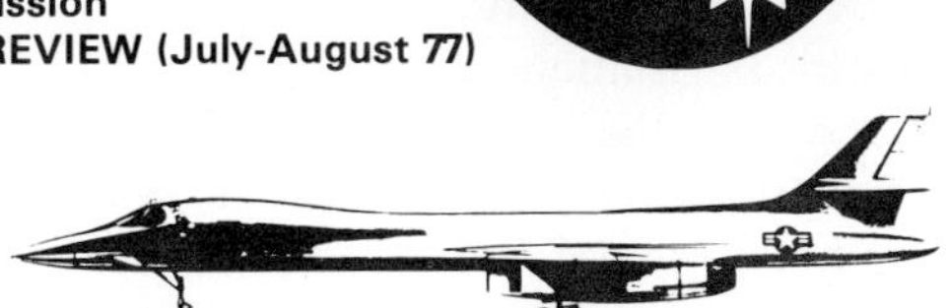

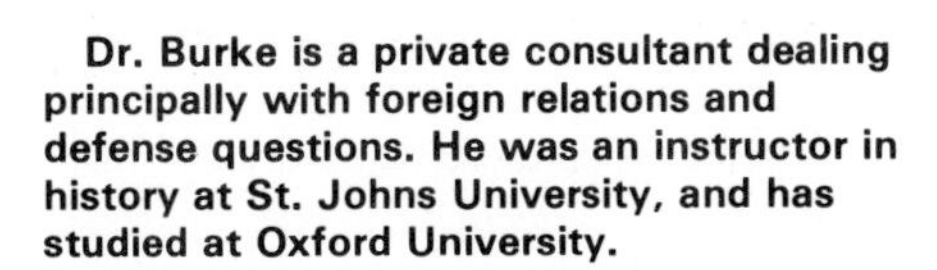

Dr. Burke is a private consultant dealing principally with foreign relations and defense questions. He was an instructor in history at St. Johns University, and has studied at Oxford University.

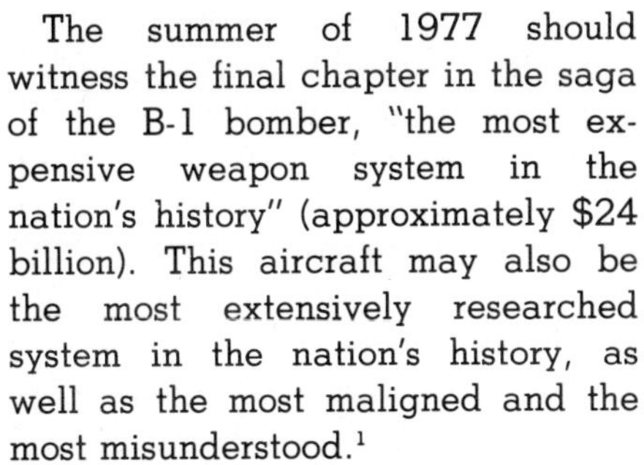

The summer of 1977 should witness the final chapter in the saga of the B-1 bomber, "the most expensive weapon system in the nation's history" (approximately $24 billion). This aircraft may also be the most extensively researched system in the nation's history, as well as the most maligned and the most misunderstood.[1]

Possibly the difficulties surrounding this controversial system are more apparent than real. If the decision to procure the projected force is made, that force will constitute the backbone of one leg of the strategic Triad for a period of some thirty years, commencing around 1986.

In theory the Triad, composed of the land-and sea-based missile forces in addition to the manned bombers, provides the nation with the capability to inflict unacceptable levels of damage on any power or group of powers capable of threatening the survival of our nation.[2] In recent years it has also been considered sagacious to include a redundant capability, a capability beyond assured destruction, for inflicting lower levels of damage (counterforce) aimed at targets other than a potential adversary's urban and industrial centers (countervalue).

In the future, this counterforce capability will continue to be necessary because in its absence the nation would be perilously exposed should a foe choose to strike at weapons rather than populations. In such a circumstance, the national leadership would find itself in the position of having to reply to a limited strike aimed at isolated weapon systems with massive countervalue retaliation or capitulate to the adversary's demands. Since the first alternative would invite the foe to reply in kind, assuming the adversary still possessed active weapons, the situation created could be harrowing. The latter alternative is unthinkable.

In the face of the dual nature of this strategic counterforce/countervalue challenge, the question becomes what level of force is necessary? Sizing the force is not easy because realities differ from nation to nation, and no man is able to predict with high confidence what the exact circumstances of some unnamed future crisis will be.

Nevertheless, the best analysis available, that made by then Secretary of Defense Robert S. McNamara twelve years ago, estimated that 400 one-megaton equivalents (OME) optimally delivered upon the Soviet Union would be sufficient to destroy 30 percent of the civil population and 76 percent of the industry. It was further estimated that this megatonnage would effectively terminate civilization in the U.S.S.R.[3]

Considerable doubt has arisen over the continued validity of this analysis. The question hinges largely on one's interpretation of four vital caveats. *First, the use of OME as a unit of measurement for nuclear destruction has been criticized as being obsolete.* OME is calculated by raising the explosive force of the warhead to the two-thirds power. Hence, a 27 megaton (MT) warhead equals 9 OME, or has a burst equal to nine one-megaton explosives.[4]

Recent developments have improved the sophistication of the model. This upgrading was made necessary by the fact that most targets are small, and many are not circular, as is the arc of the blast. By raising the number to the three tenths power for American warheads and to the four tenths power for those of the U.S.S.R., more of the wasted energy released by a typical burst is accounted for. Within the framework of this model, a 27 MT warhead would equal 2.691 adjusted one megaton equivalents (AOME)[5].

Second, and most disturbing, *for some ten years the Soviet Union has been systematically engaged in hardening its society.*

These efforts would appear to have invalidated analyses based on 400 OME or even 400 AOME.[6]

Third, the 400 OME or AOME must be optimally delivered. Prior to the enemy's initiation of hostilities, the most vital targets would be covered by at least one weapon apiece. But afterwards, in the confusion and destruction that will inevitably accompany any thermonuclear counterforce strike aimed at weapon systems, this may not be true. Some targets may still be covered by more than one weapon, others by none. The following rule then prevails: If prompt response is desired, more than 400 OME/AOME must survive in order to deliver 400 OME/AOME optimally.[7]

Fourth, the 400 OME figure fails entirely to account for the redundant force necessary to reply to lower-level strikes. When all is considered, a conservative estimate of the force necessary to deter the Soviet Union across the entire counterforce/-countervalue spectrum must be increased by an order of magnitude, with particular emphasis on weapons programmed to offset Soviet "societal hardening" measures.[8] A conservative, responsible estimate would indicate that the original McNamara estimate is now inaccurate by several orders of magnitude.[9]

The role of each arm of the Triad in producing this force level might next attract the attention and consideration of the analyst. The most heralded of the arms is the ballistic missile submarine because of its invulnerability to detection and destruction in terms of one swift strike. Given weeks or months, portions of the ballistic missile fleet could be detected and destroyed.

A second difficulty results from the relatively small and inaccurate RVs that current and projected sea-launched ballistic missiles (SLBMs) are armed with: they are ill-equipped to attack hardened or dispersed industrial and population targets.[10]

Of more significance, small warheads produce small quantities of fallout, which is one aspect of a nuclear burst that is directly proportional to the size of the explosion (approximately 100 pounds to a one-megaton detonation). Since large quantities of fallout may be necessary to secure the assured destruction of an evacuated populace dispersed in hastily constructed shelters, small submarine-launched RVs may be expected to have limited value in attacking it.[11]

The following should also be observed:

(1) Attacks on submarines at sea produce little or only limited collateral damage. This system invites attacks upon itself to a far greater degree than either of the other two arms of the Triad.

(2) Ten Trident boats will cost an estimated $15 billion. To procure a force approaching the level required for assured destruction would impose an unendurable fiscal burden on the nation.

(3) A sudden technological breakthrough could wipe out the entire deterrent if it were committed to a single medium.

(4) Severe communication problems may exist in relation to the ballistic missile submarine. The boat may possess more survivability than the means to communicate with it.

(5) Limited strikes from a ballistic missile submarine present serious difficulties. Once a missile is launched the boat has disclosed its position and would in some instances be in danger of destruction.

Conclusion: Though a key portion of the strategic Triad, the ballistic missile submarine lacks the cost-effective survivability to cover the entire counterforce/countervalue spectrum alone.[12]

The second arm of the deterrent is the land-based missile force. This arm of the Triad is currently passing through a stage of uncertainty. Careful analysis indicates that if current trends continue, by the late 1980s a Soviet land-based missile force of 313 SS-18 and approximately 65 SS-19 ICBMs will be sufficient to destroy about 95 percent of the current American land-based missile force.[13] Since the Vladivostok *aide-memoire* permits each superpower to procure 1320 missiles with multiple warheads, the situation could assume critical dimensions.

The logical response to such a threat would appear to be to move at once in the direction of one of the mobile basing alternatives available. Eventually this may become necessary, but at present powerful forces argue against the adoption of such a course. Three factors are paramount: (1) The inability to detect the number of missiles present in either the covered trenches or the multiple aim points of the two most-often-proposed land mobile systems could mean the end of SALT negotiations and serve as a powerful catalyst to nuclear proliferation. (2) The existing silo force is relatively inexpensive to maintain. This would not be true of land mobile force, which will be 1.5-2.0 times as costly.[14] (3) The greatest problem may lie in creating cost-effective firepower.[15]

Conclusion: With regard to the land-based missile force, follow current force planning goals. Introduce modifications to the current silo-based Minuteman force, withhold land mobile basing and large-scale MX alternatives pending the outcome of U.S.-Soviet negotiations.

Manned bombers constitute the third and final arm of the strategic Triad. They are the least understood of the three systems comprising the Triad, but paradoxically they are also the most lethal. The maximum payload per unit is enormous. Today a single B-52G/H optimally equipped with four gravity bombs (10 MT X 2 + 5 MT X 2) and six 200 KT short-range attack missiles (SRAM) disposes a payload that due to its enormous weight is optimal for attacking widely dispersed targets. In terms of megatonnage, an optimally armed B-52 would notionally dispose 30 MT as compared to a Trident submarine's 19.[17]

From another standpoint, the extreme accuracy of air-delivered gravity bombs renders the manned bomber the ideal candidate to attack such portions of Soviet society that have been hardened to withstand nuclear assault. In fact, witnesses before the Senate Armed Services Committee have testified that some targets in the single integrated operations plan (SIOP) cannot be attacked cost effectively by any other means. (For example, the Hoover Dam may require up to 10,000 psi.)[18]

Considering that the prime mission of our strategic forces is deterrence, the manned bomber force is the only leg of the Triad that can be flexed in time of crisis. Neither ICBMs nor SLBMs can be recalled once fired. It is unthinkable that either of them would ever be launched under any conditions other than general hostilities. But the bomber force can be launched on warning and recalled at any time short of reaching the target. In short, the bomber force gives our national command authorities a strategic option between all or nothing when hostilities appear imminent.

As with the other arms of the Triad, several caveats need to be observed. First, a single B-52 with ten weapons embarked could strike at no more than ten targets, while the Trident class submarine could theoretically (if implausibly) strike at up to 192. Second, in terms of a large target structure, one comprised of many small noncircular sites, the manned bomber fares poorly when contrasted to the submarine. Across some profiles the submarine would be superior.

But it must be remembered that the bomber is a reusable platform, and no other system has this capability in cost-efficient terms. When this capability is factored into the equation, it invalidates most other measurements. Invariably, the manned bomber is undervalued because its real-time capability for multiple strikes is ignored. This becomes more apparent when it is remembered that the Minuteman 3 and Poseidon missiles have a current price of approximately $9 million per unit. The cost of reloading missile silos or submarines would be prohibitive. Of equal significance, many bombers may be purchased for the price of one submarine. The proposed force of 244 B-1 bombers has an estimated cost of $24 billion. A mere ten Trident-class submarines have been cost estimated at $15 billion. Conclusion: No high-volume force capable of coping with current Soviet civil defense measures is able to be developed in a cost-effective manner without a high volume, high accuracy manned bomber.[19]

If it is assumed that a manned bomber is a viable portion of the Triad, the next consideration should involve the configuration of the aircraft. Under this heading few serious analysts question the need to procure a replacement for the aging B-52 fleet. Built to perform over a 5000 hour flight profile, the average B-52 has logged over 8000 hours with many exceeding 11,000; built to perform for ten years, the last B-52/H was delivered in October 1962. By 1990 these veterans will no longer be able to perform a first-line mission. If they are not supplemented by some alternative system, it is doubtful that they will be able to perform at all.[20]

In terms of the future, two candidates have been proposed to augment/replace the aging B-52 fleet: the Air Force B-1 manned, penetrating bomber and the Brookings Institution cruise missile armed, wide-body, stand-off bomber.[21]

Regarding launch survivability, two factors become relevant: (1) In a "worst possible case" scenario, eight minutes of vacillation by the national command authorities in the face of a dedicated attack by depressed trajectory SS-N-8 missiles could lead to the complete loss of *any* bomber force. Conclusion: *Any* bomber force is vulnerable even to small human error. No bomber force possesses a sufficiently high level of survivability to be entrusted with the whole of the nation's strategic defense.[22] (2) In "worst plausible case" scenarios, crisis preparations combined with the anticipated timely decision-making at the national level will be able to invalidate any foreseeable enemy attack aimed at preempting the bombers on their runways. Conclusion: In most plausible crises, the manned bomber is a valid system and remains a vital portion of the strategic Triad.

This conclusion is most readily grasped by examining a typical crisis. Whatever the origin (possibly involving Europe), the administration would have adequate warning of the impending collision, would be eager to control the level of tension, but at the same time would find it advantageous to optimize its force level should negotiations collapse. From the standpoint of the manned bomber, this state could be most readily achieved by deploying the available number of bombers at sparsely inhabited points where a preemptive strike could not reach them prior to their successful escape. Though it should be observed that at present the Soviets have not

developed depressed trajectory capability for their modern naval missiles (SS-N-6 and SS-N-8), they have tested it on the enormous land-based SS-9 system. Inasmuch as the next generation of bombers is expected to have a life-cycle span of thirty years, the possibility that this "within the state-of-the-art" concept may be deployed is difficult to ignore.[23]

Were it to be deployed, its impact may be illustrated by observing that the total escape time for B-1 is approximately 240 seconds. The Brookings wide-body cannot perform the same mission profile in less than 330 seconds and is only able to perform it that well if rocket-assisted takeoff is provided. In addition, B-1 is smaller and hence is able to take off more swiftly than the wide-body, one every 7.5 seconds as opposed to one every 15 seconds. Finally, B-1 is able to operate from short (7500 foot) runways, as contrasted to the approximately 10,500 foot runways needed for the wide-body. The result of this phenomenon is that only Rapid City, South Dakota, has the right combination of runway length and distance from the sea to enable the wide-body to escape attack from dedicated SLBMs. In contraposition, the four largest municipal airports in the state of Wyoming themselves possess a notional capability to launch 49 percent of the proposed B-1 crisis force of 210 in the face of the same assault. Conclusion: Of a total force procurement of 244, generating a crisis alert force of 210, B-1 produces 210 survivors operating either from existing airstrips or easily modified airstrips, the wide-body fewer than ten.[24]

In light of this harsh reality, a force of wide-bodies would have one of two choices: either to risk preemptive destruction or proceed to airborne alert. The latter move would preserve that portion of the force so deployed, but the gambit has its drawbacks. Primary among these is that it requires a very large force on the ground to sustain a very small force in the air. It is estimated that in the early 1960s the United States possessed a capability to sustain 12.5 percent of the then existing bomber force in the air for a period of one year. For crises of shorter duration improved percentages should be possible, but the implications are obvious.[25]

A second problem encountered in scenarios involving airborne alert touches on the delicate question of crisis management. While in some situations it might be deemed desirable to place bombers on airborne alert to signal resolve, in others it might not. It should be understood in advance that placing bombers on airborne alert has ominous escalatory overtones. Even a quick deployment to deep interior bases, away from vulnerable coasts, would not have the same impact. Conclusion: Airborne alert is not necessarily a desirable state.

Provided the bombers survive attempts aimed at preemptive destruction, they must then possess the capability to penetrate to their targets. Twenty years ago Albert Wohlstetter estimated that on a typical mission an individual bomber possessed between .5—.9 chance of survivability, depending on the state of the defense and the skill and execution of the offense. Under present conditions, analysis would indicate that the wide-body is near the lower end of that profile, while the B-1 is at the upper end.[26]

The wide-body, presumed to be in the 747 class, is vulnerable to a wide range of criticisms. Among them is the fact that the 1500 nautical-mile-range cruise missile that the Brookings experts proposed as the irreducible minimum does not exist. Proposed air-launched cruise missiles (ALCMs) have notional ranges of between 1000-1200 nautical miles. And while a longer-range ALCM could be developed from the Navy's sea-launched cruise missile (SLCM) program, to achieve 1500 nautical miles in the airborne mode would entail additional expenditure.[27]

Of more significance, the aircraft and its missile (whatever the range) are vulnerable to enemy air defense. The Soviet Union has the world's largest and most modern air defense system. It currently lists in its inventory 5000 air surveillance radars, over 2500 interceptors, and some 12,000 surface-to-air missile (SAM) launchers. It is constantly being upgraded. This air defense would have a wide range of options to deploy against the handful of surviving lumbering wide-bodies that could be maintained on airborne alert.[28]

First, the Soviets could attack them at extreme range with a combination of airborne warning and control system (AWACS) aircraft and transports (possibly of the Il-76 class) armed with air-to-air missiles. The proportion that this problem is apt to assume may be best illustrated by observing that analysis indicates that even the current primitive Soviet AWACS (NATO code-named Moss) works acceptably over water, the very medium above which the wide-body would be expected to launch its ALCMs. Future AWACS should be far more effective, especially when they are combined with the high-altitude profile the lumbering wide-body would be flying and its vast radar cross section.[29]

Second, in regions where the wide-body would have to approach nearer to the coast to attack its targets, less costly and sophisticated measures should suffice. Current or projected interceptors supported by AWACS and, in some instances, by in-flight refueling should be able to exact a considerable toll among the lumbering wide-bodies. It is possible that the mortality rate would approach the 50 percent mark.[30]

Third, in addition to the bomber, the ALCM itself may be attacked. It is not often noted, but all projected cruise missiles fly a portion of their profile at very high altitudes (as much as 45,000 feet). At such heights, even primitive interceptors with standard air-to-air weapons should be lethal. Of perhaps more importance, during the next 30 years interceptors equippped with look-down-shoot-down systems should be developed and deployed. Brookings experts estimate that 600 such interceptors, each equipped with six missiles capable of a 50 percent intercept/reliability, could inflict up to 1000 kills on a notional cruise missile force.[31]

Fourth, sophisticated terminal air defenses should be able to engage and kill several hundred more slow-moving, subsonic cruise missiles. The effectiveness of these defenses hinges largely on the number of sites to be defended, the number of rounds each launcher is able to fire at an approaching ALCM concentration, and, most significant, how many sites are able to be avoided by skillfully preplanning ALCM flight patterns. Taking these factors into consideration, one conservative estimate yields approximately 300 additional kills.[32]

Finally, it would appear that the wide-body/cruise missile system is a vulnerable weapon. It would also appear that the Brookings experts were cognizant of this. To overcome the weaknesses inherent in the wide-body system, they proposed that a path for the ALCMs be cleared through the terminal defenses with air-launched ballistic missiles (ALBMs). This system does not exist in any form today, and the expense of developing it might be exorbitant.[33] Conclusion: Without ALBM support the wide-body system is highly vulnerable and, in addition, is subject to the earlier elaborated caveats concerning the difficulty of attacking dispersed populations with low fallout producing explosives (ALCM warhead = 250 KT).[34]

The strategic picture involving B-1 is different. B-1 combines a relatively small radar cross section with an electronic counter-measures suite that will prove difficult for many enemy sensors to penetrate. The aircraft is capable of near sonic speed at heights as low as 200 feet above the ground. This renders tail chase by any foreseeable interceptor highly implausible. Should the enemy improve his sensors or electronic counter-countermeasures, B-1 has room to grow and will be fully capable of accepting advanced systems such as the short-range ballistic defense missile (SRBDM) and the advanced strategic air-launched missile (ASALM). The former would be used to protect the bomber from air-to-air missiles; the

latter will be capable of nuclear engagement against air or land-based targets and will combine SRAM speed (Mach 2.5-3.0) with ALCM range (650 NM). If their deployment becomes necessary, they should provide an acceptable answer to an advanced Soviet AWACS and look-down-shoot-down interceptors.[35]

Above all B-1 will penetrate to the target with its weapon mix of SRAMs (as many as 24), or gravity bombs, or ALCMs, or SRBDMs, or ASALMs, providing unrivaled flexibility of payload, extraordinary accuracy of delivery, and even some immediate reconnaissance of the target area. The fact that this system is manned optimizes system survivability by providing the maximum number of options for defense suppression, ranging from jamming, to avoidance, to destruction. Unlike the ALCM it will not be bound to a set, slow-moving flight profile devoid of alternatives.

The lethality of the system is best gauged by observing the performance of the aircraft under conditions depicting "the worst-plausible case." This scenario envisions the national command authorities' failing to disperse exposed aircraft to secure inland sites. Thus, only those bombers deep based at Minot and Grand Forks, North Dakota, and at Rapid City would survive preemption. If Rapid City were outfitted with a double squadron wing, some 34 B-1s could escape a dedicated attack by depressed trajectory SLBMs with a notional capability to travel 1100 NM in 45 seconds and have operational access to Hudson Bay.[36]

The amount of firepower deliverable by such a force in a single strike would vary with the payload carried. But if .9 of those able to escape preemption proved to be mechanically reliable and .85 survived enemy defenses (as the USAF has hypothesized), then 26 B-1s should reach their goals armed with as many as 24 SRAMs per bomber (if each proved to have a .9 reliability that would equal almost

600 weapons delivered on target) or a far smaller number of heavy gravity bombs.[37]

Under similar conditions, a mere six wide-bodies would escape from their only safe haven, Rapid City. And it is questionable that any weapons would be delivered on target by this handful of survivors if the rest of the equation included: .9 mechanical reliability, .5 survivability, and serious degradation to the 50 ALCMs (.9 reliability) embarked aboard each aircraft from both interceptors and terminal defenses.

In the period of the late 1980s the concept of the strategic Triad may be discarded, and revolutionary strategies may develop. However, until such plans are reasonably formulated, the classical model will necessarily have to be followed. If it is, and if it is decided that a high-volume, high-accuracy payload is a desirable feature for the strategic forces of the United States, then there would appear to be little cost-effective alternative to the B-1 bomber system.

Notes

1. Francis P. Hoeber, "The B-1: A National Imperative," *Strategic Review* (Summer 1976), pp. 111-17; John W. Finney, "Who Needs the B-1?" *New York Times Magazine* (July 25, 1976), p. 7.
2. U.S., Congress, Senate, *Hearings Before The Committee On Armed Services, Part 1 Authorizations*, 94th Cong., 1st sess., February 5, 1975, p. 50.
3. Geoffrey Kemp, "Nuclear Forces for Medium Powers: Part 1; Targets and Weapons Systems," *Adelphi Papers*, No. 106 (Autumn 1974), p. 26.
4. Ian Bellany, "The Essential Arithmetic of Deterrence," *The Royal United Services Institute for Defence Studies* (March 1973), pp. 28-34.
5. Thomas J. Downey, "How to Avoid Monad — and Disaster," *Foreign Policy* (Fall 1976), pp. 172-201.
6. Editorial, "The Erosion of the U.S. Deterrent: The Real Intelligence Crisis," *Strategic Review* (Summer 1976), pp 4-5; John W. Finney, "Stronger U.S. Civil Defense Effort Urged by an Industrial Study Group," *New York Times* (November 18, 1976, p. 16; "Intensified Soviet Civil Defense Seen Tilting Strategic Balance,"*Aviation Week & Space Technology* (November 22, 1976), p. 17; "Strategic Defensive Systems Emphasized," *Aviation Week & Space Technology* (September 20, 1976), p. 49.
7. Robert J. Carlin, "A 400 Megaton Misunderstanding," *Military Review* (November 1974), pp. 3-12.
8. Arthur A. Broyles and Eugene P. Wigner, "The Case For Civil Defence," *Survival* (September/October 1976), pp. 217-20.
9. Kemp, p. 5; P. J. McGeelan and D. C. Twitchett, editors, *The Times Atlas of China* (New York: Quadrangle/The New York Times Book Co., 1974), p. xvi; Alan Golenpaul, editor, *Information Please Almanac Atlas and Yearbook* (New York: Dan Golenpaul Associates, 1975), p. 708.
10. For Currie statement see U.S., Congress, Senate, *Hearings Before The Committee On Armed Services, Part 6 Research and Development*, 94th Cong., 1st sess., March 7, 11, 17, 19, 21, and April 25, 1975, p. 2828. Clarence A. Robinson, Jr., "New Propellant for Trident Second Stage," *Aviation Week & Space Technology* (October 13, 1975), pp. 15-19; Bellany, loc. cit.; "Study Finds Joint MX/Trident Impractical," *Aviation Week & Space Technology* (October 13, 1975), p. 17; Downey, *loc. cit.*; All SSKP calculations were done with a General Electric Missile Effectiveness Calculator.
11. Henry A. Kissinger, *Nuclear Weapons and Foreign Policy* (New York: Harper Brothers, 1957), p. 74.
12. U.S., Congress, Senate, *Hearings Before The Committee On Armed Services, Part 1 Authorizations*, 94th Cong., 1st sess., February 5, 1975, p. 74.
13. Lynn Etheridge Davis and Warner R. Schilling, "All You Ever Wanted To Know about MIRV and ICBM Calculations But Were Not Cleared To Ask,"*The Journal of Conflict Resolution* (June 1973), pp. 207-42; John W. Finney, "Soviet Deploying 2 New Missiles,"

New York Times (January 15, 1975), p. 1; Paul H. Nitze,"Assuring Strategic Stability in an Era of Detente," *Foreign Affairs* (January 1976), pp. 207-32; Thomas A. Brown, "Missile Accuracy and Strategic Lethality," *Survival* (March/April 1976), pp. 52-59.

14. General William J. Evans, "The Impact of Technology on U.S. Deterrent Forces," *Strategic Review* (Summer 1976), pp. 40-47; "B-1 Bomber Crux of SAC Plans," *Aviation Week & Space Technology* (May 10, 1976), pp. 39-45.

15. U.S., Congress, Senate, Subcommittee On Arms Control, International Law And Organization Of The Committee On Foreign Relations, *U.S. — USSR Strategic Policies*, 93rd Cong., 2d sess., March 4, 1974, p. 19; U.S., Congress, Senate, *Hearings Before The Committee On Armed Services United States Senate, Part 6 Research and Development*, 94th Cong., 1st sess., March 7, 11, 17, 19, 21, & *25, 1975, pp. 2804-5.*

17. "Weapon Advances Raise B-52 Capability," Aviation Week & Space Technology (May 10, 1976), pp. 132-35; *The Military Balance* 1972-1973 (London: International Institute For Strategic Studies, 1972), pp. 66-67, 85-86; *Jane's All The World's Aircraft 1975-76* (London: St. Giles House, 1975); Kemp, p. 6; Bellany, *loc. cit.*

18. U.S., Congress, Senate, *Hearings Before The Committee On Armed Services, Part 7 Research and Development*, 93rd Cong., 2d sess., April 4, 5, 12, 16, 23, 25, 26, and May 2, 1974, pp. 35-39; U.S., Congress, Senate, *Hearings Before the Committee On Armed Services, Part 4 Research and Development,* 94th Cong., 1st sess., March 4 and April 5, 1975, p. 2122.

19. Geoffrey Kemp, "Nuclear Forces for Medium Powers: Parts 2 and 3: Strategic Requirements and Options," *Adelphi Papers,* No. 107 (Autumn 1974), p. 19; Clarence A. Robinson, Jr., "Minuteman Production Defended," *Aviation Week & Space Technology* (January 19, 1976), pp. 12-15.

20. "B-52 Lifetime Extension Effort Pushed," *Aviation Week & Space Technology* (May 10, 1976), pp. 140-42; "Policies Altered to Stretch Funds," *Aviation Week & Space Technology* (May 10, 1976), p. 151; "SAC Tests Consolidation of Maintenance," *Aviation Week & Space Technology* (May 10, 1976), pp. 152-53.

21. Some critics have suggested that the B-52 might be replaced by a stretched FB-111 or an improved version of the B-52 itself. At this writing neither suggestion has attracted much bipartisan support.

22. Alton H. Quanbeck and Archie L. Wood, *Modernizing the Strategic Bomber Force* (Washington: The Brookings Institution, 1976), p. 44.

23. *Jane's Weapons Systems 1973-1974* (London: St. Giles House, 1973).

24. *Low Altitude Instrumental Approach Procedures*, Volumes 1-9 (St. Louis: The Defense Mapping Agency Aerospace Center, 1976), Quanbeck and Wood, pp. 47, 52, 109; U.S., Congress, Senate, *Hearings Before The Committee On Armed Services, Part 7 Research and Development*, 93rd Cong., 2d

sess., April 4, 5, 16, 23, 25, 26, and May 2, 1974, pp. 3873-78: *Jane's All the World's Aircraft 1975-76, loc. cit.*; Craig Covault, "FB-111's Effectiveness Increased," *Aviation Week & Space Technology* (May 10, 1976), pp. 103-13.

25. Quanbeck and Wood, p. 23.

26. A. J. Wohlstetter, F. S. Hoffman, R. J. Lutz, and H. S. Rowen, *Selection and Use of Strategic Air Bases* (Santa Monica: The Rand Corporation, 1954), p. 20.

27. Clarence A. Robinson, Jr., "Strategic Programs Scrutinized," *Aviation Week & Space Technology* (March 31, 1975), pp. 12-13; Clarence A. Robinson, Jr., "Tentative SALT Decision Made," *Aviation Week & Space Technology* (February 16, 1976), pp. 12-14.

28. *The Military Balance 1975-1976* (London: International Institute For Strategic Studies, 1975), p. 8.

29. Senate, *Armed Services Committee*, February 5, 1975, p. 224.

30. Hoeber, *loc. cit.*

31. Quanbeck and Wood, p. 73; Robinson, *Aviation Week & Space Technology* (March 31, 1975), *loc. cit.*

32. The equation is: Total surface-to-air missile launchers: 800, Shots per launcher at ALCM concentration: 1, Percentage of launchers successfully effecting engagement with ALCM concentration: .46, Overall reliability of surface-to-air missiles: .81, Kills: 300.

33. Donald E. Fink, "Minuteman Experiences Aiding MX," *Aviation Week & Space Technology* (July 19, 1976), pp. 113-20.

34. Clarence A. Robinson, Jr., "Tomahawk Clears Crucial Test," *Aviation Week & Space Technology* (November 22, 1976), pp. 14-16.

35. U.S., Congress, Senate, *Hearings Before The Committee On Armed Services, Part 4 Research and Development*, 94th Cong., 1st sess., February 25, 27, March 4 and 5, 1975, p. 1995.

36. *Low Altitude Instrumental Approach Procedures, loc. cit.*; Quanbeck and Wood, pp. 44-48.

37. Wohlstetter, pp. 94-95; Quanbeck and Wood, p. 65.

THE B-1: CANCELLED CARTER SAYS 'NO'

Reprinted by Permission
AIR PROGRESS (October 77)

Calling it "one of the most difficult decisions I've made," President Jimmy Carter last July decided to halt production of the Rockwell International B-1 bomber and rely instead on the cruise missile as the nation's weapon of the future.

The decision is expected to have a major impact on U.S. military strategy for years to come. Reaction to the move was predictable — foes of the much-discussed bomber were elated while supporters of the B-1 were caught by surprise and expressed shock at the President's action.

Carter told a nationally-televised press conference that development of the cruise missile — small pilotless jet aircraft with a high degree of accuracy in reaching a designated target — would continue while B-52s would be used as launch platforms for the missile.

"During the coming months, we will also be able to assess the progress toward agreements on strategic arms limitations, in order to determine the need for any additional investments in nuclear weapons delivery systems," Carter said, "in the meantime, we should begin deployment of cruise missiles, using air launch platforms such as our B-52s, modernized as necessary."

Carter said he wanted some type of research and development of the B-1 to continue, but the extent of such activity had not been settled as this issue went to press. It was believed, however, that some flight testing of the first three prototypes at Edwards Air Force Base in California would continue.

Top Air Force officials and government "insiders" had believed that Carter was ready to approve some type of production of the Rockwell bomber and the President's announcement caught many off guard.

"They're breaking open the vodka bottles in Moscow," Rep. Robert Dornan was quoted as saying after the announcement was made. Dornan, whose district includes the Rockwell plant where much of the B-1 work would have been centered, said the decision would cost 68,000 jobs in California alone.

Cost of the B-1 production termination could not be determined immediately as Rockwell and the various major subcontractors on the program began studying the alternatives to full-scale production. These alternatives ranged from a complete halt in the B-1 program, to the completion and flight testing of the fourth B-1 prototype which currently is partially assembled at Rockwell's Palmdale, California, facility. The fourth B-1 was to have been the test vehicle for the electronic countermeasures system.

In explaining his decision to continue some form of research and development, Carter said, "The existing testing and development program now under way for the B-1 should continue to provide us with the needed technical base in the unlikely event that more cost-effective alternatives should run into difficulty. Continued efforts at the research and development stage will give us better answers about the cost effectiveness of the bomber and support systems, including electronic countermeasures techniques."

The cost of the controversial swing-wing aircraft was one of the reasons for the production cancellation, according to officials. Although the President had been a strong opponent of the aircraft during his campaign, some Defense Department officials believed President Carter had changed his stand on the B-1 in the months since his election.

Officials cited several other reasons for the B-1 cancellation. It is believed the B-1 may be more vulnerable — even at low altitudes — when the Russians develop "look down" radar which may be able to pick the B-1 out of the "ground clutter" as it flies at nearly tree-top level at near the speed of sound.

The development of the cruise missile also was a key element in the President's decision. Use of cruise missiles from B-52s some distance away from their targets could make the aging Boeing bomber an effective "stand off" weapons delivery system which would not need the speed or low altitude capability of the B-1 which would penetrate enemy territory.

Following the President's decision, defense officials began discussing means of converting wide-body aircraft into cruise missile carriers to supplement the aging B-52 fleet. There even was some talk of using a simplified B-1 as a cruise missile carrier.

While B-1 work at Rockwell's facilities was winding down, the pace at two other aerospace companies was increasing. Boeing, which is building the Air Force's Air Launched Cruise Missile — or ALCM — and General Dynamics, builder of the Navy's Tomahawk cruise missile, are gearing up for the expected increased demand for their missiles.

Boeing is proceeding with full-scale development of both the "A" and "B" models of the ALCM. The "A" model, which was successfully tested in 1976, has a range of approximately 750 nautical miles while the "B" version's range would be doubled. First assembly of a production ALCM-B presently is planned for the autumn of 1978, according to Boeing, with flight tests to begin after that year. Operational missiles are expected to be available in 1980.

Defense officials said consideration was being given to establish a competitive flyoff between the Tomahawk and ALCM-B for the contract for future air-launched missile production.

B-1 TEAM AWARDED COLLIER TROPHY FOR '76

by YEARBOOK Staff

The Air Force-aerospace industry team responsible for the development of the B-1 Strategic Aircraft System was awarded the 1976 Collier Trophy in ceremonies in Washington, D.C., in May, 1977. General David C. Jones, Chief of Staff of the Air Force, and Mr. Robert Anderson, president of Rockwell International, received the award as representatives of the service-industry team.

The Collier Trophy was established in 1912 by the pioneer aviation enthusiast and publisher, Robert J. Collier. The award, "for the greatest achievement in aeronautics or astronautics in America, with respect to improving the performance, efficiency or safety of air or space vehicles, the value of which has been thoroughly demonstrated by actual use during the preceding year," is presented annually by the National Aeronautic Association.

Headed by Arthur F. Kelly, Chairman of the Board of Western Airlines, the selection committee of 28 noted aerospace authorities and leaders unanimously awarded the Trophy to the B-1 team for the successful design, development, management and flight testing of the strategic bomber.

The team, in addition to the Air Force as user and Rockwell International as prime contractor, included over 3,000 associate contractors, subcontractors, and suppliers, including Boeing, Cutler Hammer, Inc. (Airborne Instruments Laboratory Division), and General Electric.

The first of the "fly before buy" projects of the Department of Defense, the B-1 program met or exceeded all mission requirements specified by the Air Force.

The advances in technology achieved by the B-1 program will provide a major impact on the design of any future large turbo-fan aircraft, particularly in the areas of safety, performance, and efficiency.

According to the Department of Defense, the three development aircraft currently flying had "fully demonstrated the B-1's operational capability" in more than 440 hours of flying.

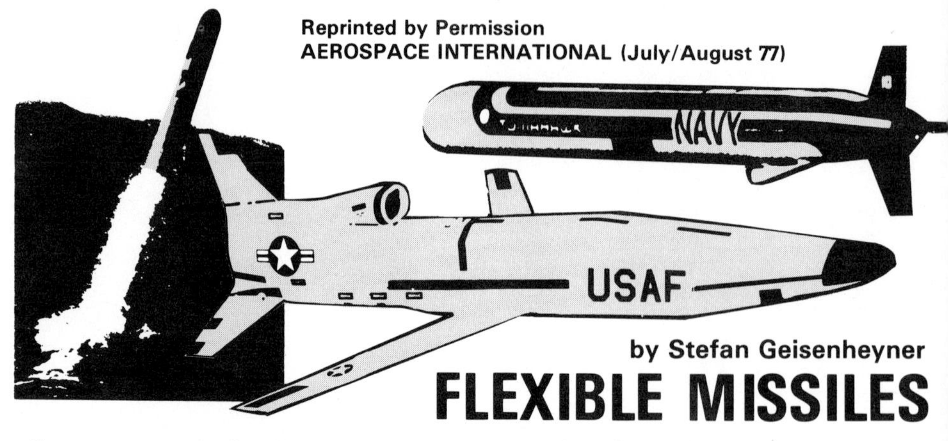

Reprinted by Permission
AEROSPACE INTERNATIONAL (July/August 77)

by Stefan Geisenheyner

FLEXIBLE MISSILES

Long-range missiles bring to mind the vision of gigantic rockets being launched into space where the warhead separates and enters the atmosphere at incredibly high speed and plunges towards its target where it would obliterate with its multi-megaton nuclear charge everything for tens or hundreds of miles around. This vision is quite correct, as are the assumptions that such missiles are extremely costly, unwieldly, vulnerable and basically inefficient, for what they lack in accuracy they have to make up with explosive power. True, they have helped greatly — and to some military philosophers they are primarily responsible — that no war has taken place among the superpowers of this world. But their efficiency and deterrence value are shrinking day by day. The MIRV (Multiple Independent Reentry Vehicles) system was invented which allows several warheads to be loaded onto one missile and then to be targeted separately. Still, the possibilities of some missiles of the retaliatory force of the attacked nation remaining intact after an attack for a devastating counterstrike diminishes, and with this the deterrence value of the missile forces in general.

The solutions suggested so far center around "hiding" the missiles by keeping them mobile. For years this has been done by mounting them on submarines, and it is now being contemplated to place land-based missiles on continuously moving trains on the nation's railroad net, or in vast tunnel systems where they would be kept constantly on the move. Other ideas center on placing them on large aircraft such as DC-10s or B747s. From there they would be extracted by parachute and then launched. All these measures must be developed and tested, and massive amounts of hardware have to be constructed in the end. The whole process is expected to cost so much that it is bound to cause severe budget problems.

Alternative approaches to retain the deterrence value of the nuclear missile force at the highest possible level have also been tried. They include improved hardening of the launch silos, "MIRV-ing" the warheads and providing new guidance packages. These measures, indeed, lengthen the life span of the ICBM concept, but the day is inexorably arriving when they will have to go the way of all weapons, i.e. become obsolete as novel systems and ideas replace them.

This process seems to be in its initial stages now and the serious competitor of the megaton ICBM is rapidly becoming the — by comparison tiny, cheap, flexible and highly precise — cruise missile. Such missiles are by no means based on new ideas, but the most critical part of the weapon system is namely an accurate, largely ground-independently working guidance package which could not be supplied by the available technology before 1970. The key to the "smart" weapons of today, and the cruise missile in particular, is the microprocessor — or in other words, the minicomputer — which allows a large volume of data and commands to be stored in very limited space on board the weapons. The cruise missile thus becomes a typical child of the dawning electronic age which now seems to be superseding the industrial age. As an example, products based on microprocessors are already beginning to destroy long-entrenched industries. Such objects as the electronic wrist watch, the pocket calculator and many other systems for industrial use are replacing straight-forward mechanical machines which are much more difficult and costly to produce. The impact of these developments on the traditional social structure of our world will be substantial and will be as far-reaching as the introduction of industry into the farming societies of the past century.

The cruise missile is bound to accelerate this process as a defense against it can only be mounted by using fully automated systems incorporating microprocessors. And because self-preservation, i.e. defense, is one of the basic mechanisms of humanity, enormous amounts of money and effort will be invested to diminish the effectiveness of the weapon. This will again increase the knowledge in electronics, a knowledge which will find its way into the civil market in the form of more advanced electronics and processes for daily and industrial uses. Thus, the cruise missile stands at the dawn of a new era which applies to weaponry as well as society.

WHAT IS A CRUISE MISSILE?

The term cruise missile is rather ambiguous and hard to define as this weapon incorporates many features from other specific weapon categories. It has a very close relationship with the RPV (remotely piloted vehicle), but shows much similarity with typical sea-skimming ship-to-ship missiles as well. Much of its guidance comes in miniaturized form from manned aircraft and former long-range air-breathing unmanned weapon systems. It utilizes aerodynamics and electronics from "smart" short-range missiles. On the whole, it is as such not a new weapon but a combination of the best features of many.

The cruise missiles which are under development today are unique in that they are low-radar-cross-sectional aerodynamic vehicles capable of flying at very low altitude for hundreds of miles according to a preplanned course to deliver a warhead with high accuracy to a target. The cruise missile can be launched from under water, sea surface, land or aircraft and is therefore the most flexible weapon system developed so far. The guidance is self-contained and can be completely passive from launch to impact, though it can be assumed that certain position values needed for an updating of the system are occasionally entered into it. It is claimed that guidance is so accurate that a terminal homing phase is unnecessary. This lightens the missile considerably because a radar or IR homing head is not exactly lightweight.

As air-breathing missiles, they utilize light low-cost and one-way turbofans which have an engine life of up to six to ten hours at the most. Their fuel consumption is very low as the engine could be optimized for one specific speed and one altitude level only. Special fuel is not necessary and this reduces the weapon handling hazards considerably if compared to liquid or solid fuel rockets. In fact, the cruise missile can be handled like a small aircraft. It can be carried aloft with empty fuel tanks for an air launch and can be fueled from the launcher aircraft. This allows more cruise missiles to be carried on wing pylons or weapon bays.

The warhead can be conventional or nuclear and there is little doubt that only in extreme national emergencies the nuclear option will

Cruise missiles are small. Seen here is an ALCM in comparison to a person.

be used as the accuracy claimed is so high that nuclear charges are not necessary. However, targeting will not be simple and the prime difficulty will be pinpointing the desired impact point on the map so that the guidance system can be "educated" accordingly. Here only satellite photography will help and there is little doubt that observation from space will be a decisive factor for the success of cruise missile targeting. But this problem will be explained in depth in the course of this article.

THE GENESIS

The cruise missile actually came on the weapon scene by accident, as nobody had thought of replacing or augmenting presently used strategic weaponry by a novel weapon. The weapon concept started out as a simple penetration aid and radar decoy for intercontinental bombers.

In 1972 the United States Air Force initiated a program to replace the aging and obsolescent ADM-20C Quail which since the late fifties has been carried on the B-52s of the Strategic Air Command as radar decoy. It has a range of some hundred miles and simulates by electronic means speed, size and flight patterns of the big bomber. Basically it is a cruise missile, but it cannot be termed "smart"; in fact, it is "stupid", and its flight behavior might designate it to enemy radar as just what it is: a decoy. Thus, a replacement became mandatory and the SCAD (Subsonic Cruise Armed Decoy) program was born.

The new missile design was tailored to counterbalance the steadily increasing number of Soviet high-performance interceptors, which eventually would have a look-down, shoot-down capability, either through a sophisticated onboard radar system or by being teamed with the airborne warning and control system MOSS, the Soviet answer to the AWACS effort. Such upgrading of Soviet air defense capabilities might center on the Mach 3.2 MIG-25 (Foxbat) high altitude interceptor and its successors, and could represent a long-term threat to the B-52 through the interceptor's ability to detect and track low-flying bombers.

SCAD was designed to negate this potential threat. Its mission is to confuse, dilute and, if possible, saturate the enemy's area air defenses. The fourteen-foot long (4.27 m) missile was to be fully maneuverable and to simulate the B-52 in an "active" as well as a "passive" manner. SCAD could be carried by both the G and H models of the B-52 in quantities up to twenty missiles per aircraft. This subsonic decoy would not only fly the same flight profile as the bomber and passively reflect the same radar "signature" to the enemy's intercept radar, but it would also generate active electronic emissions similar to those of the B-52. In addition, the new weapon was supposed to be capable of evasive maneuvers similar to those a manned bomber might perform.

Were the enemy to succeed in developing new systems that could break SCAD's "code", i.e. differentiate between the decoy and a B-52 under some circumstances, SCAD could be equipped with a small nuclear warhead. This would

An ALCM is launched from a B-52's bomb bay.

restore this cruise missile's basic effectiveness by forcing the enemy to expend his own defense weapons against SCAD, just as he would against a penetrating bomber.

Advances in radar technology brought about just this much earlier than expected, and the decision was made to continue with the program with a different purpose, namely to use the weapon primarly as carrier of warheads with a very secondary role as decoy. As a weapon system it cooperates closely with the SRAM (Short-Range Attack Missile) which has the same size, but is a typical penetration aid built to eliminate air defense installations blocking the paths of the attacking bomber. Subsequently, the SCAD was renamed ALCM (Air Launched Cruise Missile) and given the military designation AGM-86, but the basic vehicle remained the same as SCAD.

The missile is fourteen ft (4.27 m) in length, twenty-one inches (53 cm) in diameter and weighs about 1,900 lbs (855 kg) empty. A fuel load of about 500 lbs (225 kg) is carried, giving the missile an excellent range which at present is estimated at 600 nautical miles (1,110 km). The weapon can be transported either on pylons or in the internal rotary launcher of the B-52. While stored aboard the aircraft, the ALCM's eight-foot (2.44 m) wings are retracted, switchblade fashion, and its horizontal stabilizer is folded over to conform to the launch facilities it shares with SRAM. The air scoop of the engine is likewise folded down and snaps up after launch. Primary power is provided by a low-bypass, 600-lb (270 kp) thrust turbofan engine. Elevons, fins and inlet surfaces are activated by small gas generators. The elevons are powered in flight by electric motors, which are an integral part of the flight control system. Mission will be preprogrammed in a manner similar to that of SRAM, and it shares some of the on/board equipment with the latter, including the inertial measurement unit, a key element in precise guidance. But the key to the precise guidance is found in the ALCM's terrain comparison system (TERCOM) which will be described later.

ALCMs have been designed to be carried by the B-52 bomber and the supersonic B-1, the latter of which is now being flight tested. A B-52 could carry 12 missiles on its wings and eight internally; the B-1 could carry 24, all internally. ALCMs transported on B-52 wings could be outfitted with auxiliary fuel tanks for greater range. The Boeing Aerospace Company is prime contractor for airframe and integration. The propulsion unit is provided by Williams Research Corporation and guidance by McDonnell Douglas.

This cruise missile is presently under full development testing by Boeing and the USAF. Boeing holds a contract for the construction of a total of 19 missiles of which a number have already been flown. The success — in spite of some failures — has been so substantial that the USAF has decided to order the development of a longer-range ALCM, the AGM-86B, which would fly up to 1,300 nautical miles (2,400 km). Without any substantial increase in size (so as not to jeopardize the compatibility with the B-52 and B-1 bomb bays) the needed additional fuel tanks can be housed. The first launch of a B version missile is scheduled for 1979.

ENTER THE SLCM

There is little doubt about the awesome combat potential of such a missile which flies at altitudes of 50 to 100 ft (15 to 30 m) over ranges of

more than 1,000 miles (1,600 km) and is able to hit a target with high accuracy. A defense with presently employed means such as radar and attached missile and gun installations is virtually impossible, because there are in the case of war not only one or a dozen cruise missiles on the "war path" flying totally unpredictable courses, but literally hundreds trying to penetrate the defense supported by SRAMs. The air defense of a nation can be fully saturated and this raises the spectre of a new type of warfare against which no defense is possible with the present means. When this fact began to make an impact shortly after the first studies on the cruise missile's military potential, a new type of cruise missile was projected by the US Navy which would surpass in flexibilty and range the ALCM. Whereas the latter cannot hide its ancestry, namely that of having once been a penetration aid and armed decoy designed to support the manned bomber, the US Navy's SLCM (Sea Launched Cruise Missile) is a true attack missile having just one purpose and that being to hit and destroy a target over any area reaching from 60 miles (97 km) to 2,000 miles (3,200 km) or more.

The US Department of Defense's approach to the development of the two cruise missiles has been to pursue the two separate programs with subsystem commonality. By insisting on commonality at the costly subsystem level but accommodating the peculiar launch platform and environment with different optimized airframe designs, the US DoD is benefiting from top performance while reducing overall costs.

The SLCM, known as Tomahawk or the BGM-109, the latter being the US Navy's designation, is being produced and developed by General Dynamics. It has been designed primarily to be launched from the torpedo tube of an SSN (submarine, nuclear). The large torpedo tube dimensions enable the Tomahawk design to take advantage of the large volume available for fuel storage; thus, the Tomahawk is capable of very long ranges of up to 2,000 miles (3,200 km). The Tomahawk also has an extremely low radar cross section and cruises at a speed of 550 mph (880 km/h) at low altitude to deliver its warhead to target with an accuracy similar to that of the ALCM. Furthermore, the Tomahawk is readily adaptable to launches from surface ships and from land vehicles. With appropriate modifications, the Tomahawk can also be launched from the B-52. The US Navy has demonstrated successfully the concept of underwater launch of a Tomahawk from submerged torpedo tubes.

The Tomahawk and the ALCM use the same terrain contour matching (TERCOM) guidance system, the same Williams turbofan engine, and the same nuclear warhead. Thus, the most costly components of the cruise missiles will be essentially identical in the two cruise missiles under development. The only difference is in the airframes which, as mentioned previously, are designed for specific carrier platforms and environment.

To emphasize further the efforts being made to achieve maximum commonality, the warheads for the

Tomahawk has been successfully air-launched on numerous occasions.

ALCM, Tomahawk and for the SRAM have been melded into a single requirement. The development and use of a single warhead for these three systems should result in significant savings.

Equipped with a conventional warhead, a shorter-range variant of the nuclear Tomahawk is also under concurrent development for the anti-ship mission. Like the nuclear Tomahawk, the conventional Tomahawk can be launched from a variety of platforms. It will have the same airframe and underwater rocket booster as the strategic Sea Launched Cruise Missile. Its guidance is derived from that of the McDonnell Douglas Harpoon and its conventional warhead is that of the Bullpup B now in the inventory. Its range will be in the 200- to 600-mile (320 to 960 km) class.

The missile's dimensions are dictated by the size of the standard submarine torpedo tube of the US Navy which has a diameter of 20 in (51 cm) and a length of 19 ft 3 in (5.9 m). Thus, for underwater launch the missile, including its protective steel container which is jettisoned when the surface is reached, is expelled like a torpedo. While rising to the surface a rocket booster cuts in and speeds the weapon upwards. After the surface is penetrated wings, tail and air scoops unfold. Then the engine starts and the weapon is on its way. Nuclear attack submarines have up to six torpedo tubes and can thus fire salvos at six different targets simultaneously. The tubes can naturally be reloaded.

A high-power booster-equipped version of the Tomahawk has been test fired from land vehicles with full success and air launches have been performed. The latter possibility raised the question as to whether it might not be possible to cancel the further development of the ALCM and replace it with the Tomahawk because this would save the considerable expenses incurred by creating basically duplicated weapon systems. However, in order to fit into the bomb bays of the B-52 and B-1, the weapon would have to be shortened by at least 24 in (60 cm) which would lower the range considerably as space for fuel would have to be sacrificed.

Hence, a side-by-side development will continue for the foreseeable future. But this holds certain benefits for cruise missile technology as such because the present funding does not allow the construction of larger numbers of test vehicles of both missiles. Since their guidance and engine are alike test experiences made with the ALCM can be applied to the SLCM and vice versa.

THE GUIDANCE SYSTEM

The key to the desired efficiency of both cruise missiles is obviously the guidance system. Originally SCAD was to have a rather conventional inertial system with Doppler updating (or vice versa), but then E-Systems Inc. generated a revolutionary idea which is today probably the most advanced guidance method in existence anywhere. The novel method — called Tercom, which stands for Terrain Contour Matching (or Terrain Comparison) — has been brought to perfection by McDonnell Douglas Astronautics and equips both cruise weapons today.

Tercom makes use of the surprising fact that the elevation contour of every five square miles (or less) of land surface (including even the flattest plain) of the whole earth is different from that of any other location of the same size. Thus, the elevations and depressions of these terrain blocs are as different from each other as the fingerprints of all human beings are.

Since the whole land surface of the earth has been well mapped by aerial survey and stereoscopic satellite photography, it has been relatively simple to compile an exact tridimensional and comprehensive world map featuring each and every elevation and depression with near absolute accuracy. Modern computer technology permits this contour map to be transformed into digital form for storage in an electronic processor from which any desired part can be recalled at a moment's notice.

The modern microprocessor of small dimensions is capable of storing a substantial amount of these contour map data and of making selected parts available to the user immediately. The Tercom system is based on this capability and the availability of exact contour maps. Tercom's simple principle is that the cruise missile's guidance system compares the stored contour map data of its preselected desired flight track with altitude measurements taken of the topography to be flown over. Thus, a continuous updating of position and track can be achieved which in turn allows the highest navigational precision to be attained.

Tercom is by no means a stand-alone system. In the cruise missiles it cooperates with a high-precision inertial system which serves as the supporting element. The missile's inertial navigation tells it where it is in relation to a given starting point. Once a cruise missile is launched, its geographic location is fed into its computer. Armed with this starting reference and a preprogrammed flight path, inertial guidance provides the reference to bring the missile to its first checkpoint. At this checkpoint the guidance system uses Tercom to correct any deviations from the intended flight path, and sends the weapon toward the next checkpoint. By comparing the geographic features on the preplanned course with the features seen by the missile's sensors, the navigation/guidance equipment knows where the missile is, and can update the inertial guidance unit at each checkpoint. Any necessary course corrections are then made. After passing several checkpoints, the inertial system's drift can be predicted and compensated for. The missile thus becomes more and more accurate as it approaches its destination. It is obvious that Tercom will not work over water and the inertial system is a must here.

But as soon as a landfall is made Tercom begins to operate.

The sensor supplying the altitude measurements is a radar altimeter working straight downwards. Naturally these measurements are relative only to the distance between ground and missile and do not provide the needed reference base for the altitude of terrain and missile above sea level. This reference base is provided by barometric pressure, and by subtracting the radar altitude from it, the missile's computer arrives at the true altitude of the terrain flown over.

It would not be economical and would increase the size of the computer, which at present weighs about 90 lbs (40 kg), if the whole flight track was to be stored in the Tercom system in the form of contour maps. Thus, for the less critical parts the inertial system is used predominantly. But for the target run — the last 200 to 300 miles (320 to 480 km) — Tercom takes over completely. The missile flies at lowest level anywhere between 50 and 100 ft. (15-30 m) at ever-changing headings. It might even make a 360° turn. This becomes possible because with the contour map the guidance system has a built-in terrain avoidance mode as all elevations which might block the flight path are in the computer's memory and can be avoided.

As good as Tercom may be, it will not provide the navigational precision needed to hit a target with pinpoint accuracy. For this another system, called SMAC (Scene Matching Area Correlation), is used which compares a digitalized photo of the target area with infrared images continuously produced by the missile's optics during flight in the target region. This method yields excellent results for highly accurate terminal guidance and has been tested successfully.

The guidance system is largely passive as the downward-pointing radar altimeter beam will rush by the ground observer at 550 mph

Exploded view of Boeing ALCM

(880 km/h) and is therefore hard to pick up. Reportedly the Tomahawk missile as such has the radar signature of a sea gull and should thus escape even the most sophisticated and automated radar defense systems.

It is obvious that the combination of these guidance systems and the Tomahawk or ALCM airframe and engine offers an unheard of combat potential, and that therefore cruise missiles play a major role in international politics. However, even if the strategic long-range cruise missile should be banned by international accord, it stands to reason that short-range versions will be built under a different label or designation.

The technology exists now and cannot be just swept under the rug. Moreover, it will be very simple to increase on shortest notice the combat range of the short-range cruise missile to any desirable figure. In addition, the new weapons are basically simple machines, easy to construct and — when considering their military importance — are extremely cheap at $500,000 per missile plus warhead. Even if the ban against the weapon goes into effect on the basis of a treaty between the superpowers, it can be expected that other nations will create their own cruise missiles. For instance, it has been rumored that the Anglo-French Martel ship-to-ship weapon might be fitted with a small turbofan engine and a new guidance system. This would already make it a cruise missile. The Oto Melara/ Matra Otomat is basically one and it will be simple to give it a better range than 200 nautical miles (370 km) it features now. The Soviet Union has for years employed cruise missiles as armament for their navy and there is no reason to believe that the Russian scientists and technicians are not able to create a Tercom-type system. Without any doubt, the world will again have to get used to living with a new weapon system against which there is presently no defense.

This situation is well known to the military historian, as it arose each and every time a radically new weapon entered the combat scene. It never took too long before proper countermeasures were developed. This pattern will not change with the cruise missile. However, until a true defense is discovered the new weapon will help to retain the status quo in military might among the superpowers and thus will help to prevent the outbreak of conflicts.

AEW NIMROD IS GO

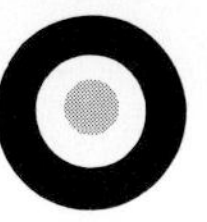

Reprinted by Permission
FLIGHT INTERNATIONAL
(April 9, 1977)

"In the light of the continued uncertainty about the procurement of the AWACS system by NATO we have decided that we must now go ahead with the Nimrod system, which, subject to the successful conclusion of the necessary contract negotiations, will now proceed to full development." Thus British Defence Secretary Fred Mulley announced the go-ahead for an airborne early warning version of the Hawker Siddeley Nimrod on March 31.

The move comes less than a week after the latest NATO Defence Planning Committee meeting in Brussels, called by Mr Mulley to consider new proposals on E-3A cost-sharing arrangements. After the meeting it was announced that the British Government reserved its right to decide on the best way of contributing to an AEW system for the alliance as a whole.

Now, according to Mulley, the Nimrod go-ahead "will give the Alliance an urgently needed and modern airborne early warning capability to replace the Shackletons now operating in the UK Air Defence Region and in the (NATO) Eastern Atlantic and Channel areas." He talks of "making a contribution to NATO AEW in kind rather than in cash," pointing out that European countries could still decide to buy a joint force of aircraft to supplement the Nimrods. He hopes that they will, noting that the existence of an 11-aircraft RAF early-warning force will allow NATO to buy fewer than the 27 planned E-3As at a much reduced overall cost. These aircraft could do without certain specifically maritime enhancements since the RAF's Nimrods will be responsible for virtually all overwater surveillance.

Mulley says that it is still almost impossible to give a reliable estimate of the cost of the Nimrod programme because negotiations on the full development contract are incomplete. He admits, however, that it will be more expensive than the British contribution to a NATO E-3A buy would have been. *Flight* understands that the latest Defence Ministry estimate is in fact about £ 300 million, although industry is claiming it can do the job for nearer £ 200 million. The proposed UK share in a NATO programme was about £ 230 million.

The British Government is going ahead on AEW Nimrod mainly because a substantial increase in the rate of expenditure is now necessary if further delay is to be avoided. The nine months of abortive AWACS negotiations have almost certainly put service-entry date for the British aircraft back from the unofficial 1981-82 projected in June 1976.

NATO headquarters reaction to the news has been predictable. "It is a great disappointment," said one spokesman. Mulley expects reactions in the United States to be mixed, but he thinks that the US administration understands the British position.

The United States Air Force is rediscovering the art of close-in air combat and has based a specialist squadron in Europe to train fighter crews in their potential wartime environment. CHARLES GILSON takes on the Aggressors.

TIGER ON YOUR TAIL

Reprinted by Permission
FLIGHT INTERNATIONAL (March 12, 1977)
Edited for Space

Heading 040°, Flight Level 170, airspeed 420kt, and all is quiet on the Western Front. Or is it? Two miles behind, 2000ft above and in our seven o'clock, the merest fleck of dark grey in an otherwise virgin sheet of foolscap cloud accelerates after us. We've been jumped. The options for my unarmed RF-4C Phantom pilot are limited, and slow or indecisive reactions are not among them. Within three or four seconds we roll hard left, hit full afterburner, point ourselves 60° downhill and put on about 3g. With luck we should be able to out-range his guns and missiles, or at least keep him in sight to know when to try to out-turn the missile if we see it leaving the rail.

Heads straining back and over left shoulders, we keep turning for a handful of seconds. So far, so good — the dark-grey fleck is still out there but he's nowhere near gun range and looks as if he could be having difficulty getting "into parameters" for his missiles. We have to risk it. Ease off the turn, momentarily down to 1/2g to let those burners really get us moving out; that character could have had 200kt advantage over us before we even saw him. Which reminds me, what are we up to? Someone told me (probably during a simulator ride, which was never like this) that we are q-limited at 750kt, so several seconds of diving in full afterburner causes me to take my eyes off the joker behind us and steal a glimpse at the airspeed.

I just have time to see 720kt going through the little window before life suddenly gets very serious. The mind's eye of Maj Don Pickard, USAF, South-East Asia reconnaissance veteran and at this moment making the most of his RF-4C experience, spots a tell-tale flash and smoke-puff from a wingtip about a mile behind. Amid the sounds of effort from the front seat, and as my head sags towards the radar tube at the insistence of 6g which also pins both forearms to the side instrument consoles, two disconcerting words rattle sharply around the inside of my helmet: "Atoll! Atoll!"

Exclamations indicating that a Russian infra-red missile was homing on my exhaust plume might normally have been expected to come from a worried pilot three feet in front. The fact that oblivion was on this occasion short-lived, and that Don had time and breath only for the evasion tactic, proves otherwise. The words came from a Capt Karl Whittenberg of the 527th Tactical Fighter Training Aggressor Squadron, who had just "fired" the missile from his Northrop F-5E Tiger II. The location for the exercise — not, as it might have been and certainly felt like, Central Europe in the opening hours of a hot conflict — was NATO's supersonic air-combat training area over the North Sea.

The not-very-happy record of American fighters in air combat during the 1960s in South-East Asia is fairly well known and documented. Air-defence aircraft, and the F-4 in particular, had for a number of years been developed to exploit at least medium-range missiles and an associated interception radar. Engagements were planned to take place as far away as possible from the point or area being defended, and reliance was to be placed on identification equipment (IFF) to an extent which would virtually preclude the visual, dogfight engagement.

A number of unforeseen circumstances conspired to make this approach to air combat less than entirely successful when it was put to the test. American pilots were constrained for political reasons to firing only after a positive visual identification, which meant that all air-to-air combat was at relatively close range. Greater agility than was originally designed into the F-4 became necessary, hence the later adoption of leading-edge slats. The enemy's aircraft, particularly MiG-21s, were considerably more agile at most dogfighting speeds, as well as smaller and thus much more difficult to see, identify and track.

Even given the initial position of advantage, the RF-4C overshoots or "spills out" badly when confronted with the F-5E's first tight turn, in the second engagement. Having lost some energy in this turn, the Aggressor is still able to better the Phantom's move into the vertical, roll over the top and get a firing position at maximum range after the Phantom has decided to dive and accelerate out of trouble. If the Tiger II had pulled even harder over the top, he might have been able to fire a missile earlier and at shorter range

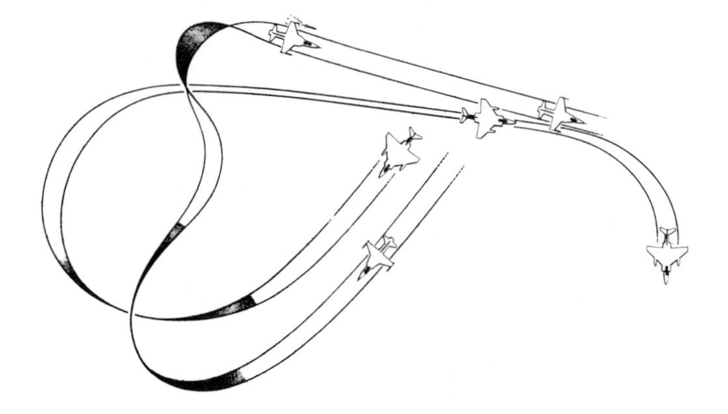

US missile reliability often fell far short of expectations. And perhaps above all, the crews had insufficient experience and training in this close-in form of combat. The result, until about 1972 was kill:loss ratios of little better than 2:1 for both the USAF and USN, compared with nearly 10:1 in similar circumstances during the Korean War.

Better results were achieved in South-East Asia only towards the end of the conflict after the introduction of a new (to the American Services) concept called Dissimilar Air Combat Training (DACT) using, at first, the T-38.

It was realized from the outset that the F-5E would make a very much better imitator of the MiG-21 as well as a more realistic fighting adversary regardless of whether the Fishbed continued to pose the threat it posed in South-East Asia. The Tiger II has, by a considerable margin, a better turn rate and radius than the T-38. It is capable of speeds up to about Mach 1.6, has a computing gunsight, ranging radar, camera gun and the ability to simulte Atoll missile launches as well as to carry a transponder pod to work with the air-combat manoeuvering instrumentation equipment now being introduced.

When the USA withdrew finally from Vietnam, and the North subsequently took over the South, what had earlier seemed to be a major market for the F-5E ceased to exist, so the opportunity was taken by the US Air Force to acquire a number of new-production Tigers which no longer had customers. There are now four Aggressor squadrons in the USAF, two at Nellis, one with the Pacific Air Forces in the Philippines and the 527th, based at RAF Alconbury, England. All but the PACAF unit, which still operates T-38s, are equipped with the F-5E.

The primary mission of the 527th is "to improve the air-to-air capability of USAF Europe by realistic simulation of the enemy threat." For this purpose it is assigned 18 F-5Es and 20 pilots. All the aircraft were delivered in three C-5A Galaxy loads last summer and re-assembled on the spot.

The *dramatis personae* in this production, where even the coffee mugs in the crew room hang on hooks arranged round a red star, have a varied background. The least experienced pilot has well over 1,000hr flying time, nearly all in fighters. All pilots assigned to the squadron have flown the F-4 at some time in their careers, though other experience ranges from T-33s through T-38s, F-5s, F-100s and F-106s to A-7s.

Aggressor squadrons are unique in that they have their own, integral GCI (ground-controlled interception) radar operators and an intelligence section. Air-to-air combat training done before formation of the squadron was conducted against similar aircraft and using USAF tactics. Not only did the pilots need to try fighting against a dissimilar type, they needed the "enemy" aircraft to be flown by pilots who had studied the real adversary's tactics. The GCI operators thus are of very high value in the Aggressor squadrons because the Soviet tactical philosophy calls for close radar control of fighters on virtually every mission.

In the case of the 527th, there are six CGI experts, all but one of whom have come from Aerospace Defence Command — the odd man out is from Tactical Air Command. They learn everything the USA has managed to find out about Soviet interception tactics.

"When a course of students comes here," says 527th commander Lt. Col.Bruce MacLennan, "we start with a briefing on the local area, procedures, techniques and so on; then we move on to about ten hours of academics." Briefings are given on the Soviet pilot, his aircraft, avionics, armament, formations, tactics, weapon training and the potential battle arena.

"Having acquainted him with why he is here," says Lt Col MacLennan, "we have to tell the student about our own very particular mode of operation. The most important part is probably the debriefing after a mission, which may take anything up to two hours." Tape recorders are carried by the Aggressor pilots on every flight and apply a crucial role in the debriefing, as do gun-camera and, in the case of F-4Cs and Ds, radarscope film.

Three main training areas are used by the Aggressors: one in the Wash, one farther out over the middle of the North Sea and the other off RAF Bentwaters in Suffolk. The first is subsonic only, for environmental reasons, but the second is supersonic and therefore more widely used, particularly when disengagement manoeuvres are being taught. It used to be the Northern Air Training Area but, with the rather special requirements of the Aggressors, has been renamed the Air Combat Training Area (ACTA) and is occupied for the majority of the time by the 527th. Its only slight disadvantage, apart from being over some of winter's most inhospitable water, is its distance from the home base, 125 n.m. Ideally a closer supersonic area was desirable to allow maximum "playing" time for the F-5Es, which always fly clean, the average mission lasting perhaps 45min. The Bentwaters area satisfies the need well, and is now quite extensively used by the Aggressors.

We brief at 0800hr. Lots of time to build up a churning stomach before we even get airborne, planned for 1100hr. The twinkle in Karl Whittenberg's eye is decidedly off-putting. I cannot help feeling that, not only has he laid down the rules for the exercise, he also is bound to know the best way to use them. The feeling of "being trained" is inescapable.

The reason for the long pre-flight briefing quickly becomes clear. The weather, though of crucial importance, is one of the least time-consuming items. It is categorised as Delta Sierra (equals messy). The temperature is -1° C at Alconbury and a lot lower at our combat altitude; the cloud starts at 800ft and visibility is a mile, until 1200hr when the 8-10kt wind will bring in 8/8ths stratus at 600ft and almost certainly a considerable amount of snow. To add to the fun, there are isolated thunderstorms over the North Sea and 4/8ths alto-cumulus between 8,000ft and 14,000ft, where we shall be spending a considerable amount of time during our three planned engagements. Of some importance to tactics, particularly Karl's, contrail level is from FL220 up to FL450. Remember that for engagement number three, I tell myself.

A great deal of the point of dissimilar air-combat training is to acquaint the fighter pilot with adversary aircraft which are a lot smaller and less visible than the Phantom. In South-East Asia, many American aircraft were shot down by a totally undetected enemy. Even when a MiG was seen, crucial errors were often made in visual estimates of range, which resulted in certain manoeuvres being made either at the wrong time or not at all. In transit from Alconbury to the training area, Karl will allow me to do some eyeball calibration by performing a series of "snatch-back" manoeuvres from about two miles, pulling out from loose formation, accelerating to about 400kt, then rolling up and in for a simulated gun attack while calling ranges over the radio down to about 1,000ft.

Amidst all the talk and aura of air combat, a deeply held concern for air safety makes a welcome and, rightly, sobering incursion into the adrenalin stream. Everyone knows the basics, but the Rules of Engagement are stated in full before every mission. Above 10,000ft, the Tiger is not allowed to fly at less than 100kt (except momentarily) and Don Pickard must keep our Phantom at more than 250kt. We shall be carrying two underwing fuel tanks, a load which limits us structurally to 6g. Karl could go to 7.33g in the F-5E but will try to

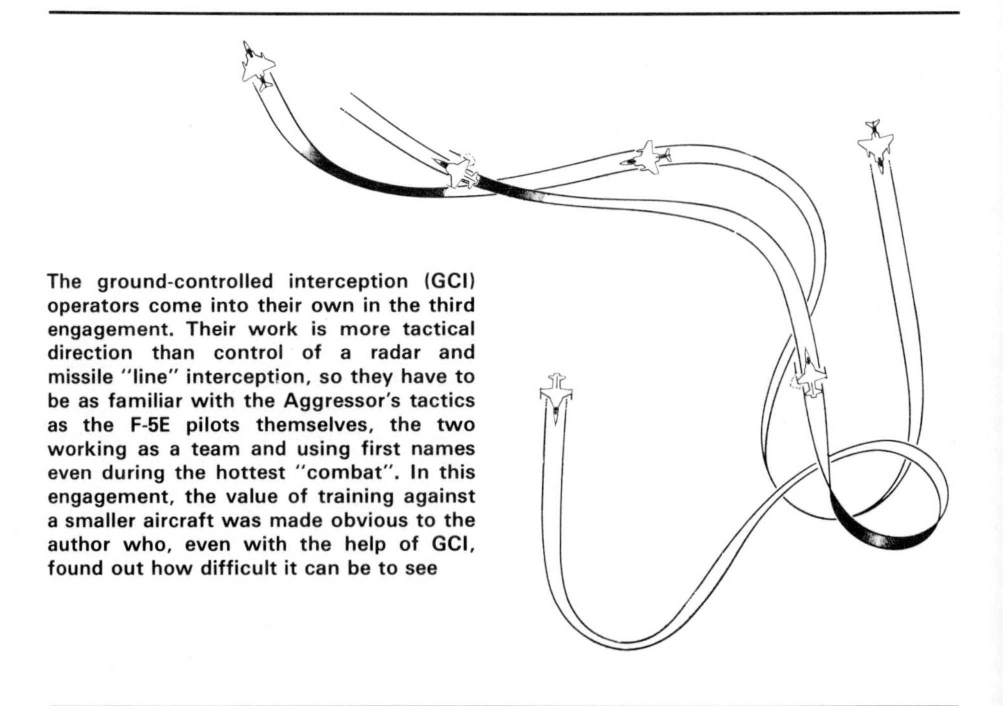

The ground-controlled interception (GCI) operators come into their own in the third engagement. Their work is more tactical direction than control of a radar and missile "line" interception, so they have to be as familiar with the Aggressor's tactics as the F-5E pilots themselves, the two working as a team and using first names even during the hottest "combat". In this engagement, the value of training against a smaller aircraft was made obvious to the author who, even with the help of GCI, found out how difficult it can be to see

keep it to 6.5 for the sake of this particular exercise — with that look on his face again, I will believe it only if I see his g-meter after the sortie. Minimum altitude under any circumstances is 5,000ft a.g.l.; flight must be maintained 2,000ft from cloud and in 5 n.m. visibility.

MINIMUM GUN RANGE

Gun range is taken to be 1,000ft minimum, and a "bubble" of that radius round each aircraft must not be penetrated. There are to be no front-quarter gun attacks more than 135° off boresight — if the two aircraft are approaching head-on, both will clear to the right unless that causes their flight-paths to cross. The same applies in the vertical sense: whichever aircraft has the higher nose attitude has the right to stay that way and the pilot should tell his opponent to go low. Generally speaking, this will naturally give him an advantage in the engagement, but the object is training, not necessarily just winning, and safety must always come first. If the defender takes the attacker into the sun, it is then his responsibility to maintain clearance.

About all that remains on the safety score is to know when and how to "knock it off," or bring the engagement to a complete stop. A knock-it-off radio call is made if the dogfight strays out of the training area, if visual contact is lost, if either crew thinks he cannot maintain the required separation or if the desired learning objective has been achieved. Likewise if the defender rocks his wings, radio contact is lost, the 1,000ft gun "bubble" is penetrated or the engagement is stalemated. Either aircraft reaching Bingo fuel state (the amount required to get him home from his current position, including reserves) brings the fight to a close, as does "loss of awareness" on the part of any crewman — I am understandably reminded that that includes me: "Don't hesitate to holler if you are no longer aware." Yes, sir.

Partly because of rules such as these, the safety record of the Aggressor squadrons is almost unblemished. No aircraft has ever been lost during an actual engagement, although there was one fatal collision when a relatively inexperienced pilot under training turned the wrong way after an engagement had been knocked off.

We taxi out and the Tiger lines up on the left-hand side of the runway for a staggered take-off at 1100hr on the dot. With a struggle I take up those extra couple of inches of play which are always there in the lap straps, even when you can feel them biting through g-suit, flying overalls, immersion suit, Nomex and all — legs, among other things, will probably go numb anyway, and it all helps to keep blood nearer to the data processor when it is under strain. Two afterburners light up the rear of the Tiger as Karl heads off and, within seconds, I am being J79-boosted into what is euphemistically called the high-g environment. Don cleans up the RF-4C, sets up 350kt on 050° and makes for our cruising altitude, where we join up in loose battle formation. Having been down at the bottom of the ejection seat's travel for take-off, I now jack it up a few inches until my helmet is almost touching the canopy — disregarding all thoughts of stiff necks to come, I want every single degree of peripheral visibility that I can get. Karl proceeds to prove the point during his ranging manoeuvres. There is not much trouble seeing him in plan view two miles beyond one wing tip, but anything approaching head-on aspect as he passes across to our rear verges on the impossible.

We have briefed for the first engagement to demonstrate cancellation of a threat and then evasion, a typical manoeuvre for the unarmed RF-4C but not necessarily so for all Phantoms, or indeed other combat aircraft. A short radio call confirms that we are in the training area, at our briefed altitude and ready to "play." Karl has set himself up on a so-called perch, about two miles behind, 2,000ft higher and accelerating. That first missile is itching to go.

The first shock of manoeuvring catches me with my neck muscles relaxed and throws my head down on to the left shoulder — not an auspicious start if I am expected to be of any assistance in telling Don where that Tiger is. But no time to think, get a grip on the canopy sills and look where he has to be. Don has the difficult task, cancelling the threat. Classically, we turn into the threat area, throwing the cone of exhaust heat outwards to make it as difficult as possible for that infra-red missile. Its performance also degrades significantly at low level, so we head for the imaginary ground, building up speed and precious energy.

The Atoll call comes at longish range and brings a fierce reaction from Don. We are already up to 750kt and practically at "ground" level but we must try to break missile lock-on with some of the spare energy. I spend five excruciating seconds jacknifed over the radar tube, incapable of defying Newton's discovery. After that time we should know whether or not we have been hit — Don, assuming success, rightly decides that the loss of about 50kt in the 5-6g turn needs remedying if we are to continue our escape. Loss of airspeed is loss of energy is loss of manoeuvrability, which cannot be afforded with a Tiger on your tail. We ease off and start "extending," or accelerating out of range again. Ease off is hardly the expression, since it involves the almost instantaneous transition from plus 6g to minus 1-1/2g, snapping me back upright in my seat like a rag doll.

The good news is that I can regain my purchase on the canopy sills and look back just over the wing to see the F-5E. Not so encouraging is the fact that I get visual contact just as a second Atoll is called. This time there is no doubt about it. Don knocks it off without hesitation. At the debrief, Karl points out that, being so close to the ground, we had pulled slightly up as well as in during the evasion manoeuvre. This caused us to lose even more airspeed and allowed him to cut off the corner of the turn and close in. Immediately we started extending for the second time, we turned our heat cone just where he wanted it. Lesson number one: try not to turn your six o'clock directly at the adversary. Mitigating factors? At extreme low altitude, the MiG-21 has a lower q limit than the F-4, though not very much lower: and the Atoll has quite severe g limits at low level. But don't count on it. That first "cheap shot" missile, fired for effect, brought the exact response required of it and lined us up for the kill.

All our engagements come under the general heading of Basic Fighter Manoeuvres (BFM). In the second one we reverse the roles and take up an offensive position on a perch behind the Aggressor, who is back at FL170. In the precious seconds before battle commences, I recall the all-important pre-flight briefing and manage to summon up a mental picture of F-5 and F-4 manoeuvring envelopes. The term corner velocity (CV) describes the point on the graph at which the aircraft can achieve maximum turn rate and minimum turn radius. This takes account of all the important parameters except energy, and the disadvantage of being at that point in the envelope is that airspeed can decay very quickly indeed — tens of knots per second. At 17,000ft, CV for the F-5E at 6-1/2g is about 360-370kt. For the F-4 at 6g it is about 420kt; at 8g it is nearer 470kt.

The bigger Phantom, particularly the lightly loaded and unslatted RF-4C with a good thrust/weight ratio, clearly gains advantage compared with the tighter-tuning F-5E if it is able to stay in the high-energy part of the envelope. If it goes simply for high g, it loses airspeed too quickly and thus loses turn rate as well, witness our first engagement. For this reason, the F-4 pilot will often go for best-energy Mach number rather than the slightly lower CV, keeping in an area of the envelope where he can generate the maximum g for the minimum loss of energy or airspeed. In training, fuel and other restrictions tend to limit the use of best-energy Mach, but we have settled for 400kt starting speed in the second engagement, just where Karl likes it best, and we have to struggle a bit or accelerate.

"I will be trying to counter your attack," warns Karl, "and then force a mistake. You can do everything right but the F-5 will still make some angles on you." He means it, as can be seen in the diagram. The tape recorder carries the message, punctuated by the panting and groaning involved in heaving a control column back into the tightened pit of the stomach: "I want to put a hell of a turn on him here" is recorded for the debrief, followed by a few seconds of silence as Karl causes us to overshoot his first counter-turn by hundreds of yards.

Don decides to take the fight into the vertical, raising the nose slightly while still turning hard — must try to keep the energy level up so don't climb too fast in the turn. We are in full afterburner all the way round. As we start the climb to regain some of our lost advantage, Karl reverses his turn to see where we have got to. After the first hard pull to the right his energy is relatively low, but he can still get his nose up faster than we can to start making the angle for a rear-quarter attack. With about 60° climb angle and the altimeter racing to keep up, Don decides he cannot get the required nose-tail separation angle to put us behind the F-5E. We roll hard right and dive, extending with maximum acceleration in a straight line.

Karl has already tape-recorded his delight at seeing us trying to mix it in the vertical, but is now faced with a very slippery Phantom accelerating across and underneath him at a time when he could use some more airspeed. He solves his problem by doing a variation of the Derry-turn theme, rolling over the top and turning out in the opposite direction to dive steeply after us. He loses some more energy going over the top but manages to get a radar lock-on at extreme range, about 8,000ft. Since we are still accelerating away from him it is almost too late, but he calls an Atoll at maximum range and we are forced to counter, turning right and bringing the g-suits back into biting action once more. Don realises any consequent loss of airspeed will allow Karl to cut off the corner and close in again, so knocks it off, "the required learning objective" having been achieved.

At the debrief, a moral victory is chalked up for the pink aeroplane on the blackboard, but discussion centres round what might have been. The cause of our problems was the first overshoot. To regain any advantage, we had to climb, but in doing so lost some life-blood energy. If we had rolled out farther to the right, facing south-east rather than east, we could have forced Karl to lose a lot more airspeed rolling over the top, and then accelerated our way out of trouble.

And so to Round Three. The two GCI operators at Coltishall are contacted and the two aircraft split up. Don is allocated the band of airspace in the training area between 15,000ft and 20,000ft. Karl, who is to have the advantage of full GCI on his own frequency throughout the engagement, must remain outside that band until either he has visual contact or the separation is 5 n.m. We will have partial GCI until 5 n.m. separation or visual contact, and then the full service thereafter. We have to counter the attack and escape, the engagement continuing until there is a result or, as it is the final one of the sortie, separation fuel state is reached by either aircraft.

"First sight, first shot," goes the fighter pilot's saying, which is why two sets of eyes should be an advantage over one. On the other hand, the high speeds and big manoeuvres of the F-4 compared with those of the F-5 mean that it is relatively easier for the Phantom crew to lose sight of the enemy, who may then be able to take the opportunity of, for example, slipping in behind and below for a gun attack from the blind spot. In terms of visibility, the advantage clearly lies with the smaller aircraft, even though he still needs a positive identification before firing. I am about to find out how inexperienced fighter pilots can mis-estimate range of an Aggressor by as much as 50 percent — throughout engagement number three, in spite of GCI, I do not see the F-5E once.

USING THE SUN

On the ground, the GCI operators are marking a sun symbol at 170° on their radar screens — it all helps with the tactics. Don flies a wide circuit at 19,500ft (Karl is most likely to be high rather than low, so we do not want to give him any unnecessary height advantage — give-away contrails start at 22,000ft). We roll out on 170° and head south into the sun as a head-on attack is least likely. Karl, 10 or 12 miles to the south-west of us, is flying on 180° above our height band until GCI tells him to turn left to meet us. At 7 n.m. and on a heading of 020° he calls tally-ho to his GCI for visual contact, and continues to turn in towards us.

The ground radar in use was unfortunately not designed for this sort of exercise, and an antenna rotating only once every 15sec or so results inevitably in less than continuous GCI instructions. We are aware of Karl's general area but, search as we might, cannot see him. As the 5 n.m. separation mark is passed, GCI tells us to turn right into the oncoming attack. We do so maintaining a shallow dive, wanting to keep speed up for any imminent hard manoeuvring. All in vain. Karl has continued to turn right and now climbs steeply into a tight wingover to keep us in sight all the way round the turn. Only at the top does he decide, correctly, that we have not yet seen him.

A GCI instruction to turn hard left on to 270° gives us another clue to where the bogey Aggressor is. Under the now-familiar buffeting of 6g I struggle to rotate my head manually through a totally unnatural further ten degrees. Tunnel vision makes the inordinate effort practically worthless and the call of "Atoll! Atoll!" rings in my helmet once again. If there is any blood left that far up my body, my ears must be pink with frustration. Karl is in our six o'clock, diving, cutting off the corner and trying to get into gun range as we extend to the west. He loses out on acceleration at high airspeed compared with the RF-4C, fails to get close enough for gun-camera film and so calls a final Atoll as we hit 700kt and start a sweep round to the north. Thre is little left to do except knock it off.

The decision to base an Aggressor squadron in Europe is seen as crucial to the proper training of USAFE fighter crews, in the air-space and weather (even if un-fortunately not the correct altitude) in which they would have to operate in war. Lt Col MacLennan is en-thusiastic about the location of the squadron: "The UK is one of the best if not the best, radar environments in which the military can work."

"What we must teach pilots is to know their aircraft and the par-ticular problems of the adversary, that is his size, turning ability and so on," says MacLennan. "We do not so much simulate a particular type of aircraft, but we must expose our crews to different aeroplanes. The emphasis, and the key to this whole business, is on dissimilar flying." It certainly concentrates the mind wonderfully.

SIKORSKY BEGINS UTTAS PRODUCTION

Reprinted by Permission
ROTOR & WING (February 77)

Sikorsky Aircraft, Stratford, Conn., is gearing up to begin production on the first 15 Utility Tactical Transport Aircraft System (UTTAS) helicopters ordered under the U.S. Army contract announced Dec. 23, (1976).

The first 15 ships were ordered for procurement in fiscal year 1977, with options on another 353 aircraft for delivery over the next three fiscal years. Sikorsky expects to deliver the first production UH-60A to the Army in August 1978.

The UTTAS is to serve as the Army's principal combat assault helicopter for at least the next 20 years and Sikorsky expects to receive contracts for production of more than 1,100 helicopters between now and 1985. Including foreign military sales, the UTTAS contract is expected to bring Sikorsky in excess of $3.4 billion.

Army sources said the UTTAS award went to Sikorsky because the UH-60A offered the lower overall life cycle costs, including investment, operations, and maintenance over the 20-year expected lifetime; the lower technical risk; and a greater maturity of design than its competitor, the UH-61A built by Boeing Vertol, Philadelphia, Pa.

According to Gerald J. Tobias, Sikorsky president, the contract will have far reaching effects, not only in Connecticut, but in many other states where Sikorsky has subcontractors. For Stratford, it will mean stabilization of Sikorsky's present work force followed by a gradual increase. The rate of buildup will depend solely on the rate of production established by the Army for fiscal years 1977 through 1980. Sikorsky's present employment is about 6,100 and industry sources have estimated that it could go up by as much as 2,000 workers. The company has already begun advertising for engineers and designers.

In a message of congratulation to Sikorsky employees after the UTTAS announcement, Tobias said that he believed the Army selected the UH-60A "not only for its outstanding performance and maintenance qualities, but because you designed and built a truly production-ready helicopter."

The UH-60A marked Sikorsky's first application of advances in rotary-wing technology such as corrosion-proof, ballistically tolerant main-rotor blades with pressurized titanium spars, fiberglass skin, and aft-swept tips; elastomeric main-rotor bearings that need no lubrication or seals; a maintenance-free bearingless tail rotor; and a completely modularized transmission system.

Other features include Sikorsky's self-tuning main-rotor vibration absorber and energy-absorbing landing gear with a tailwheel to protect the tailboom and rotor in high flare-angle landings.

Tobias stated that these advances make the UH-60A "a simple, easy-to-produce helicopter that offers a combination of high performance, excellent maneuverability, increased reliability, reduced maintenance, and high survivability."

Sikorsky has incorporated many of these new developments into its S-76 — a twin-turbine helicopter for the off-shore-oil and executive markets, the first prototype of which was rolled out Jan. 11.

A big question in the industry now concerns the fate of Sikorsky's civil version of the UTTAS helicopter, the S-78, a subject which has drawn no comment so far from company officials.

On the losing side of the UTTAS announcement, Boeing Vertol announced that it planned to lay off some 600 employees by the end of January. Spokesmen for the company expressed optimism for the future of its UH-61A in the upcoming U.S. Navy LAMPS selection. The Navy plans to buy some 200 or more helicopters for its Light Airborne Multi-Purpose System. Manufacturers in competition for the LAMPS include Boeing Vertol, Sikorsky, and Great Britain's Westland.

LAMPS Update: On September 1, the U.S. Navy announced the winner of its Light Airborne Multi-Purpose System (LAMPS) helicopter competition to be the Sikorsky S-70L. The S-70L is an application of technology first developed for the Army's UH-60A Black Hawk (UTTAS) helicopter.

— JSAY

SEEING BOTH SIDES: THE SPEER -EAKER INTERVIEW

Reprinted by Permission
AIR FORCE MAGAZINE (April 77)

On October 21, 1976, retired Air Force Lt. Gen. Ira C. Eaker, who commanded both VIII Bomber Command and the Eighth Air Force in 1942-43, and Dr. Arthur G. B. Metcalf, Chairman of the Board of the United States Strategic Institute, met with Albert Speer, Hitler's minister of armaments production, at Mr. Speer's home in Heidelberg. These highlights of their discussion concerning the effects of Allied airpower on German production have been made available to AIR FORCE Magazine by General Eaker and Dr. Metcalf. The insights that were revealed in the conversations are a significant contribution to understanding the development of strategic airpower and its contribution to victory in World War II.

EAKER: Mr. Speer, it seems we worked at cross-purposes in the last war. It was your mission to supply the weapons for the Nazi land, sea, and air forces. It was my job to prevent your accomplishing that by bombing your munitions factories and their supporting systems — oil, ball bearings, power, and transportation.

If I had had a more accurate estimate of your problems, it would have improved our chances of accomplishing our mission.

Now, more than thirty years after Allied bomber operations began in World War II, there is a renewal of interest in airpower operations in that war. One of the major current interests concerns this question: Which hurt you more, the RAF night bombing or the American daylight bombing? Or was the combination, called "round-the-clock bombing," the most effective Allied strategy?

SPEER: At first, of course, it was the British night bombing. We had that to deal with a year before the American daylight raids began, and a year and a half before you made significant attacks with a hundred or more of your daylight bombers.

After the British night bombing raids on our industry in the Ruhr, and especially their heavy raids on coastal cities like Bremen and Hamburg, I was directed to concentrate on night-fighter production. Eventually, we began to take heavy toll of the British night bomber force as a result of devising tactics and techniques and developing equipment to deal with the night bombing effort.

I often wondered why the RAF Bomber Command did not continue their thousand-plane raids on our cities. Had they been able to do so, the morale of the German population and the German labor force might have been significantly weakened.

Of course, one reason why the burning of Bremen did not hurt the morale of our people more was because they did not know at the time the full measure of that catastrophe. Hilter's Propaganda Ministry had full control over all communications. Naturally they did not play up bad news. I, myself, did not know the full extent of the fire bombing of Bremen, the horrible loss of civilian life, until much later.

Later on, when American bombers came in daylight in ever-increasing numbers, attacking our munition factories very effectively, our military leaders repeatedly told Hitler that unless the daylight bombers could be stopped, the end of the war was clearly in sight. So I was ordered to concentrate on day-fighter production. For a time we held our own, often causing your raids heavy losses, as at Schweinfurt and Regensburg on August 17, 1943, but eventually you over-whelmed us. So I should suppose that it was the combined air effort that destroyed our means to wage war, and eventually the will and resources to continue.

You will note that in my book *Spandau* I pointed out that you in fact had started a second front long before you crossed the Channel with ground forces in June 1944. Air Marshal Milch told me that your combined air effort forced us to keep 900,000 men tied down on the so-called "West Wall" -to defend against your bombers. This, of course, included the fighter defenses, the antiaircraft artillery people, and the fire fighters, as well as a large number of workmen needed for repairing damaged factories. There was also the large number of artillery pieces required all over Germany because we never knew which of our industrial cities you would attack next. It was your freedom of target choice and our uncertainty that enabled a limited number of bombers to tie down such tremendous numbers of people and equipment in our defense effort.

I suspect that well over a million Germans were ultimately engaged in antiaircraft defenses, as well as 10,000 or more antiaircraft guns. Without this great drain on our manpower, logistics, and weapons, we might well have knocked Russia out of the war before your invasion of France.

EAKER: Your view of the bomber offensive as constituting a second front is one I have never seen advanced elsewhere. I know you called it to the attention of Sir Arthur Harris, and he quoted it in an address he delivered last September.

(The summary referred to by Speer appears on page 339 of the English translation of *Spandau, The Secret Diaries*. An excerpt follows:

> August 12, 1959. Recently a book was smuggled into my cell, *The Army Air Forces in World War II*, a semiofficial history by Craven and Cate. . . It seems to me the book misses the decisive point. . . It places its

emphasis on the destruction the air raids inflicted on German industrial potential and thus upon armaments production. . . The real importance of the air war consisted in the fact that it opened a second front long before the invasion of Europe. That front was the skies over Germany. . .)

EAKER: Which of the target systems — shipbuilding, fighter plane and engine factories, oil, ball bearings, or transportation — was most decisive?

SPEER: It was the combination. At first I was most worried about ball bearings. If you had repeated your bombing attacks and destroyed our ball-bearing industry, the war would have been over a year earlier. Your failure to do so enabled us to get bearings from Sweden and other sources and to move our damaged ball-bearing machines to dispersed localities.

EAKER: There were several reasons why we did not repeat our attacks on Schweinfurt immediately. In the first place, the strike photos showed great damage. Secondly, we sent out 376 bombers that day against Schweinfurt and Regensburg and lost sixty. No air force can sustain that loss rate. We always tried to hold our operational losses below the programmed number of replacement bombers and crews. I was determined that our bomber force should always be a growing force.

In addition, we had other target systems of high priority, such as aircraft production, oil, transportation, etc. If we had continued all our effort against one of these systems, you would have concentrated your defenses around that system, and our resulting losses would have been unacceptable. Further, we always endeavored to send our daylight bombers against a high-priority target, which was for that particular day free of cloud cover. All these conditions naturally diversified our bombing attacks.

SPEER: You are quite right. Ball bearings were not our only critical weapons production system. Your attacks on our petroleum supply, for example, were also decisive in our pilot training program. After your successive raids had severely damaged Romanian oil sources, you followed up by mining the Danube and by constant attacks on locks and barges so that eventually our supply of gasoline and oil from natural sources was greatly diminished. Then you turned, quite logically, to our synthetic oil production. By that time you had such overwhelming air superiority that your long-range fighters were not all required to protect your bombers, but began very disastrous attacks on fighter planes on our airdromes.

Your air attacks on our transportation system were also very effective. They not only interfered with transport of troops and their equipment, but also disrupted my weapons production system. We often were producing engines and planes in required numbers, but we could not get them together from our dispersed factory sites. This was particularly true with respect to rail and barge transportation throughout Germany, especially in critical locations like the steelmaking Ruhr, which also supplied coal and coke to other critical industries.

The Allied air attacks on our shipping did much more damage than you apparently realized at the time — not only the destruction of the shipbuilding facilities in our coastal cities, but the attacks on our submarine pens in the occupied Channel ports as well. And, of course, it was your long-range air reconnaissance over the Atlantic sea lanes that eventually reduced our submarine effectiveness and enabled the Americans adequately to supply those vast invasion forces. Sir Arthur Harris undoubtedly was correct in his contention that the so-called Combined Bomber Offensive was critical, perhaps decisive, in the three great campaigns he described: land, sea, and air.

EAKER: Aside from the bombing of German industry, a very high priority with the Allies was the destruction of the Luftwaffe. Since the Luftwaffe did not show on June 6, 1944, when that great naval armada appeared off the three French invasion beaches, we thought we had positive evidence that our Allied air offensive had largely destroyed the Luftwaffe.

SPEER: I think your surmise was essentially correct. I was still turning out the required number of fighter planes, but by that time we were out of experienced pilots. We were so short of fuel that we could give the incoming pilots in our flying schools only 3-1/2 hours flying training per week. These poorly trained and inexperienced Luftwaffe pilots, by that time, were suffering heavy losses. A pilot only survived for a maximum of seven missions against your bombers and their accompanying long-range fighters in 1944 and '45. This was very discouraging to German pilots. It represented an attrition of fourteen percent for each mission. I do remember Hitler had ordered that 1,000 fighters take to the air on the day of the invasion. I do not know the reason for their not showing up. Perhaps General Galland (chief of German fighters) could tell you.

METCALF: Do you believe, as some do, that the Luftwaffe was misused?

SPEER: Yes, I do. First of all, the performance of our fighters and bombers, which had been developed well before the war, was inferior to your military aircraft. Hitler insisted that the Me-262, the twin-jet fighter we developed, be converted to a bomber, since Hitler was interested only in *offensive* weapons. It was a great mistake. I believe that as a fighter, it would have offered much more serious opposition to your bombers than the fighters we did use. When we removed the guns, ammunition, and other fighter armament from the Me-262, it was capable of carrying only a single 500-pound bomb, which was hardly worthwhile. Also, the shift of our aircraft industry from the production of bombers to the production of fighters and then back to the production of bombers was a nightmare. This disruption was hardly conducive to producing the aircraft we needed with which to fight the war.

METCALF: Was Göring's leadership of the Luftwaffe bad?

SPEER: One would have to say yes. After all, he spent most of his time at Karinhall, his country estate,

dressed in long, exotic robes, heavily bejeweled, etc. As you know, he was on drugs for a time. At the time of the Nuremberg trials Göring was, of course, off the drugs and he had lost a great deal of his excess weight. At that time he behaved like a new person and exhibited many qualities of leadership and clearheadedness. It was quite a surprising transformation.

METCALF: Was the German failure to execute the cross-channel invasion of England ("Sea Lion") due to your inability to gain command of the air over Britian?

SPEER: Yes. And here again, the need was for a superior fighter capable of knocking down the Royal Air Force, which would have played havoc with our invasion flotilla and our troops on invasion barges during the long passage across the Channel.

METCALF: Was it a mistake to interrupt your campaign against the Royal Air Force, whose fighters were having such telling effects on the Luftwaffe during the Battle of Britain, in order to bomb population centers? That shift in strategy gave the RAF a breather — a chance to recover from the systematic attrition of its fighter forces.

SPEER: Yes, it was. Here again was seen the influence of Adolf Hitler.

EAKER: As I remember, you were charged at the Nuremberg trials with the use and abuse of a so-called slave labor force of some 6,000,000 conquered people.

SPEER: The foreign labor force was guarded, housed, fed, and under the general supervision of Himmler. I only made requisitions and was alloted the labor required in our factories. In hindsight, I should have been more concerned about the treatment of this labor force. My factory managers complained about the training problem resulting from the frequent loss of labor, probably due in part to lack of proper housing, feeding, and care.

This labor force had some distinct limitations. As you probably know, the loss of our code machine, which enabled your Ultra process to intercept (and decode) our radio communications, was due to this labor. There were many factory fires that probably were set by the laborers, and continual reports of sabotage.

How much wiser you were to bring your women into the labor force. Had we done that initially, as you did, it could well have affected the whole course of the war. We would have found out, as you did, that women were equally effective and, for some skills, better than male labor. We never did, despite our hard-pressed munitions production in the late years of the war, make use of this great potential.

METCALF: Was foreign labor worth the number of occupation troops you had to use to combat local resistance activities that were heightened by taking those workers out of the countries.

SPEER: We had an expression then: "Sauckel (Fritz Sauckel, Gauleiter of Thuringia, who was in charge of all foreign labor) was the greatest ally of the French Maquis," whose activities pinned down large numbers of military manpower. On balance, I guess it was not worthwhile. It also was a management problem within our own country to guard these people to prevent sabotage, etc. It was through (Polish workers) that the cryptographic machines for Ultra were handed over to the enemy. No, I don't think the foreign labor program did as much good as it did harm.

EAKER: In your book you refer often to the unity of effort of the whole German people behind Hitler and his war effort. Would you anticipate that the people of West Germany would be equally unified under their present government if the Warsaw Pact countries attacked across the NATO line?

SPEER: Your premise that the German people were all united behind Hitler I do not believe to be entirely valid. You will recall, there were many attempts to assassinate him. As the dreary war years wound on, there was great disaffection about various phases of his leadership. Undoubtedly Hitler's early successes in the Low Countries and in France gave our people hope that all Germany would again be reunited, that all the territories lost in the First World War would be recovered. Also, as you may remember, we had been suffering great economic depression and deprivation with many people out of work and with the tragic depreciation of the mark. With the Second World War, all that changed, of course. This undoubtedly made a tremendous impression on our people, and I can see where you, on the other side, would get the idea of our united effort.

There was great doubt about the wisdom of attacking Russia. I believe most of our military leaders and knowledgeable civilians doubted the wisdom of fighting on two fronts. After 1944, we frequently heard of Churchill's remark that Hitler was the Allies' secret weapon, and that was probably true.

I have little doubt that the German people will support their NATO commitment and will fight with their accustomed valor against any invasions of our Homeland. The great difficulty NATO faces, in my judgment, is that it is composed of fourteen separate nations. It must be very difficult to get concerted action and quick decisions from such a conglomerate.

Now I would like to ask some questions about the Allied air effort in World War II. I have often wondered why you began your bombing attacks with such limited forces. Would it not have been better to have waited until you had several hundred, perhaps a thousand, bombers available?

EAKER: We did not have that option, for several reasons. After Pearl Harbor, there was great pressure, both at the political level and among the military leaders, to send all our bombers against the Japanese. If we had not begun operations against the Nazis, according to our prewar plan, this Pacific deployment would have taken place. The RAF bomber force would then have been left to deal alone with the Luftwaffe and German weapons production. It was only by demonstrating, as early as possible, that the daylight bombing offensive against Germany was feasible and productive that we were able to sustain our bomber buildup for operations out of Britain, as originally planned.

We learned during those limited early operations how to operate bomber forces under the conditions that then prevailed. If we had waited for the arrival of a thousand bombers before making attacks on German-occupied Europe, it probably would have been a tragic disaster. We learned how to deal with the weather, what kind of training we would have to give our combat crews, what types of formations to fly, and what communications we would require. We also learned that significant changes would be required in our aircraft.

Here is another consideration you may not have taken fully into account. Armies and navies have clashed for centuries, and their battles, strategies, and tactics have been recorded, studied, and analyzed by historians and war colleges of many nations. Prior to World War II, airpower had never had similar experience. Although Lord Trenchard of Britain, General Douhet of Italy, and Gen. William Mitchell of the US had prophesied that strategic airpower could exercise a decisive influence on warfare, those theories had never been tested.

The airplane was less than fifty years old. Flying machines with the power and capacity to test the visions of Trenchard, Douhet, and Mitchell had not been developed. For the first time, the US Eighth Air Force, operating out of Britain, and Britain's own Royal Air Force were to be given the resources to test those theories of the use of strategic airpower.

Gen. H. H. Arnold, head of the US Army Air Forces, was a dedicated Mitchell disciple. His instructions to Gen. Carl Spaatz and to me were clear-cut, specific, unmistakable. We were to take the heavy bombers General Arnold would send us and demonstrate what airpower could do. Could it, as he hoped and believed, exercise a decisive influence on warfare by destroying the weapons-making capacity of an industrial nation like Germany?

General Spaatz was diverted from the test temporarily when he was ordered, in October 1942, to accompany General Eisenhower to Africa to conduct the campaign against Rommel and to seize North Africa. I moved up from leading VIII Bomber Command to be Eighth Air Force Commander. Air Marshal Arthur Harris had been RAF Bomber Commander for six months. This responsibility for the vital test of airpower fell upon us for the next two critical years.

So, during 1942 and '43, this process continued, cooperatively, out of Britain — the RAF by night, the US Eighth Air Force by day.

SPEER: Why did you not attack our sources of electrical power upon which our weapon production so largely depended? We were always apprehensive about the vulnerability of our dams, our transformers, and our electric grid, so essential to continued war production.

EAKER: Our target planners had suggested electric power as one of the critical Nazi targets. However, the operational people, including myself, pointed out that the bomber was not an effective weapon against electric power production and distribution. We had no bombs available of a size and characteristic needed to destroy your dams, and thus interrupt your water power. Transformers could not be seen at night, or even in daylight from bomber altitudes, and they were much too small to be attacked successfully. The power lines were discernible, but any bomb damage could be quickly repaired, and we realized you undoubtedly had provided for quick repairs of lines and transformers.

You will recall that the British spent a great deal of effort in the development of a bomb large enough to damage your dams. But the work of the dam-busters, though spectacular, did not accomplish decisive results.

As late as the Vietnam War, with the great technical advances that had been made in the meantime, the North Vietnamese powerplants, transformers, and electric grid did not become especially lucrative targets until the smart bombs were available. Of course, with nuclear weapons, power sources of the enemy would be productive, perhaps decisive, targets.

SPEER: Why did you not join the British in attacking civilian industrial centers and our labor force?

EAKER: Airpower pioneers, including Lord Trenchard, General Douhet, and General Mitchell, had long believed that bombardment aviation might be able to reduce the will of civilian populations to resist. Our own doctrine held that the way to reduce civilian morale was not by killing people, but by depriving them of the resources for further resistance.

The US airpower doctrine, which covered the employment of the Eighth Air Force out of Britain, never contemplated attack on civilian populations, other than that incidental to attacking munition factories. A letter I wrote to General Spaatz in 1943 contained this often-quoted observation: "We must never allow the record of this war to convict us of throwing the strategic bomber at the man in the street."

I do not imply any criticism of the Royal Air Force bomber effort. Their position was entirely different. German planes had brutally attacked London, Coventry, and other cities, inflicting heavy loss of life. When the RAF began to retaliate with the limited resources available, all they could do with their night operations was to hit German industrial areas. As the bomber force grew, they were able, as you have said, to effect considerable destruction of your war effort by bombing German industrialized areas.

METCALF: At what time in the war did you feel that the Allied bombing was becoming unbearable to the German people?

SPEER: The best answer I can give is that the gradual buildup of your bombing attacks permitted the German people to become accustomed to and fortified against the great increase in destruction. So it is difficult to say at what point the tolerance of the population may have shown signs of being exceeded. Of couse, the fire bombing of Hamburg, Dresden, and the like, were great disasters locally. It would have been better if you had been able early in the war to have abruptly increased the size and weight of these bombing raids.

EAKER: I believe you have expressed some surprise that there was not closer cooperation between the British night bomber and American daylight operations. It was realized early that the British and American bombers had differing characteristics and limitations and crews with different training and experience. This made it advisable for each to be assigned the distinctive air task that each was best qualified to perform. Occasionally there was close collaboration. The RAF attacked targets we had hit and set afire in daylight, bombing on our fires. We in turn made daylight attacks on installations they had hit at night and which were discernible, even in bad weather, by the fire and smoke.

There was close cooperation in the exchange of target data, operational data, and in logistics and communications. Seldom, if ever, have two national military forces cooperated as effectively as did the RAF and the US Eighth Air Force in the war years.

Lt. Gen. Ira C. Eaker, USAS (Ret.), completed pilot training in 1918. Prior to World War II, he served as Executive Assistant to the Chief of the Air Corps and participated as a pilot in many pioneering flights., including the Question Mark endurance flight and the Pan-American flight of 1926. During the war, he commanded successively VIII Bomber Command, Eighth Air Force, and Mediterranean Allied Air Forces. General Eaker flew on the first heavy bombing raid against Occupied Europe and the first shuttle bombing mission to bases in Russia. Retired since 1947, he writes a syndicated column on defense affairs, and has been a frequent contributor to AIR FORCE Magazine.

Dr. Arthur G. B. Metcalf is the Chairman of the Board and President of Electronics Corp. of America, the founder and Chairman of the US Strategic Institute, and Strategic Studies Editor of Strategic Review. A former faculty member at MIT and Harvard, Dr. Metcalf has been a test pilot and was a pioneer in the field of aircraft control and stability. During World War II, he served as a lieutenant colonel. He is the author of many articles in the fields of mathematics, aerodynamics, and strategy and doctrine.

CHEAP ENERGY POSSIBLE FROM SPACE

by Al Scholin

Solar reflectors, power relay stations, power generating stations, and improved applications of technology are but a few of the possible advances available in space when it is viewed as a field for human investment.

These are predictions of 60-year-old Dr. Krafft Ehricke, a colleague of the late Wernher von Braun at Peenemunde in World War II and now chief scientist for space systems at Rockwell International, which is building the space shuttle for the National Aeronautics and Space Administration.

We spoke with Dr. Ehricke at the Aviation/Space Writers Association meeting in San Francisco in May, 1977, and have just received from him some of the many papers he has written on space applications.

"The age of cheap energy is not over," Dr. Ehricke said. "Before our fossil fuel resources run out 25 to 30 years from now — if, indeed, such predictions are accurate — we can have energy-producing stations in space that will provide power not only to all parts of the earth but also to manned stations in space and on the moon."

If that sounds a little visionary, it's because you haven't met the very persuasive and brilliantly imaginative Dr. Ehricke.

He has no patience with those who argue for a slowdown in technological advances in order to conserve our dwindling natural resources.

"The greatest danger to our civilization is no technological advance," he said. "That route leads only to disaster, for as resources grow scarcer men would fight one another to survive. Instead technology will enable us to grow and prosper."

For Dr. Ehricke the path of advancement leads into space, to the moon, and beyond. The space shuttle represents the first rung on the ladder.

As the initial step to conserve energy, he suggests the construction in space of huge solar reflectors that would throw a beam 50 times brighter than the moon during hours of darkness.

The reflectors would be carried in sections into near earth orbit aboard the shuttle. There they would be unfolded to form a thin screen a third of a mile in diameter — "the rim could be a sort of bicycle tube," says Dr. Ehricke — and hauled by a space tug into geosynchronous orbit at the desired points above the earth.

The reflected sunlight would eliminate the need for street lighting in urban areas. Reflectors could be shifted to Arctic regions in winter to overcome the long hours of darkness — "it would have saved lots of time and money in building the pipeline" — or be used to illuminate areas hit by earthquakes, hurricanes, or other disasters to speed rescue and recovery.

The second step would be to position power relay satellites in space to relay power from points on earth where it is generated to points where it is needed.

"There are many untapped sources of energy on earth." he said. "The Middle East is an excellent source of solar energy as well as oil. There are rivers in New Zealand that could be harnessed to produce electricity — far more than New Zealand could ever use.

"These resources are being wasted because we have no economically feasible way to transfer the energy to potential users.

"With power relay satellites in space, we can produce the energy at the source, send it by microwave beam back to earth where the power is needed."

The power loss in transmission would be less than in conventional overland lines, he said, not to mention the savings in copper wire and metal for towers.

But the real successors to oil as an energy source, in Dr Ehricke's view, are power generating stations in space. These would employ nuclear fusion reactors to produce power that would be beamed by microwave to stations on earth and reconverted to electrical energy.

"Nuclear fusion reaction should be possible by 1981," he estimates. "In many ways it will be simpler to achieve fusion in space than on earth. We should be able to have power stations operating in space by the end of the century."

Ehricke said the vacuum and germ-free environment of space offers many advantages in industrial and biological applications. Factories and hospitals could be assembled in space, drawing power for their operations from the orbiting power stations.

He also sees the moon as the first "exoterrestrial" colony for humans.

Television pictures of the moon relayed to earth by NASA astronauts do not show the moon's potential as a colony, he declares.

"With the availability of nuclear power, it will be possible to construct homes and factories on the moon and create a 'shirtsleeve' living environment there," he says.

"The moon has no water, but along with many other valuable resources it has enormous quantities of oxygen. Hydrogen imported from earth can be combined with the oxygen through underground nuclear explosion to produce tons of water."

To Ehricke, the moon is first only because it's nearest. It will also provide an excellent springboard for more distant ventures.

Though his mind ranges throughout the universe, Dr.

Ehricke's feet are firmly on the ground.

"Peoples and governments everywhere will judge the contributory potential of space activities not by the brawn of unbridled imagination, but by the skill with which we temper imagination with responsiveness to human needs and economic realities," he told a congressional committee.

"Their support of space activities will be in proportion to the degree to which we can distinguish between 'spending on' and 'investing in' space."

SPORT AVIATION

AVIATION YEARBOOK

INTRODUCTION

It's easy to understand sport aviation if you understand sport in general. There is no lack of definitions, mind you, but the concept can be somewhat elusive. My favorite definition says that sport is that which makes mirth, pastime or amusement. And when you carry m, p and a into the air they take on many faces.

Actually, this year saw the conclusion of a vast international MPA project. Not mirth, pastime and amusement, but Man Powered Aircraft. Ever since 1959, when British industrialist Henry Kremer offered 50,000 pounds sterling for the first human powered airplane, pedal-powered aircraft have wobbled and tottered into near-flight. This year, on August 23rd, Bryan Allen of Bakersfield, California powered the "Gossamer Condor" through seven minutes and twenty seconds of flight to claim the $85,000 prize. The 70-pound craft, made of tubing, balsa, plastic, and tape was the twelfth MPA design by Paul McCready, a PhD in aerodynamics from Pasadena. Speed seekers remained unimpressed with the Condor's top speed of 11 mph.

It all depends on your personal choice of mirth, pastime and amusement. Race pilot Darryl Greenamyer has spent his past ten years painstakingly assembling an F-104 Starfighter from surplussed and scavenged parts. Despite military and political opposition, Greenamyer sought to break the 15 year jet-powered speed record of 903 mph, and he did it in spades. Streaking across the Nevada desert at a mere 100 feet of altitude, the resurrected Starfighter ticked off a neat 1,010 mph. Still, ten years of pastime and amusement produced little mirth for the hapless Greenamyer as technical problems with the filming of his historic flight prevented any official recognition. A second attempt on October 24, 1977 was successful; establishing a record speed of 996.99 mph.

While Allen pedalled through the sky on shear sweat, and Greenamyer thundered across Nevada on souped-up kerosene, a more conventional airplane circled the globe on a single jar of homemade jam. Well, almost.

On September 30th, Don Taylor landed his 180 hp Thorp T-18 at Wittman Field in Oshkosh, Wisconsin becoming the first person to circumnavigate mother earth in a homebuilt aircraft. Taylor's epoch flight was substantially financed by a friend's generous bid of $8,000 for a jar of his wife's jam. Navigating through Greenland fjords, Middle Eastern deserts and incredible red tape, Taylor churned eastward at 140 kts for 171.5 grueling hours. He tells his own mirthful story, leg by leg, in the excellent feature, "Today The World, Tomorrow??".

Mirth, amusement and pastime. Ed Yost had lots of pastime as his balloon, *Silver Fox* drifted west on autumn winds in an attempt to conquer the Atlantic Ocean. Buoyed by a modest 60,000 cubic feet of helium, Yost floated from the coast of Maine to an open-ocean ditching 700 miles short of Portugal. Although that transatlantic goal remains an elusive one, the Silver Fox set new records of 107 hours and 2,475 miles. It is a milestone which will surely contribute to the inevitable successful crossing.

And speaking of balloons. This year's annual conclave grew into a full fledged, 5-day trade show in Reno, NV. Three hundred sixty registered participants gathered to pass the time in mirthful amusement. This sport is growing with over 500 licensed balloons and 1,000 active pilots, and is becoming a treasured national resource in the war against boredom. Even Billy Carter, the President's peripatetic brother, took off in a race to raise $50,000 for muscular dystrophy. A crowd of 3,000 watched Carter and fellow beer drinker Ed McMahon compete in hot air floaters. Carter claimed the victory on the unique grounds that his suit had the fewest smudges. "I had the cleanest landings," he said.

One new kick is parachuting or gliding from balloons. In Torrance, California a new giant balloon carries up to eight hang gliders aloft to 800 feet for the short glide back to earth. Over in Colorado they've started another trend with balloon jumping. Three or four sports (mirth, pastime, amusement — remember?) float up and parachute down in a round trip vote of confidence for ripstop nylon.

There have always been some interesting parallels between parachuting and hang gliding and this year there are even more. Skydivers have recently developed square canopies, inflated by ram air pressure, which look and act like wings. Now that glide back to earth is a bit more like flying.

And speaking of hang gliders, 103 pilots migrated to Heavener, Oklahoma to compete in the 1977 hang gliding championships. Flying unusual craft with improbable names like Dinger Wing Whizzer and Easy Rider, they competed for eighteen prizes in three individual classes. Hang gliding, however, is still somewhat of a step-child. The Woodstock atmosphere which prevailed at Heavener may have obscured the inevitable establishment of meaningful rules which could enhance the development and safety of this interesting sport.

Traditional gliding, or soaring as it is more accurately called, continues its active growth. The 1977 Smirnoff Derby, a long distance race from Los Angeles to Washington was won by Ingo Renner, a 37 year old Austrian. Another, even more significant record was set by Karl Striedieck, an Air National Guard pilot from Port Matilda, Pa., who soared a record round trip of 1,015 miles.

Actually there is no end to it. Sport aviation by its very nature defies any limitation because it involves imagination, originality, whim and fancy. It is wherever you find it and you'll find a lot in this 1978 edition. Three cheers for m, p and a!

Dan Manningham
December 1977

MacCREADY'S GOSSAMER CONDOR WINS KREMER PRIZE

Reprinted by Permission
NATIONAL AERONAUTIC ASSOCIATION NEWS (September 77)

Persistence pays for designer Dr. Paul MacCready, whose "Gossamer Condor" manpowered aircraft on August 23 completed the arduous course to claim the elusive Kremer Competition prize of 50,000 British pounds — at today's de-rated British pound, some US $86,000. The prize was originally established in 1957 by Henry Kremer, a British industrialist who deposited the prize money with the Royal Aeronautical Society. Though there have been numerous actual flights of man-powered aircraft, none have successfully completed the extremely demanding Kremer requirement of clearing a start-finish line ten feet high, then flying a figure-eight around two pylons one-half mile apart and completing the flight over the same ten foot high start-finish line.

It's been "back to the drawing board" numerous times for MacCready, an aeronautical engineer and former national and international champion soaring pilot,

Cruising at only nine miles per hour, the American Gossamer Condor may be the world's slowest man-carrying airplane.

in his determination to design and construct the successful aircraft, which finally evolved as a transparent vehicle made of corrugated cardboard, balsa wood, paper-thin aluminum, pianowire, Styrofoam, scotch tape and cellophane. The end result: a craft with a wingspan of 96 feet (more than that of a DC9) but weighing 77 pounds (a DC9 grosses up to 116,000 pounds). As an "engine", as MacCready describes it, he selected Bryan Allen, a young but experienced hang-glider and bicycle racer — an ideal combination of flying skill and "leg-power" — to transfer human energy through bicycle pedals to a fragile propellor. It worked!

MacCready's claim is subject to final approval by the Royal Aeronautical Society based on submission of documentation of the flight acceptable to the Society.

NAA has by request been assisting the Royal Aeronautical Society for a number of years, acting as a point of contact for anyone in this country interested in entering the Kremer Competition. Though numerous copies of the Competition Rules have been distributed, only two other vehicles have been constructed and actually flown in the U.S.: one on the same concept as the Gossamer Condor, one-man-powered monoplane, and one on the concept of a multi-man-powered triplane — the triplane to shorten the wing span theoretically to minimize the problem of low altitude turns — by a group of aeronautical engineering students at MIT.

The prize to the winner of the Kremer competition is the largest purse ever offered for a single aeronautical achievement.

THE GOSSAMER CONDOR: THE FIRST REALLY PROMISING U.S. MPA

by Paul Wahl

EDITORIAL NOTE: Because of varying magazine and newsletter lead times, this article was actually printed before the NAA article. — JSAY

Even for a man-powered aircraft, the Gossamer Condor is a strange bird. Big and spindly, it looks like a cross between an indoor model airplane and a rigid-wing hang glider. It has the wingspan of a Boeing 737, yet weights only 77 pounds. And it flies tail-first at a lazy *nine* miles per hour.

Someday, perhaps soon, this sky cycle may earn a whopping $86,000 for less than 10 minutes of pedaling by winning the Kremer Prize, aviation's richest purse. It will go to the first man-powered aircraft to fly a figure-eight course around two markers a half-mile apart, crossing both starting and finishing lines at an altitude of at least 10 feet.

Established in 1959 with the sponsorship of British industrialist Henry Kremer, this competition has inspired the building of a number of man-powered aircraft in the United States. But, until this one came along, the few that ever got off the ground have been disappointing performers.

After some 400 flights since last fall — in itself an impressive record — the American Gossamer Condor has now become the number one contender for the Kremer Prize. It has outdone its sole rival, the Stork by staying aloft for five minutes 15 seconds, to better the Japanese plane's four minutes 28 seconds. Much more significant, the Gossamer Condor has flown figure eights, something that no other man-powered aircraft has accomplished.

To put this feat in perspective: Of the 20-odd pedal planes that have flown since 1961 (besides the Gossamer Condor), only the British Puffin II more than a decade ago and the Stork in 1976 have managed even 180-degree turns.

According to recent reports from Japan, the Stork has had difficulties in making turns (perhaps in attempting figure eights). The plane has been in downtime lately while the wing was rebuilt to improve handling during such maneuvers. Also, it has been fitted with electric motors to relieve the pilot of pedaling during practice sessions, permitting him to concentrate on the problems of turning.

With its complex airframe, painstakingly built of spruce and balsa, the Stork is basically an updated and scaled-down version of the British Jupiter, a classic design of the same Sixties generation as the Puffin II. It represents the best of the old, while the Gossamer Condor is the first of a breed of simple, efficient sky cycles.

Designer-builder of the plane is Dr. Paul B. MacCready Jr., well-known aerospace scientist and president of AeroVironment Inc., a Pasadena, Calif., firm described as "providing services and research in the atmospheric environment." A veteran pilot of both airplanes and sailplanes, he is a former world and national soaring champion, and flies hang gliders, too.

"I was thinking of hang gliders and about the indoor model airplanes that I built many years ago," Paul MacCready told me, "and I got this idea for a man-powered aircraft that could win the Kremer Prize." That brainstorm resulted in Gossamer Condor.

The optimized airfoil designs used in the craft were developed in computer studies by the noted aerodynamicist, Dr. Peter B. S. Lissaman. Formerly professor of aeronautics at Caltech, he is now vice-president of AeroVironment Inc.

Another key member of MacCready's team is engineering consultant Jack Lambie, who is famous in hang-gliding circles as a pioneer of the sport and designer of the popular Hang Loose glider. He also built and flew replicas of the Wright 1902 glider and 1903 Flyer for the 1971 NET-TV documentary on the famous brothers.

The Gossamer Condor represents an innovative approach to design and construction. A parasol-type monoplane, this machine has its wing suspended 10 feet above the fuselage, while the horizontal stabilizer (which also provides lift) is out in front.

Overall, the aircraft is 30 feet long and stands 18 feet tall. Its wing spans 96 feet, has a root chord of 10 feet and a tip chord of five feet. The stabilizer has a span of 24 feet and a chord of five feet, is mounted on an 18-foot boom. Empty weight of the plane is 77 pounds. With a 140-pound pilot aboard, the wing loading is about four ounces per square foot.

The pilot sits in a semi-reclining position with his seat at about pedal level. A long drive chain, twisted 90°, transmits the pedal power to the 12-foot pusher propeller, the blades of which are similar to model-airplane wings and are adjustable to vary pitch. There is no driven ground wheel, a feature of most man-powered aircraft. Landing gear consists of two five-inch toy wheels fore and aft.

The prop turns at 110 rpm and the power required to fly a 140-pound pilot is about 0.37 hp. Ergometer studies, using a specially modified exercise bicycle that provides a readout in horsepower, indicate that a champion cyclist can sustain an output of 0.5 hp for 10 minutes. That's long enough for the Gossamer Condor to fly the Kremer Course.

The control wheel is mounted on a pivoting shaft. Turning it rolls the stabilizer to control yaw. Pushing the wheel up or down tilts the stabilizer vertically to bring the nose down or up. For turning, the wing is warped by pushing or pulling a separate lever.

Although man-powered aircraft typically are tricky to pilot, MacCready says that the Gossamer Condor is a docile bird, so easy to fly that anyone can learn to handle it in a few sessions. However, its very light wing loading, large wing, and unusually low airspeed make the machine highly sensitive to wind. Although it has been flown in winds up to seven mph, turns are not feasible if wind velocity exceeds two mph. It will need almost a dead calm to fly the Kremer Course successfully.

The airframe, suggested by hang-glider construction, is built of aluminum tube, chemically milled for lightness, in diameters from 1/4 - to 3/8-inch for the ribs, to two inches for the wing spar, keel, and vertical post. Piano wire provides bracing. The double-surface wing and stabilizer are covered with 1/2-mil Mylar, material also used in the streamlined fairing that surrounds the pilot.

Because of this simple construction, the Gossamer Condor originally was built in less than eight weeks at a materials cost of under $2000. The concept has made possible numerous modifications, including six major redesigns in as many months. With most other man-powered aircraft, such work would have taken years. Repairs can be made quickly; no damage has required more than 24 hours to fix.

It looks as if the Gossamer Condor concept may evolve into a generation of home-built sport sky cycles. Equipped with model-airplane engines, some could be "flying mopeds." Plans for the Kremer Competition Gossamer Condor — and maybe a more compact version for fun flying — will soon be on sale. For details, send a stamped, self-addressed envelope to Gossamer Condor, AeroVironment Inc., 145 Vista, Pasadena 91107.

OH TO BE ABLE TO SOAR

by Hope Hines

Photographs by Bettina Gray

Reprinted by Permission
SWIFT'S AIR (January-February 77)

You can hardly drive along the California coast these days without noticing those tiny, multicolored specks in the sky. At a distance, they might be mistaken for many things; but past sightings remind you that some man or woman, hanging from a few square yards of cloth affixed to a frame of metal tubing, is out for a gentle ride on nature's toll-free airstreams.

There is an intangible bond that binds these free-spirited performers. They fly by wit, courage and dexterity, climbing and maneuvering above rocky mountains and yawning canyons. Hang gliding probably comes closest to making man's timeless dream of personal flight a reality.

To be a bird must have appealed to earliest man as he witnessed an eagle with its prey clutched firmly in its claws ascend to its nest high atop a mountain.

Over the centuries, numerous stories have told of man's first birdman. Greek mythology tells of Daedalus and his son Icarus who found themselves imprisoned on the isle of Crete. Daedalus fashioned two pairs of large, plumed wings which attached to the shoulders. The fate of Icarus is well known. Once aloft, he began to feel so exhilarated and powerful that he flew higher and higher. The sun melted the wax holding his feathers on, and Icarus found himself cooling his heels in a watery grave.

But it was the genius of Leonardo Da Vinci that blazed the scientific trail to human flight. By studying birds and falling leaves, Da Vinci learned about airflow and drag. He invented a parachute of linen which a man could use to descend from any height without injury. Da Vinci sketches of flying machines included a model in which the pilot flew prone, flapping his wings with leg power and activating his elevator with a harness attached to his head. His contraptions never flew, but they showed the way for others who would follow.

Many did follow. In the nineteenth century, two brothers, Otto and Gustav Lilienthal, dreamt of flight. At age 13 and 14, they slipped silently from their home in northern Germany one summer night and strapped on clumsy wings made of beech veneer. Imitating storks, the brothers ran for hours flapping their wings until they fell exhausted. They did not fly that night, but from their meager efforts came history's first knowledge of aerodynamics.

On August 28, 1883, John Montgomery and his brother, James, trekked to a hill near the Mexican border just south of San Diego to fly their handmade glider. It had two wings, each about ten feet long and four and a half feet in width. The tail was horizontal and could be elevated or depressed by a system of cords and pulleys. Hanging beneath the frame was a bicycle seat for the pilot.

Just before daylight all was ready. James held a long tow rope and ran down the hill. As he felt the rope slacken, he looked upward to see John soaring 50 feet above the ground. It's estimated that John glided about 600 feet.

The problem with this story, as with many others, is that no records were kept of the flight, and there were no witnesses. Those who accept the word of the Montgomery brothers consider John's exploits to be the first hang glider flight in this country.

The modern day hang glider is the offspring of space technology. What makes any glider successful is the wing. The wing considered standard by the thousands of 1970's pilots is the Rogallo. It is an A-shaped frame of aluminum tubing fitted with Dacron sailcloth. The wing span goes up to 26 feet with a nose angle of 80 to 90 degrees. The glider weighs 35 to 40 pounds and is maneuvered by a control bar. The

pilot flies sitting in a swing seat or prone in a harness.

The sport of hang gliding has grown at a phenomenal rate. Probably no other sport has grown so fast from scratch. Prior to 1970, there were fewer than a hundred hang gliders in existence. The sport originated in Southern California but has spread rapidly to Europe, South America, the Far East, Mexico and Canada.

Several months ago, I went through a hang glider school to learn to fly and to film my efforts for a series to be used on my daily sports show of KFMB-TV in San Diego. A gentleman by the name of Bill Armstrong, a member of Ultralite Flyers Organization of San Diego, called me and asked if I would like to go through one of his hang glider schools. He noted that there had been a number of tragedies in and around San Diego and that he wanted to stress the safety value of the sport and educate the public, using television as the medium and me as the student.

Thinking myself to be a daring individual, I said yes. Several years before back in Tennessee, I had attempted to fly a tow kite and had met with immediate and dangerous results. Tow flying is much the same as hang gliding except the pilot is pulled aloft by a boat if on water or by a truck if on land. Both, I might add, are extremely dangerous. On my first attempt, I tangled with the rope. On the second, I got up smoothly but the nose of the kite tipped forward suddenly as we sped along at 30 m.p.h. I crashed into the water hitting the control bar with my nose. By the time the boat reached me, I was regaining consciousness with blurred vision and a broken nose. That was the end of tow kiting for me.

Here I was again, I was to go through a week of training to become a hang glider pilot. As Bill's only student, he began by explaining the make-up and design of the kite I was to eventually fly. I learned such terms as: keel, sail area, aspect ratio, span, nose angle, wind loading, kingpost and wash out. At the end of my first lesson, Bill gave me several books to take home and study. I had yet to feel the weight of the kite on my back.

My next series of lessons were spent in ground school. Assuring him I had read the books he had given me and after a question and answer session to prove I had, I was ready. Bill had instructed me to wear a long sleeved shirt, blue jeans and heavy boots which I did. I strapped on the pilot's helmet, and Bill harnessed me to the kite. Since I was learning to fly in a sitting position, the harness is much the same as a parachutist's. The seat is attached to the top of the kite by clips. The rather small horizontal seat is fitted with straps through which you put your legs. The straps fit snugly around the top part of your thighs and the seat rests under your buttocks. Your hands are placed on the control bar.

The kite rests on the control bar, nose down, to keep the kite from blowing over. The pilot, harnessed in, simply lifts the bar and brings the kite up with it. Usually, you need someone to steady the kite, especially if the wind is blowing.

With the kite lifted, I experience my first feel of the kite. After this, Bill instructs me to simply walk forward as he holds the nose. After several minutes, Bill walks away; and I try to maneuver the glider by myself. On this particular day, the wind is blowing briskly (as it always does in San Diego), and I find maintaining the balance of the kite somewhat difficult.

After learning to walk with the kite, my instructions are to run with it. Bill tells me to push out on the control bar as I run to allow the wind to push the nose upward. With enough speed and wind, the glider will lift. This is the most difficult and exhausting part of ground school. To run is torture enough, but to run with the weight of 30 to 40 pounds of aluminum and cloth is grueling.

The third lesson is a continuation of the first and second, but now Bill talks about safety. And safety can't be emphasized enough in hang gliding. It begins when the pilot assembles his kite. Assembling a glider requires the same care as packing a parachute. Do it properly, and you'll have no problems. Do it hurriedly, and it could be fatal. You can never be too safe. It's only your life at stake.

Having learned the basics, lesson four is spent watching experienced pilots. Bill takes me to Torrey Pines, a favorite launch site. I watch and talk about hang gliding. Because of the obvious dangers involved, experienced pilots are concerned and go out of their way to help the person just beginning.

Lesson five takes place in Cantamar, Mexico, about 40 miles south of Tijuana. It is a beautiful spot. Sand dunes run for miles along the beach, and it's off the dunes that hump upwards of 50 to 60 feet that a beginner learns to fly. The height is right; and, more importantly, the landing is soft.

It is a Saturday. The sun is hot, and the cool breezes off the ocean are refreshing. We have assembled the kite, and I am ready for my first airborne venture. As I stand atop the sand dune, Bill goes over his instructions for the last time. I hang on his every word. Sixty feet may not sound daring, but climb up sixty feet and look down sometime!

I could feel the adrenalin racing through my body. My hands were sweaty on the control bar. Bill told me to run as hard as I could down the dune, and when I was airborne, to push out on the control bar. If I didn't, the nose would slip forward; and I would go crashing into the ground.

When the wind was just right, Bill said "go," and I began running with all my might. Suddenly, I forgot all he told me. All I could think of was the ground under me and where I would fall. I pushed out hard on the bar. The kite lifted, and I felt as if I were falling backwards. I pulled back on the bar, and the kite stabilized.

For the first time I realized my feet were not on the sand. I was flying. My god, what a sensation! It was the most unnatural natural thing I had ever experienced.

I looked down and saw the sand rushing beneath me. I could hear cheers behind me and felt an accomplishment of greatness. Then, I was crashing into the on-rushing tide.

Unlike the Montgomery brothers in 1883, I did have witnesses. It wasn't the greatest first flight ever, but there were others to follow — more durable, longer lasting and higher. But none will ever compare to the first.

Like the first haircut, the first day in school, the first time you get away with something you know you shouldn't — you remember them all. But the first. . .well, it's kinda special.

1977 U.S. HANG GLIDING CHAMPIONSHIPS

by Pork

Photographs by Bettina Gray

Reprinted by Permission
HANG GLIDING (September 77)

There were a hundred and three pilots listed in the program, but only 94 actually began flying at this year's USHGA-sponsored/sanctioned championships. Few of them remained to watch the eighteen winners in three classes receive their awards when it was over. Some left early because of their low standings in the scores, some because they couldn't abide the officiating, and no doubt some left because they were tired of constantly being soaked with rain or sweat.

Mount Poteau (the natives pronounce it po-do), was chosen for the nationals because it was the right size, in the right geographical location, with three launches and a large landing area, and predominantly favorable winds. Those favorable winds blew only infrequently however, during the two weeks of competition, and the much anticipated cross-country task which would have sent pilots out over a ten-mile course, was not attempted. A few days before the meet, though, Chris Perkins, winner of the hotly-contested class two in Southern California, launched from the southwest crest of Mt. Poteau into an eighteen mile an hour wind. A few minutes later he found himself two thousand feet above the top in a dandy thermal generated by the wet heat, so Chris turned downwind over the top and proceeded fifteen miles in fifty-five minutes. He went five miles past Sugarloaf, a large hill to the north, and landed in the state of Arkansas. He made cloud base at 4500' agl several times.

After Perkins' flight there were no other remarkable flights for distance or altitude during the rest of the meet, although hopefuls organized a cross-country contest at Buffalo Mountain, a larger hill some forty miles away. The longest flight over ten miles was to win $500. No one could collect. Rain and thunderstorms kept the meet itself from getting started until Friday, July 21. During the rain on Thursday contestants and others kept themselves amused at Sky Unlimited, the local shop, by playing ping pong, throwing soft-nosed air darts, and hanging from the flight simulator in a new harness made by Lorin Ellsworth of San Diego, the Ultimate Hi prone harness, which supports the pilot's weight through a length of tubing. Although not in the contest, Bob Trampanau of Santa Barbara was there with his latest Sensor glider, the Sensor IV, a strutted, double-surface kite with a keel pocket extending aft of the flying surface, supported by battens. Car freaks admired Jim Jaworski's Polish War Wagon, a custom-built lengthened late-model Chevrolet in the form of a Blazer or topless Suburban. The War Wagon performed nearly flawlessly through the hundred degree heat which was nearly matched by the humidity rating. And on Thursday night, frustration reached a minor summit when a disgruntled pilot burned his glider on the training hill near the landing area. Charlie Gillespie maintained that the machine was divergent, that the manufacturer couldn't fix it and wouldn't take it back. "I'm a practicing pyromaniac," said Charlie. "I have another glider."

The Nationals got underway on Friday, July 22. Contestant Steve Perry, on a Whizzer made by Dinger Wings, flew as always with his eight-year-old green wing macaw, Chico. The bird rides Perry's control bar, on the uprights, but Perry said that on long flights Chico steps down to the base bar and flaps his clipped wings, pretending to soar. "He's had a few hundred flights," said Perry, "mostly in the Sacramento area." He said the bird often got ready for flight in the set-up area by clambering over the bar and squawking. Late that afternoon many contestants drove across the nearby state line to swim in Arkansas, in a cooling, clear length of water called "the Quarry."

Earlier, while we watched from the steep southwest launch, contestants flew in five-man heats while overdevelopment occurred; cumulus clouds shot up towering chimneys. A motorized Easy Riser biplane flew over Mt. Poteau and cut his engine.

A double-headed eagle graces the sail of this glider.

Reminiscent of dust bowl days, high winds from nearby thunderstorms put an end to this day's competition.

Little thermals switched around on the launch, at times holding up the meet. Chico the macaw climbed the windsock, but came down after an hour or so to get under the shade of an awning. He sat on the back of a chair that Dennis Pagen was dozing on, and shredded a piece of bark on Dennis' sweating shoulders. And later that night two bands played in a field nearby, where iced three-two beer was being hustled, and shrimp, and Mexican food. Many were unable or unwilling to leave the field that night, and the next morning I saw Chico the macaw waking up, perched on a glider on a truck near the empty stage.

Competition flying continued through the weekend. Some who had registered for the fly-in were unhappy. They had thought most of the meet would be open to them, for free-flying, but this was not the case. At the pilots' meeting Sunday morning several announced they were going home. Joe Greblo said that George Worthington had flown his ASG-21 ninety-five miles in a repeat of his earlier effort on the Mitchell Wing. Competitor Gary Wilson left for Canada, for the paying meets at Fernie and Swansea and Grouse Mountain.

"Do you have any comments I could print?" I asked Gary.

"No," he said, "but plenty of them that you can't."

The *Sunday Oklahoman* had predictions for another hundred-degree day. For the first time, the bullseye began to count in the scoring, and pilots started to hit it with some regularity, though thermal action on the landing field made it hard at times. J. C. Brown hooked a thermal for a two hundred foot gain, the first one witnessed during competition, and got back up to launch level. Tom Vayda blew a crosswing launch with his Fledgling and suffered abrasions on an arm and a hand. About one o'clock a shaded thermometer in the landing zone reached 110 degrees, and spectators could see the overdeveloping clouds punch through an upper inversion, sending streamers down their white sides. Thunderheads began to develop on all sides, virga or rain curtains beneath them, trailing especially heavily out west near Wister lake, where there was noise and electrical flashing as the biggest cumulonimbus approached. As the westerly winds rose, official flying was called off. A couple of dozen pilots tried soaring in the gusty pre-frontal winds. One of them, Mike Zarracina, rose two or three hundred feet above Poteau Mountain, and was looking like he might get up and go somewhere else when the wind picked up to a non-penetrable speed and Zarracina descended vertically into the landing zone, whereupon he reached for his flying wires at touchdown. He was the last pilot down.

Monday morning brought clearing but hazy skies, and winds from the southeast at ten miles an hour. It was the commencement of the second week of flying. A 60 per cent chance of rain was predicted and a garbage truck worked in the spectator area, where the weekend's business had been good. National Champion Keith Nichols, wearing the complimentary number one, made the day's first launch, cleanly and into the wind.

Around eleven, the wind showed a north factor, and the air over the landing zone became turbulent. Some high-performance gliders acted like standards and couldn't penetrate to do the pylon task or even make the target. Their disgusted pilots walked under the downed machines toward the judge's stand and dismantling area. Then, while the meet was stopped, it got soarable. Tom Goodman and Steve Moyes flew for several minutes, up to 200′ above the hill, and Moyes landed on top with a smooth and unpracticed move that brought applause from the competition pilots. Gary Harkins of East Lansing made the classic mistake — he forgot to hook in. Gary was lucky enough, though, and suffered only a sprained ankle and a few scratches. Conditions became a big factor. Many of the leaders had poor flights in sinking air. A decision was reached, to allow free-flying after all pilots had finished their second heat of the day. Some scores were posted showing the results of the first five flights in class II:

Jokes were being made among the pilots as to whether the next champion would be flying a Phoenix or a Phoenix. Bill Bennett certainly seemed to have a clear field. In fact, when the qualifying was over, thirteen of the twenty-one qualifiers were Phoenix pilots.

Chico, the macaw, has his feathers ruffled by a tailwind, while his owner Steve Perry answers an interviewer's question.

Brad White launches a Mitchell Wing into first place, Class III.

After a day's bad weather, flying commenced again on Thursday, July 28th. I watched the competition begin under gray luminescent clouds, and noticed Keith Nichols wasn't limping as much as he had been. Keith had hurt his foot on a rock while launching himself into the Quarry from a trapeze. There were still lots of short landings being made. Dave Beardslee failed to make the foul line and dropped far from the third place he'd been occupying. Young Dave Braddock did the same thing, though he'd been leading the standard class.

"I flared too early and didn't make it across," he said mournfully. "I really rookied out."

Come rain or shine, come judge's rightful or wrongful decisions, come understanding or confusion, come ready or not, up air or down, the finals approached. Verbal and written protests were as thick as flies, and many pilots had already gone home by the time the last flights in heat number seven were completed on Friday, July 29th. That day began with a big thunderstorm at dawn that flooded the streets of Heavener (that's Heavener, not heaven), and turned the landing area into a mudhole. Competition began after ten, and so did the usual overdevelopment. At 3:04 that afternoon the cut was made for the finals, the top twenty-one pilots in class two. Dave Beardslee was relieved — he was number twenty-one.

And so on Saturday, July 30, the finals for the 1977 National Championships of Hang Gliding were held. There were three flight tasks for the day. They were pictured on a sheet given to the pilots.

The night before, half a dozen hang glider pilots performed in the Heavener Rodeo. Steve Perry and Joe Greblo were among those who rode calves, and reports had it that

Sirocco pilot, Hank Syjut, flares for a landing as judges rush up to mark the spot.

Greblo, with gymnast's training, had been backflipped, dismounted by his steed. The citizenry loved it.

At 10:07 Saturday morning, Jim Braddock made the first competition flight on his Moyes glider, just as he'd made the first flight at Thunder Bay a month earlier. Jim had a good one, as they say, and got the bullseye. Dave Beardslee went second, and then Dick Reynolds. All of them hit the target. Then conditions began to change rapidly. I went up to the top to watch, and arrived there as the wind was becoming soarable, from seven to fifteen miles an hour out of the southwest. Chris Perkins remarked that it was a day not unlike the one when he got his fifteen mile flight. Many pilots felt that this should be the day the long-awaited cross-country task was begun. A young elk watched the launches from the hill's lip a hundred feet below. Several elk had been planted on the mountain by Mr. Ward, the landowner, who was also there watching. Ray Leonard made a low launch on his Eagleman Spirit. "If there'd been a rattlesnake there it'd have bit him," said Mr. Ward.

On his next flight, Jim Braddock stayed level with the launch all the way out to the course. Once there, he worked thermals with his Moyes Maxi to gain as much as 400′, effectively "blowing out" his competitors Tom Peghiny and Dennis Pagen. In these final heats there were only three pilots. "That's my dad, the forty year old maniac," said Dave Braddock as he watched his father soar the course. "The toughest kid in the family."

The toughest kid in the family turned out to be Jim's brother Henry Braddock, who beat Jim out for first place, in the hotly-contested class two. Next toughest might have been young Dave himself, who hoisted his position back up to number two in the standard class. That class was won again by Robert Reed, the first time a champion has repeated in the nationals.

The unexpected and perhaps unprecedented sweep by the Braddocks and their Moyes gliders left the Michigan family jubilant. Their jubilation was to be short-lived, however. The following day, Sunday, saw one of their prototype Moyes Maxi gliders dive from the sky carrying Ed Vasquez, a competition pilot. Witnesses said the glider fell from an easy left turn into a slipping dive from two to four hundred feet, crashing into an oak tree at the edge of the landing area. Vasquez was taken by ambulance, conscious and aware, to the Poteau hospital. Later that afternoon the ambulance driver reported that Vasquez was not badly hurt and was to be released the same day. Other, less serious accidents that same day occurred to George Dyer of Lufkin, Texas, and Mark Phillips of Seattle, neither in the competition. Dyer dragged a wingtip on the west-facing cliff launch and plowed his Wills XC into trees on the hill below. He was not hurt. Mark Phillips, attempting to launch tandem with Cindy Bachman of Norman, OK, forgot to hook in. Mark fell from the CalGlider Corsair, suffering minor hurts, but leaving Cindy in the air with no previous experience. She managed to land the Corsair with no damage to the glider and only scratches on her own skin, escaping one of the most unlikely accidents possible. Motion pictures of the Phillips crash and the Vasquez crash

Robert Reed, flying a Cirrus III, won the Class I competition, and became the only pilot in the history of the Nationals to repeat the victory.

were taken by Delfin Salazar, and may prove helpful when developed.

Some pilots went away from Heavener happy. Some went away mad. Some enjoyed the flying there, and some swore they'd never return. In any case, this year's nationals will be at the center of a continuing controversy, a controversy about rules and judgements and personalities, a controversy that may never be resolved. Dean Tanji went away with the newly-created Bob Wills Award. It was voted him by the rest of the pilots, and Dean was obviously moved by its presentation to him, by the fact that his peers thought he should receive it. "I guess Bob was my hero," he said simply, blinking. It seemed safe to conclude that everyone who'd gone to Heavener was emotionally affected, that the Nationals next year would have to undergo some painful changes.

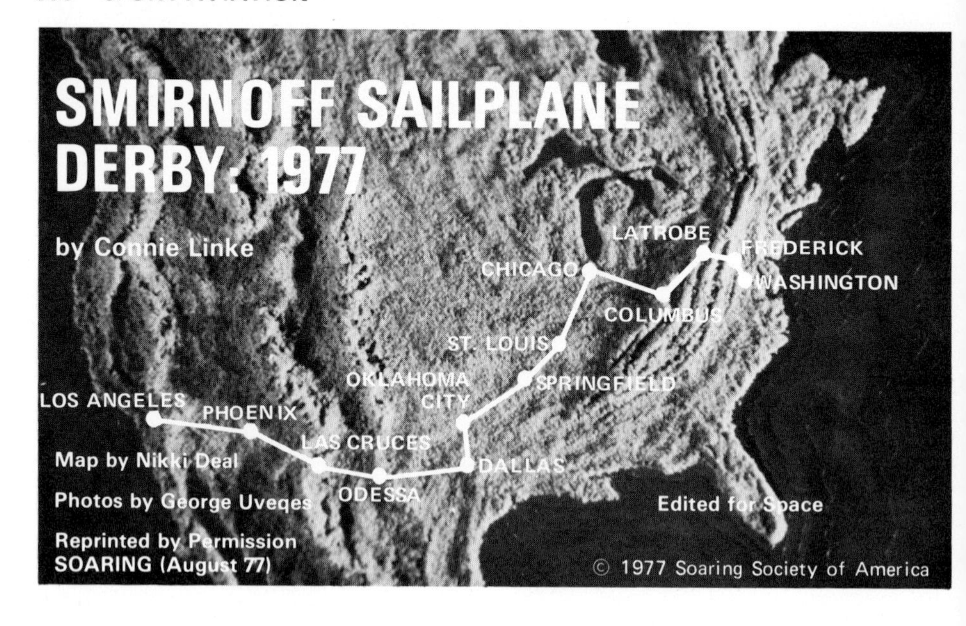

There are many ways to cross the United States. One of the most exotic ways was developed by Gregg Reynolds of Smirnoff Vodka as a publicity venture — the transcontinental Smirnoff Sailplane Derby. However, what began as a product promotion idea has grown into a world class invitational contest, with international recognition and a competitive challenge that attracts the world's best racing pilots.

"I competed as hard as I did last year in the World Comps," said Ingo Renner, current World Standard Class Champion.

This year for the sixth running of the Derby, there were three world champions, a national champion, and a world record holder — all eager to compete in this most unique soaring contest. Three PIK-20's, a I-35A, and a modified H-301 *Libelle* teamed up with their pilots; the outcome to be a measure of ship and pilot performance.

LOS ANGELES TO PHOENIX

On May 3rd, the five rivals of the Sixth Smirnoff Derby were towed out of the Los Angeles haze. In addition to preparing their ships and crews for the start of a two-week journey and attending the obligatory pilots' meeting, they spent three hours preceding their 11:30 a.m. takeoff answering questions by reporters from network and metropolitan TV, radio stations, and newspapers. Obligingly, the normal May cloudcast morning disappeared before a clear, bright, sunny day. Towing over the mountains north of Los Angeles, the five pairs of planes and gliders rendezvoused over Rabbit Dry Lake in the Mojave Desert. The sailplanes released and reformed in the same thermal at approximaely the same altitude for their "racehorse start." With all ships i position, contestant George Moffat from his PIK-20, gave the countdown, "Fifteen seconds. . .three, two, one, start!"

The first goal, Phoenix, lay on the far side of 287 miles of truly desolate desert. To one like George Lee, more accustomed to the gentle, lush, greenness of the British Isles, the first contest leg was a totally new and rather too thrilling an experience. "The landscape was awesome. I think I came in last because I was so cautious over that appalling terrain."

Five starters, five finishers, with Ingo Renner the first to cross the finish line at Deer Valley near Phoenix. The Australian who had just won his country's Nationals and the 1976 World Championships flying a PIK-20, was competing with an H-301 *Libelle* with a Schuemann wing modification.

PHOENIX TO LAS CRUCES

Las Cruces, New Mexico, was the goal for the second leg, with more frightening terrain than the previous day. After the start signal, every pilot headed out on course in a slightly different direction. Each seemed to have his own idea about how to fly the task. George Moffat: "About 150 miles out, I was working a thermal when George Lee materialized out of nowhere and came in below me. A few more turns and Ingo joined us. I hadn't seen them since Phoenix. When we left the thermal, we had separate ideas about the next area of lift, and that was the last I saw of them."

On that day Al Leffler experienced the only rough landing in the contest when he was forced down in a river bed. A four-wheel drive vehicle was procured for the retrieve, which barely managed the recovery of the PIK-20D. George

George Moffat has become a virtual soaring legend since starting in the sport 18 years ago in France. Although he started flying powered aircraft in 1935, he had been enthralled about soaring from youth. In 1962 he entered his first Nationals at El Mirage, bringing him a successful background from international level sailing. In the mid-sixties, he set several world records while developing as a threatening competitor. His first national championship was in 1969 in Marfa, Texas. In 1970 he turned in another first-place performance at Marfa, this time winning the Open Class of the World Soaring Championships. Victory seemed to suit him well as he continued to win, first the U.S. Standard Class Championship (also in 1970), the 1973 U.S. Open Contest, and finally another World Championship in 1974 in Australia.

Moffat landed on a highway, attracting considerable attention. He was impressed to learn that more than half of the people traveling on the road knew about the race! Wally Scott found a beautiful landing spot for his I-35A. With his usual nose for adventure, he landed in a mining

Ingo Renner has a German-sounding name for a very good reason since he was born in Germany. Ten years ago he emigrated to Australia largely because of the soaring opportunities it afforded, eventally becoming a naturalized citizen there. In 1968 he flew his first contest. Yugoslavia in 1972 was his first international contest and he placed sixth. At the 1974 World Comps in Waikerie, Australia, he set a good pace in the Standard Class only to have victory narrowly snatched from him by Helmut Reichmann on the last contest day.

Ingo is one of the few "professional" soaring pilots, spending Australian summers instructing at Tocumwal, and German summers training and instructing at Oerlinghausen, West Germany.

His impressive success in this year's Smirnoff Derby he modestly ascribes to "being warmed up" for competition, having recently completed the Australian contest season.

compound with locked gates and tall fences. Permission to unlock the gates took four hours. In the meantime, George Lee was finishing at the goal with the last rays of the setting sun after a grueling flight under and around frequent large masses of overcast. He was more than an hour behind the Australian, Renner. (Renner, in German, means "runner.") Two days, 2000 points, made for an auspicious beginning to a highly-touted competition.

LAS CRUCES TO ODESSA

Overcast skies and early-morning drizzle canceled the next day's scheduled departure from Las Cruces.

The following day the race continued to Odessa, Texas, 267 miles away. Once again, all five pilots finished, with only eight seconds separating Wally Scott and Ingo as Renner continued his perfect score. Wally was given only 999 points out of a thousand for his eight-second dalliance.

ODESSA TO DALLAS

The Derby flew out of Odessa the next day heading for Dallas, fortuitously evading the severe hailstorm which descended upon Odessa that afternoon. A late start precluded finishing at Dallas. Four of the pilots, Leffler, Scott, Lee, and Moffat, landed in fields and airports near the interstate highway leading to Dallas, having made distances varying from 177 miles to 233 miles. These four had been sighted frequently by their ground crews. A combination of poor radio communications and Ingo's three previous successes convinced his crew to push on to Dallas, only to learn that Ingo had landed 72 miles short but had flown farthest. Tallying up the daily scores, Inro flew 22.5 miles over second-place George Moffat.

On George Lee's outlanding that day, he was nipped by a rancher's curious canine. (This classic "Dog-Bites-Glider-Pilot" story made national and international wire services, even to a TV announcement on Iranian national television!) That day marked the first of four consecutive uncompleted tasks.

DALLAS TO OKLAHOMA CITY (NO CONTEST)

The Derby traditionally brings inclement weather to Dallas, making it a difficult place to either soar into or out of. The first two Smirnoff races spent many soggy days waiting for good weather, finally trailering out. The weather this year *seemed* cooperative until the pilots were ready to go on Monday morning, following a Sunday rest day. For three hours the temperature hovered a few degrees short of triggering temperature. As the ominous stratus overcast of early afternoon began conquering more and more of the sky, the signal was given to launch. The pilots had another problem besides weak weather: how to circumvent the Dallas TCA (Terminal Control Area), with its altitude restrictions. Of the four pilots who heeded the advice of Competition Director Hannes Linke ("Fly to the west of the TCA"), three landed less than 50 miles from takeoff, and one barely glided into Oklahoma. Scott, the sly Texas fox, flew around the TCA on the east side and came very close to completing the task. His brilliant flying became a personal triumph since only two (instead of the required three pilots) made minimum distance of 65 miles, thus resulting in the only non-contest leg of the race.

By Oklahoma City, the contestants and crews were becoming accustomed to the media interviews. Ingo Renner, with a perfect score after four contest days, was being contacted a minimum of thrice daily by the New York office of the Australian News Agency. Before the race was completed, Ingo had become front page news in the

George Lee flew a PIK-20B in this year's Derby. Born in Ireland, he is presently stationed in Scotland flying Phantom jets with the Royal Air Force. He began soaring in 1963 through an RAF gliding center "for purely mercenary reasons" he noted, because he thought that his glider training would get him selected for flight training sooner. His first soaring competition was a regional-type contest in 1970, which he won. He entered the British Nationals in 1972 because he found competitive soaring "very appealing." In 1974 he won his first Nationals and followed that by his victory in 1976 at the Internationals in Finland.

Australian papers and virtually a national hero in his sports-avid homeland.

THE OKLAHOMA CITY — SPRINGFIELD RACE

Tuesday, May 10, took on the potential aspects of an out and retrieve. Ingo made one exploratory flight and reported nothing. Anxious waiting and another probe resulted in: "No lift encountered during tow."

Anxious crews and fellow competitors watched Ingo slowly gliding down as he reported, "Still no lift." A few more minutes dragged by. "Wait, I have half-meter lift." The words were hardly out of his mouth when the next sailplane was launched with the other three following in swift succession. The stability surrounding the airport gave way to moderate lift to the east over Oklahoma City itself, continuing stronger farther on course. Crews were able to keep visual contact with their pilots for much of the course. Past Tulsa, however, conditions gradually deteriorated and progress became a matter of staying aloft. Ingo's winning streak held good and he covered 166 miles, making each mile worth about six points. By now the standard joke when someone inquired as to the day's winner was, "You have to ask?" Five contest days, and 5000 points to Renner the runner. (Ingo was asked if he had ever won a contest with a perfect score, since it seemed likely that he could win the Smirnoff Derby without breaking stride. "Yes," he replied, "at the Australian Nationals in 1973 I won every day flying a *Kestrel 19*."

Wally Scott is the hard-flying, lovable Texan, most famous for his legendary AS-W 12 and his many, many world distance and goal records. Wally began flying "way back when" and worked as a military instructor during World War II. With an offer of free lessons, he began soaring in 1961. His first competition was at McCook, Nebraska, in 1964. Although Wally is always an intense national and international competitor, he seems to have found his milieu in the Smirnoff Derby. He won the first and fifth Derbies, and also competed in the second one, making him the only four-time race participant.

SPRINGFIELD TO ST. LOUIS

Springfield, Missouri, Municipal Airport is a very busy regional airport, serviced by two airlines and many private business jets. Triggering temperatures on Wednesday occurred early accompanied by dust devils, but jet traffic delayed takeoff until past 2 p.m. The race to St. Louis saw no finishers and was yet another day of slow progress with crews afforded

Al Leffler of Ventura, California, began his flying career in 1948 and expanded to soaring in 1959. In 1963 he competed in his first soaring competition, offering "a desire to win" as his reason for entering. Al's contest number "LB" for "Little Boy" belies the not-very-diminutive pilot behind the call sign. He has always held the reputation of turning in better contest performances than his ships were seemingly capable of. Last year he won the Region 12 contest and the Unlimited Nationals.

the rare opportunity of following their pilots visually for much of the leg. Navigation claimed its first victim as George Moffat literally "took the wrong highway out of town." The lack of aerial road markers caused an expensive mistake, as he dropped to fourth overall. The fields west of St. Louis, meanwhile, were being marked by the other four ships and closest to the goal was the evidently unbeatable Renner, who flew 182 miles of the 198-mile course. Bob Fenton, a St. Louis soaring pilot, was on the road with a mobile camera unit from the TV station he works for. He arrived at Ingo's landing field simultaneously with Ingo's crew and filmed the whole retrieve for the late news. The TV station's weatherman had been prepping his viewers for a week in advance of the Derby's arrival, giving them information about soaring conditions and explaining thermal lift.

ST. LOUIS TO CHICAGO

The next goal, Chicago, was a worrisome one for the contest coordinators. The previous two Derby attempts to fly the distance had met with little success thanks to uncooperative weather. As it turned out, however, four pilots managed to complete the leg from St. Louis to Chicago. Finishing was not as easy as it seemed. There were numerous areas along the course with wet ground and strong sink. Being caught in the wrong lift-sink cycle was a critical factor, as Al Leffler discovered, much to his dismay. Hannes and Connie Linke were driving toward Chicago when they spotted one of the contest ships circling at a heart-stoppingly low altitude over the town of Lexington, Illinois. They stopped, hoping to watch Al climb out of his desperate situation. Al's superb skills and the Linkes' telepathic encouragement combined could not overcome the sink-cycle. Al's subsequent landing in a local cornfield attracted a police plane and ground unit to investigate the crash. They found, however, a dismayed California glider pilot, with his definitely heavier-than-air craft surrounded by about 50 awed local youngsters. It was quite a sight. The winner? Who else but the seemingly invincible Ingo who managed to cover the 248-mile distance at 58.69 mph — six mph faster than his closest rival.

CHICAGO TO COLUMBUS

Departure from Chicago the next day was set for noon from Howell Airport on the south side of the city. John Cody, a Chicago radio personality, called Hannes Linke at 6:30 a.m. for a live interview to broadcast to the early morning commuters. The interview was so successful that several hundred spectators showed up to see what this soaring race was all about.

The weather report didn't indicate strong conditions, but amazingly the strongest lift of the race was encountered in South Chicago. At the start of the leg to Columbus, Ohio, Ingo still had a perfect score, with 7000 points. He had more than 700 points over George Lee in second place, and Wally Scott who was a close third. But perfection in soaring is quite hard to maintain, and that day it was George Lee who made 1000 points with a winning speed of 75 mph, his speed aided by a 30+ knot tailwind. He was followed by Scott (72 mph), Leffler (71 mph), and Moffat (65 mph). Renner, surprisingly, failed to complete the task and was fifth for the day. Everyone wanted to know how Ingo had "fallen from grace," so to speak.

Over the 267-mile course from Chicago to Columbus, the barometric pressure dropped by one-half inch. Due to problems with his radio, Ingo did not call the control tower at the destination airport for a current altimeter setting. He was far ahead of his competitors at that point, but that oversight, coupled with strong sink during final glide left him three miles short, even though he could see the airport!

COLUMBUS TO LATROBE

On Sunday, May 15, several hundred Columbus residents took a Sunday drive to Ohio State University Airport to get a view of something different — a cross-country sailplane race. At the same time the airport's control tower personnel were enjoying *their* first glimpse of competition soaring. They were extremely cooperative, even assigning the towplanes and gliders their own separate runway, enabling them to launch into the wind.

Conditions were a bit weak and the rendezvous took a long time to set up. But finally a start was made at 1:36 p.m. for the 194.5-mile task to Latrobe, Pennsylvania. The middle of the course was moderately strong and fast-going, but the first and last parts were very slow. Competition Director Linke and company arrived at Latrobe Airport simultaneously with George Lee who was hotly pursued by Wally Scott. They were finishing second and third. First? none other than the little *Libelle* and its hard-pushing pilot making a successful comeback from his previous day's falter.

LATROBE TO FREDERICK

Nine contest days completed and only one to go. It hardly seemed possible that this fantastic race, with its great soaring weather (the rest of the country meanwhile alternately froze, flooded, and fried) and tremendous camaraderie, could be so close to ending. "We aren't ready to end this race," they protested. Came the morrow, Monday, May 16, and all set off for the final task: Latrobe to Frederick, Maryland, a brief hop of 123 miles.

The only close competition still left in the race was between Lee and Scott. Ninety-seven points kept Wally from second place. There was a certain grim fighting tension in these two, Scott and Lee, as they settled into their ships for the final duel. Would Scott's old ally the "Allegheny Alligator" help him again? (Wally had won last year's Derby on the final day.) No, it chose a new honoree. On this final dash, George Moffat snatched the daily victory from Ingo Renner by 45 seconds. But even this slight challenge did not take away Ingo's ultimate victory. When George Lee and Al Leffler finished at Frederick, there was no sign of the usually hard-pressing Wally Scott. Dear Wally had been deserted by his favorite lady luck and had fallen to mother earth short of the goal.

The ceremonial end of the race occurred at Dulles Airport in Washington, D.C., where the sailplanes landed in finishing order for the media reporters: Ingo, first; George Lee, second; Wally Scott, third; George Moffat, fourth; and Al Leffler, fifth. An exciting contrast this year was a British Airways Concorde jet (called *Speed Bird*), piloted by a Briton with a Silver Badge, no less, taking off on a runway parallel to the glider landing runway at Dulles. The Concorde would cross the Atlantic to another continent in less time than it had taken for most of the individual tasks on the transcontinental Derby.

STRIDIECK REPEATS HIS 1000-MILE FLIGHT

by Doug Lamont

Reprinted by Permission
SOARING (July 1977)

For the second time, Karl Striedieck has exceeded 1000 miles in an out-&-return distance flight. On May 19th last year, he became the first soaring pilot to break the "1000-mile barrier" with a thirteen and one-half hour flight from Pennsylvania to Tennessee and back. Unfortunately, his earlier turnpoint photo may not have been taken within the sector arc required for record approval by FAI rules, and homologation has lagged while the paper work has been shunted from one agency to another.

Karl evidently became restive waiting for his flight to be confirmed as a record and decided that the best solution might be to fly the course again. So on May 9th, after waiting almost a year, he set forth again in his AS-W 17 from his farm strip clearing in the forest atop Bald Eagle Ridge.

"I used a jeep tow to launch as usual, but Sue (*Karl's wife*) couldn't tow me because she is working in a metropolitan library. I got off at 5:52 a.m. and flew north along the ridge to my regular startpoint at Lock Haven. I turned south there and went through the gate at 6:07 a.m. Daylight Saving Time.

"This time, instead of the southerly wind component I faced the first time, there was a quartering tailwind blowing on the ridge at about 20 to 30° northerly. Right from the beginning I went barreling along the crests at about 500 to 1000 feet without essing or circling. My groundspeed out of Pennsylvania, across Maryland, and into Keyser, West Virginia, was over 100 mph."

Even better conditions lay ahead. He picked up a wave and climbed to 8000 feet asl. Freed from turbulence, he pushed the '17's nose down still further until the Appalachian forests were slipping by at jet speed. A hundred miles later, when the wave ended, his calculator indicated he had made 180 mph for the stretch!

He dropped back down on the ridges to take advantage of the still-continuing quartering tailwind. Approaching the state line between Virginia and Tennessee, he looked eagerly toward Clinch Mountain and more lift.

"It was only 10:30 a.m., and I had already flown five-sixths of the distance to the turnpoint at an average of 130 mph! It seemed too good to be true."

And it was. Conditions weakened abruptly. Karl dropped the AS-W 17's flaps to their number three setting and went on max L/D.

"I stopped to circle in any lift I could find; I didn't pass up anything. The convection layer wasn't very deep in Tennessee, so I didn't want to take any chances.

After the earlier speeds, I felt like I was just crawling along and it seemed to take forever to reach the turnpoint by 12:40 p.m. I took two pictures. They ought to be right this time."

Karl's caution paid off; he succeeded in returning to Clinch mountain and found that conditions to the north were still good — except that the helpful tailwind was now a headwind, of course. This situation reversed that of his flight a year earlier, and by the time he succeeded in reaching his starting point at Lock Haven, the elapsed time was also greater — 14 hours and 18 minutes as compared to 13 hours 30 minutes.

It was 8:10 p.m., and instead of attempting to return to his aerie at Bald Eagle Ridge, he landed at Lock Haven. Reasonably certain that he had secured good turnpoint photos this time, he considered his next quest: "I'm going to try for Florida," he says. "Sometime between March 15 and April 15 next year the weather should be right. I think 1000 miles straight out is possible." Hans Werner Grosse's spectacular trans-European flight of 907 miles in 1972 still stands. So far no one has made the magic 1000-mile straight-out distance. Karl would like to put his name on that one.

Karl Striedieck also flies Air Guard jets.

U.S. PILOTS SCORE WELL IN PRECISION FLIGHT CHAMPIONSHIPS

Reprinted by Permission
FLORIDA AVIATION JOURNAL (October 77)

Edited for Space

The U.S. Precision Flight Team, sponsored by the National Pilots Association, has returned from the Second World Championships of Precision Flight as winners of both the silver medal for individual performance and the fourth place team trophy. At the competition held in Wels, Austria, current U.S. National Champion, Jim Lafferty of San Jose, Calif., finished second in the individual competition and earned the silver medal for the U.S. Final team scores placed the United States fourth, following Sweden, Poland, and Austria.

Other members of the U.S. Team included Steve Schwenk, St. Louis; Richard Hoesli, Ann Arbor, Mich.; and Joe Poerschke, Miami. Team leader was William H. Ottley, Executive Director of the sponsoring association. Bruce Mazzie, also a NPA official, served as team manager.

Because precision flight competition in the U.S. is organized on an entirely non-profit basis, selection, training and expenses for the 1977 American Team were supported by voluntary donations solicited by NPA from the U.S. aviation publications.

In its second world-level competition, the U.S. Team moved from a seventh place finish (earned in 1975) up to fourth place, for which the U.S. received an "honor award" cup.

Fourteen teams competed. Final rankings were as follows:

1. Sweden
2. Poland
3. Austria
4. United States
5. Switzerland
6. Norway
7. Great Britain
8. Federal Republic of Germany
9. Denmark
10. Czechoslovakia
11. France
12. Canada
13. South Africa
14. Japan

(A Yugoslavian delegation had been present during the practice sessions, but received a message from their government before competition began ordering them to withdraw — in protest against the presence of the South Africans. The Finish team (who had finished high in the rankings two years ago) did not participate this year, for the same reason.

The U.S. Team arrived in Austria after a transatlantic flight to Frankfurt Tuesday, August 9th, to begin intensive on-site practice. The daily schedule was dawn-to-dusk, despite the fact that aviation fuel in Austria costs $2.20 per gallon and every touch-and-go landing required the payment of a separate fee. The early morning starts allowed the American pilots to beat other teams to the practice runway and flight course, which turned out to be of critical importance due to the fact that weather was marginal all but one day.

The start of official competition on Friday was delayed repeatedly because of less-than-VFR conditions. Shortly after noon the first heat set off on the navigation course, described later by Jim Lafferty as "the hardest hour and a half of work I ever did in my life." The terrain was difficult, everything "looked the same," and the weather was extremely bad. During the third heat when Hoesli and Poerschke were both aloft, lightning, and a downpour moved into the area. Conditions went to zero-zero and the two American pilots were instructed to come home. One made it through the rain; the other had to divert with radar vectors to the international airport at Linz twenty miles away.

On Saturday it went CAVU, which allowed the landing competition to run smoothly. Fine judging and excellent video-tape contributed. Lafferty received a disputed penalty call which would have thrown him out of the top scoring if upheld. A protest filed by the U.S. called for a review of the tape of the incident. The jury reversed the judge's call and Lafferty regained his second place title.

International competition in precision flight is authorized by the Federation Aeronautique Internationale and governed by rules set and approved by that organization.

Team members of the 1977 U.S. Precision Flight Team are (left to right) Joe Poerschke, Team Leader William Ottley, Steve Schwenk, Jim Lafferty, Dick Hoesli, and Team Manager Bruce Mazzie.

POWDER PUFF DERBY'S FINAL FLIGHT

by Wanda Cummings

An AVIATION YEARBOOK 1978 Special Report

Record-holding aviatrix Jackie Cochran dropped the flag at 8:00 a.m. July 1, 1977, to begin the Powder Puff Derby's 30th Anniversary Commemorative Flight. Behind her, 140 planes of all ages and horsepower were lined up on the ramp of Palm Springs' Municipal Airport. The women pilots were both veteran and new, from thirty-six states including Alaska, from the Bahamas, Australia, South Wales, and Germany. Among them were five former winners, all the former chairmen of the race, and officers of the International Organization of Licensed Women Pilots (the 99s). Three contestants were flying their 24th All-Woman Transcontinental Air Race.

The Honorary Starter's Flag was passed from Jackie to Carolyn West, who, with Bea Medes, had won the first AWTAR; to early movie flyer Clema Granger; to veteran air-race promoter Cliff Henderson, and to cartoonist Milton Caniff, who was sponsoring "Bitsy Beekman" entry Peggy Noltensmeyer of Gladstone, Missouri.

Before the final contestant departed, the temperature had soared to 130°. Former WASP Melba Beard, baking in the open cockpit of her 1928 Brunner Winkle Bird, took off, developed oil pressure problems, and went home to Phoenix.

Alverna Williams, legless since childhood, started in a 85 h.p. Ercoupe; but, discovering a gas leak, parked her plane and continued the flight as co-pilot to Susan Parish in a WW II AT6.

Last to be flagged off was Jerrie Cobb, Harmon Trophy recipient and only woman qualified for space flight, in her Aero Commander with which she flies Mercy Missions along the Amazon River.

The Derby had announced its final run in 1976, due to lack of financing and the fuel shortage. At the suggestion of the Smithsonian Institution, however, a Commemorative Flight was planned as the National Air and Space Museum's "Milestone of Flight" for July, 1977. NASM Cachet Covers were carried by the entries.

This was *not* a race, but a sentimental journey along the route of the first race in 1947: Palm Springs CA, Tucson AZ, El Paso, Midland and Dallas TX, Shreveport LA, Jackson MS, Monroeville AL, Thomasville GA, and finally ending at Peter O'Knight Airport in Tampa, Florida. There were contests of proficiency between each of the designated stops, and points awarded for each contest.

By early afternoon July 1, the majority of the planes were tied down at El Paso, where the race had stopped a record 13 times. Waiting to greet the fliers was Pulitzer Prize winner Bill Mauldin, the only man to have flown with the Derby. Bill had received permission from the AWTAR Board to fly with the Derby in 1955, when he was assigned the Powder Puff story by *Sports Illustrated*. El Paso Airport management, FBO's and airlines had reserved a section of the Terminal for the 1977 participants, and furnished frosty Margaritas and a tasty Mexican Buffet.

July 2, the Flight continued with an identification contest to Midland, and a navigation leg into Dallas, the next overnight stop. Early arrivals were given a tour of the American Airlines Flight Academy, and invited to fly their giant simulators.

The relaxed fun and games format of this Flight provided adjustment problems for veteran Powder Puff participants. Competition was not completely forgotten, for at this point, four Flight numbers were tied for first place.

July 3, everyone went picture-hunting into Shreveport. "Bitsy" was swarmed by the Press, and the Deep South's hospitality was for all. Those participants spending the night at

First Place in the Final Powder Puff Derby were (1) Patricia Udall and Nanette Gaylord, with FAA Associate Administrator of Air Traffic, William Flener.

Jane Ruehle (l.) and Jane Stevens flew this 1940 Taylorcraft from Palms Springs to Tampa.

Jackson were taken (voluntarily) in a police department bus to a private residence for cocktails and dinner, and came away saying they had been to Tara.

There was a spot landing contest at Monroeville, and none of us will ever forget C. L. Stokes on the Unicom, calmly saying to every flight,

"Have a nice flight, and come back to see us, hear?"

A golf tournament in Thomasville had left limited motel space, and when over 60 planes landed for the night, the mayor declared disaster status for the area so that mattresses could be laid out on the ballroom floor of the Holiday Inn. Additional accommodations were offered by the local mortician, who promised innerspring caskets.

Last leg of the Flight was a proficiency race, and all contestants were allowed the thrilling 200-foot-AGL "Fly-By" past Peter O'Knight's temporary tower.

Here, to join the celebration, were more of the "Greats" such as Louise Thaden, Harmon Trophy winner, first woman to win the Bendix, and winner of the first Women's Air Derby in 1929; Fran Nolde, 1948 Powder Puff winner, and founder of the Reading Air Show; Blanche Noyes, another Bendix winner, who still holds the women's speed record from East to West, and Viola Gentry, first woman to refuel in flight.

Among the 127 planes to finish the Flight by deadline July 4 was a 65 h.p. Taylorcraft which had been rebuilt by pilot Jane Stevens of Boulder, Colorado.

First place, for collecting the most over-all points, was pilot Patricia Udall, a ranger for the Navajo National Monument in Arizona, and Nannette Gaylord of Denver, Colorado, in a Cessna Cardinal. Close second was Fran Bera, 7-time winner of the Derby, and her co-pilot Joyce Failing, both of Southern California. A California sister team, Joan Paynter and Shirley Tanner, were third. Fourth place went to Thon Griffith, International Vice-President of the 99s, and co-pilot

Kay Brick, who had been chairman of the race for 13 years. They flew a Beech Baron.

In fifth place, flying their first event, were Californians Eileen Wyckoff and Linda Schreck. A Jackson MS team, Cindy Bass and Pat Brooks, were Flight 1, and came in 6th.

The Smithsonian's cachet covers were collected for final stamping at the Tampa Post Office, and returned to Washington DC, where the perpetual trophy for the world's longest and largest speed race for light aircraft is exhibited.

During its 30 year history, the Powder Puff Derby had been supported by friends from coast to coast. The Jeppesen Company had furnished charts, and Simplex had loaned time stamps. Flying Tigers had flown luggage, and a network of Amateur Radio Operators had given of their time and equipment. The FAA had arranged for flight watch and weather briefings. Countless 99s and aviation enthusiasts had, yearly, formed the ground crews so necessary to a race. Aircraft manufacturers had called us the "Sounding Board for General Aviation," for, in our cross-country "fire-walling," we had shown what their planes would really do.

Three who flew 24 Derbies: (l. to r.) Pauline Glasson; 1971 winner Gini Richardson; and Pat Gladney.

The race had inspired many to fly, and brought about the up-dating of airports. Many of the rules made for the safety of the racers were later adopted by the FAA. It has been said, "No other type of flying requires the thought, planning, and knowlege of a plane required by AWTAR."

The Powder Puff Derby covered more than 5 million race miles without a fatality. It had over 3,000 entries from 12 countries other than the U.S., and was welcomed by 41 states and 175 different cities.

The proud old race showed the world a touch of efficient glamour. I doubt it will be soon forgotten.

RESULTS OF FINAL POWDER PUFF DERBY

PILOT/CO-PILOT/PASSENGER(S)

1 PATRICIA UDALL
NAN GAYLORD

2 FRANCES S. BERA
JOYCE FAILING
KATE WEATHERED

3 JOAN PAYNTER
SHIRLEY TANNER

4 THON GRIFFITH
KAY BRICK
BARBARA EVANS
JANET GREEN

5 EILEEN HANSEN WYCKOFF
LINDA C. SCHRECK
JUDITH R. SORCE

6 CINDY BASS
LONNIE (PAT) BROOKS

7 ANN LOWELL
MARY WADINGTON

8 CAROLYN ZAPATA
BERNICE HOWELL

9 MARY R. WENHOLZ
ROSEMARY DE ANGELO

10 MARY CREASON
ELOISE M. SMITH
LISA WHITTAKER

11 BEBE RAGAZ
SUSAN RAGAZ

12 BARBARA A. GOETZ
THELMA CULL
RUTH LUMMIS
SHIRLEY KAMMEYER

13 JEAN JOHNSON
EUGENIA ROHRBERG

14 IRENE FLEWELLEN
RACHEL PRUETT

15 JERRIE COBB
BETTY LOU WRIGHT
GRACE HARRIS
G. ROSS HOWARD

16 CHRISTINE CWALINA
MARY CWALINA

17 CORALEE TUCKER
MELISSA VREELAND

18 SANDRA E. SULLIVAN
BONNIE L. KLEIN

19 JAN MILLION
NANCY SMITH
GWEN CRAWFORD
NEMA MASON

20 DOROTHY L. DUNCAN
SHARON R. M. LASH

21 PATRICIA GLADNEY
PATRICIA SHERWOOD
MARGARET STANDISH

22 MARIE HOEFER
BABS TULEY
LOUISE GUEST

23 MARTHA WOODHOUSE
JULIA SOMERVELL

24 LOUISE WHITE
LUCY MERRITT
ANNIE LEE ORR
DOROTHY M. PENNEY

25 JUNE McCORMARK
BEVERLY ALLEN
MARTY HARSTAD
LYNN BRIGGS

26 ROSEMARY ALDEN
JO BETH TINGLEY

27 ELIZABETH ENGSTROM
CLAIRE M. ELLIS
JOAN STEINBERGER

28 CHARLENE FALKENBERG
BARBARA JENNINGS

29 PHYLLIS A. PIERCE
VIRGINIA M. COLLINS
JEAN E. COLLINS

30 FRANCESCA S. DAVIS
JUDY WALTER

31 BETTY RICHARDSON
ELIZABETH SEWELL
RUTH "RONI" JOHNSON

32 LORETTA JUNE JONES
HELEN HAMILTON
ERLENE LOWE

33 DONNA BOWER
RUTH W. THOMAS

34 DEBRA ANN THOMPSON
SUSAN JANET KINGSTON

35 MARGARET RINGENBERG
PAT WILSON

36 JO ANN SIGLIN
LINDA PAUL

37 MADINE CARPENTER
HYLA BERTEA
JOYCE YOUNG

38 VALERA JOHNSON
RUTH BENEDICT

39 LINDA M. HOOKER
NORMA JO McCARRELL

40 MARGARET M. BOLTON
DIXIE LEE MILLER

41 JACQUELINE KLAUS
LOIS FREEMAN

42 MAYBELLE FLETCHER
MARY BYERS

43 SUSAN M. CLARK
CATHERINE L. HATCH
ELEANOR RICHARDSON

44 MARY JANE NORRIS
MARY HELEN BURKE

45 COLLEEN ARMSTRONG
JEANINE M. BERTRAND

46 CAROL L. LE KRON
CHARLENE KAY ALLEY

47 JEANNE NEEL
SELMA JO DOUGLASS

48 JOYCE ODOM
ALICE GUSTAFSON
JANET BONE
JOANNE MARSDEN
MARION BELLOWS

49 IMA JEAN HUFF
CHRIS McCLAIN

50 DOROTHY E. COLVARD
SHERYL C. JONES

51 PAT CLARK
LYNNE SELLERS
JOAN HRUBEC

52 BRUNHILDA K. BRADLEY
SIGRID SIKORSKY

53 SANDY ANSON
PAM MARLEY

54 MARILYN L. DICKSON
ROBERTA HUKE

55 BRANDY E. HAMER
DORI W. LOHR

56 JAN CHURCHILL
CAROLINE J. DUPONT

57 GENIE RAE O'KELLEY
JOYCE GREEN

58 LOLA RICCI
JUANITA PEEPLES

59 DR. HELEN G. ZUBROW
ANNE M. SHIELDS

60 STEPHANIE ANN WELLS
WENDY KARLEEN HURST

61 DONNA TRACY
SHIRLEY MARSHALL

62 MARILYN BURCH
BETTY DODDS

63 JANNA L. IMLAY
JUANITA JANE LOFTUS

64 PAULINE MALLARY
CAROLYN BAKER

65 VERA ARNOLD
PAT DAVIS
MARDO CRANE

66 ELLEN (PEGGY) MAYO
CINDY SCHOFIELD
LINDA HARGRAVES

67 MARY ANN HAMILTON
JOAN MANDA

68 MICKI E. THOMAS
DIANA CAGGIANO

69 IRENE WIRTSCHAFTER
MELISSA MESSNER

70 JUNE G. CUNNINGHAM
SHIRLEY L. MOORE

71 JOAN D. McARTHUR
CAROLYN J. ARNOLD

72 VICTORIA ROSS
STACY WACHS

73 EILEEN ANDERSON
JOYCE WILKINS

74 PATRICIA A. WEIR
KATIE V. CONKLING
JANE KRIHA

75 FRANCES DUNFIELD
JO ANNE MAPLE

76 ANGELA BOREN
JUANITA WADDELL

77 ALICE MARKEE
JUDY AAGESEN

78 MARY ANN JAMISON
ELEANOR TODD

79 LOUISE PRUGH
MARGIE WOOD

80 ELIZABETH McGRAW
VELTA BENN
CYNTHIA GAMBLIN

81 BARBARA SILAGI
BEATRICE SIEMON
MARJORIE B. ANDERSON
ETHEL D. SCOTT

82 JOAN E. KNAPP ENYEART
CONNIE GOULD

83 MARY ANN MOBERLY
BILLIE DUTCHER

84 CATHERINE NICKOLAISEN
PRISCILLA MORELAND

85 BETTY BOARD
FRAN LEVE

86 ANALEE HOLDEN
ANALEE H. PERICA

87 GINI RICHARDSON
JILL RICHARDSON
DEBBIE LINDEMAN

88 PHYLLIS M. BAER
ELEANOR MAY CANSDALE

89 SANDRA L. PRATT
HALLIE R. McGONIGAL

90 MARIE CHRISTENSEN
MARJE HANSON

91 SAMMY McKAY
WINNIE DUPEROW

92 PAULINE CLENDENING
VEE ST. JOHN

93 LOIS FEIGENBAUM
HAZEL JONES
BETTY M. WHARTON

94 PENNY CARRUTHERS
MARION FICKETT

95 EMMA McGUIRE
DORIS MINTER
NINA E. LAUGHBAUM

96 MARTHA PEARCE
ELIZABETH ALLEN

97 ANNE BRIDGE BADDOUR
PAT THRASHER
MARY LEE EDMONDS

98 JEANE WOLCOTT
MARY-ELLEN WEBSTER
CONSUELO HUFFMAN

99 DOROTHY F. CARROLL
JACQUELYN FREEBERG

100 ALMA HITCHINGS
MARY JANE LIBONATE
CLAIRE (KURICA) ANGELINI
NANCY STEINFIELD

101 NINA NOEL ROOKAIRD
VIVIAN HARSHBARGER

102 RENE HIRTH
MARY SWEETSER
MURIEL GUGOLZ
JUNE HIRTH

103 VERNA STEELE WEST
SUSAN NORMAN
SARAH ROHLFING

104 DIANE SHAW
MARY ROSE MYERS

105 BARBARA WELSH
JAYNE HUNTER
CAROL EWING

106 NORMA VANDERGRIFF
CAROLYN PAGE
MARJORIE HUDSON
ESTHER TALIAFERRO

107 ALMA JESCHIEN
ANNE MACY
JOANN JESCHIEN

108 JAN GAMMELL
ILOVENE POTTER

109 TOMMY JEAN HAYES
SENJA ROBEY

110 JEANNIE L. BALL
CATHERINE GROVER

111 THEOCLETE SOMMER
JEANNE MORSE

112 VIRGINIA RAINWATER
JO ANN LINDER

113 NOLA L. RHODES
VIRGINIA M. BOGGUST
JANICE N. HOFFMAN

114 PEGGY MOSLEY
HANES BURKART
JAN PERLITCH

115 HARRIET ZIMNEY
MARY FIRTH

116 PAULINE GLASSON
BARBARA D. LEWIS
AILEEN MELLOTT

117 LOIS WARD
MARY LOUISE WILLIAMS

118 V. LORRAINE NEWHOUSE
TERRY ROBERTSON
MARGARET SCHOCK

119 LYNN NEWTON
BARBARA G. WARD
BARBARA STEWART

120 BARBARA McKENNA
WANDA SCHRAMM

121 PAT FRIEDMAN
PAT GATLIN

122 LINDA FAY HAEDGE
MARY GRIFFITHS

123 BETTY GILLIES
GERTRUDE LOCKWOOD
NANCY WALTON

124 SAUNDRA K. Y. STIENMIER
KAREN JOY LINDBERG
BETTY GOSE LIGON

125 CLARICE M. BELLINO
MARGARET J. PEAKE

126 SUZANNE PARISH
ALVERNA WILLIAMS

127 JANE ORMSBY STEVENS
JANE LE FEVER RUEHLE

128 BOBBY SMITH
JOANN LANGLEY

GREENAMYER'S RENO AIR RACES

by Mark Lyon

Reprinted by Permission AVIATION NEWS (October 77)

The 14th National Championship Air Races at Reno this year gave us a new speed record, the Canadian Snowbirds, a one-day attendance record in spite of intermittent weather and the start of a new racing class.

We did not see sport biplanes (no one seemed to mind) fist-fights in the pits, enough sunshine (except on Sunday) or as many people as last year.

The air races are still one of the best shows in the west; and with the addition of more unlimited races to the exclusion of the smaller, harder-to-see biplanes, they are becoming even more entertaining.

Darryl Greenamyer flew his fastest unlimited race ever: 430.703 mph for a new competition record.

The RB-51 gave everyone a nervous stomach early in the week but hung together through the big race, Sunday, after three years of frustration and failure.

During practice on Tuesday, Greenamyer stretched the sleek red airplane to 443 mph, pulling 90 in of Mercury — considerably less than most of the stock Mustangs — when the "Red Baron" coughed a couple of times and left a thin trail of black smoke behind the aircraft and pained expressions on the faces of the Baron's pit crew.

The popular Reno National Championship Air Races will enter their 14th year this coming September 15-17. The action-packed event features the finest in air racing with a spectacular supporting air show.

"We were getting ready to qualify and I was going up on the power at a relatively low power setting," said Greenamyer. "The boost pump was intermittent. I didn't catch it and it leaned the engine out and started detonating and before I could get the power off it backfired. The intake duct which lays across the top of the engine was pooched out. Thank God it didn't come off.

"It also sheared the bolts that connect the drive shaft to the supercharger and that put the supercharger into freewheeling and that is what resulted in the engine failure — the apparent power loss.

"We had a spare engine but no spare supercharger. (The RB-51 engine is a hybrid made from a Shackleton Bomber motor and a Spitfire 2-stage blower.) That's the same supercharger that broke the gears during previous years and which ate itself up.

"We had to kind of piece parts from this and that together. We would have liked to go through the whole engine but we could only clean the parts out as best we could and check the screens and put it all back together."

On Friday afternoon during a qualifying heat Greenamyer had the opportunity to lap Gary Levitz's P-38 "Double Trouble". Darryl, sensing an opportunity to administer the *coupe de grace* with all possible showmanship and flamboyance, caught Levitz as they were coming around the number eight pylon, heading down the home stretch — the fastest part of the course.

At the distinctive sound of the Griphon Rolls-Royce honking at full power, everyone in the pits turned to watch one of the singular moments of the races. Greenamyer finally showed what the counter-rotating props and $750,000 were supposed to have been doing for the last three years.

He trounced it. He blasted by Levitz midway down the stretch so fast that he couldn't make the turn around the first pylon and had to go half a mile wide of the course line. By the time he got himself righted and back on the course "Double Trouble" had almost regained the loss.

"I got the checkered flag on that lap," said Greenamyer. "But I saw the white flag and the checkered flag. I knew that they weren't giving the checkered flag to Levitz but I figured I'd go around the course once more to make sure."

Levitz later went on to win the trophy race on Sunday at a speed almost 100 mph slower than Greenamyer's, 333.144 mph.

Everyone had seen the "Red Baron" make its move, well in excess of 450 mph. From then on it was all a question of mechanics. The aircraft left a brown misty wake when it whizzed by Levitz's plane.

In spite of all the anticipation, Sunday's final race was no contest. Greenamyer led the race from the moment Bob Hoover in the Yellow Mustang pace plane radioed, "Gentlemen, we will now go to race pawah."

He made his seventh win look easy, leading Don Whittington around the course by an easy six seconds. He did not strain to cut the pylons close nor did he fly especially low. He just flew fast.

In other years the "Red Baron" had required 90 in of manifold pressure from the 2,239 cu in engine to stay competitive. This year Darryl extracted 80 in — rated takeoff power for the Griphon — and set a record. "It should run all day at that setting. Of course on the last lap I moved the power up in case someone was hanging back. I wasn't going to make it easy on them.

"It was nice to win for Ed (Ed Browning president of Red Baron Racing). He's had so many problems with that airplane. Now he's got a real good combination that will run fast at low power. The engine still has a few little problems. After the race we found pieces of the fuel pump drive gears in one of the screens."

Don Whittington of Ft. Lauderdale, FL placed second at the finish posting a 425.701 mph speed. Clifford Cummins, Riverside, CA was third with 424.357 mph. No Bearcats made the finals, indeed only one flew at Reno this year. It was strictly a Mustang competition with two P-38's and a P-40 representing the alternatives.

In a much closer race on Sunday, Ralph Twombly snuck up on Pat Palmer at the last possible minute to eke out a .3 sec victory and break Palmer's three year winning streak in the T-6 class.

Twombly won with a speed of 209.658 mph in a T-6 he had just purchased on Monday before the race. Palmer, second with 209.505 mph was slower than his fastest winning speed of 211.350 in '74. Ralph Rina of Los Angeles, CA was third flying 206.156 mph. Twombly, a former B-26 pilot from Wellsville, NY won last year at the California National Air Races at Mojave, CA. Reno was his second win.

In the controversal IXL Formula-I class, favorite Roy Cote had a tough time when his plane "Shoestring" blew an engine on Friday and he was unable to finish his heat. He flew to San Diego for a replacement but his qualifying speed only allowed him to compete in the Silver race on Saturday. He won with a speed of 234.773 mph, slower than his '74 winning speed of 235.411 but faster than this year's IXL champion John Parker of Rancho Palos Verdes, CA who flew 226.119 mph.

Judy Wagner edged Gary Wilson for the second spot in another photo-finish, 217.523 mph to 217.413 mph for Wilson.

Parker's win makes him the first champion of the newly formed IXL class. He told *Aviation News*:

"'I wish Cody had been in the race. The victory would have been that much sweeter. But we'll settle for what we got."

Parker noted that the atmosphere of the IXL races had improved since the rules had been restructured to emphasize a positive approach to engine modifications.

"The difference is that if a guy saw something unique on my engine, he would come up to me and under the old rules say, 'You can't run something like that. It doesn't look like mine.'

"Under the new rules he'd be more likely to say, 'Hey, that's unusual. I'd like one of those on my plane.'

"So that and the absence of a few personalities has made for the best show the midgets have ever put on at Reno," Parker continued.

Jerry Duty, Reno's race director added, "The pilots who refused to come to Reno allowed this to be the most pleasant race we've had in years. Everyone got along.

"Everyone flew a good safe race and the crowd loved it."

The star performers in this year's airshow were the Snowbirds. Their graceful 30 min routine completely enthralled the crowds at the beginning of each day's program. In spite of extremely rough weather on Friday their performance was flawless.

People who watched them wore down jackets and wrapped themselves in blankets. It rained briefly, threatened to get worse and finally the wind covered the field with dust clouds and almost sent the meager turn-out of 8,000 people home.

Duty had to cancel the IXL race for the day. "When I saw the weather I wanted to throw-up. I thought it was going to be a repeat of the year before when it rained for two straight days."

Saturday was better. The sun shone. Attendance was up. Art Scholl flew an inspired routine.

Debby Gary, who had impressed early arrivals to Reno with one of the best tail slides we had ever seen during practice on Thursday, produced a nice slow-motion lomcevak during her airshow.

Bob Hoover left all the business-twin pilots agape with his Shrike performance. He spends more time in the air during the races and show than any six other pilots. He flew two performances a day, paced two unlimited races a day and if needed would fly anytime to help keep the show flowing.

The Confederate Air Force returned with its flaming orange explosions, aerial dog fights and general bedlam that continues to thrill audiences wherever they perform. Their Midway show was disappointing but Tora-Tora was the usual success.

This year they brought a Curtis SB25C Helldiver with them — perhaps the only one flying today. Of course their B-17 attempted its wounded-duck-one-wheel-draggin' landing to a chorus of OOOhhhs and AAAhhs from the crowd. The ME 108 and Zeros also drew their share of admirers.

Lefty Gardner flew a skillful show in his P-38. So did Gerald Martin in his FM-2 Wildcat. The Red Devils had trouble staying together during their overly long performance. Jim Raymond sure can fly his immaculate T-6!

By Sunday the weather had improved so much that Duty claimed a record attendance. All of the 17,800 seats were filled ". . .and we had more people than we've ever had in the pits," he said. Official estimate for the three days was 54,000, down 2,000 from last year.

The most unusual event of the weekend actually occurred after the final unlimited race on Sunday when the Mustangs were taxiing back to the pits. The crowd had started to pour onto the taxiway to look at the airplanes. John Lapham, assistant race director, stopped the small Datsun sedan he was driving on the taxiway and got out to keep people away from the still moving aircraft.

Don Whittington drove "Precious Metal" towards the pits, zig-zagging as he went so he could see ahead of the plane. As soon as he heard the first "Karraaaang" of the propeller biting into something he pulled the power. Too late, the Datson had seven slices from the trunk to the windshield and looked like someone tried to hack it up with a dull machete.

The car belonged to Jerry Duty's wife, Ernie. Duty said it was the third car he has lost during his tenure as Reno's race director. He also commented that Whittington was now an ace. His score for the year: one F-86 Sabre Jet, one B-26, 2 ME-109s and a Datsun.

Fortunately no one was hurt. Whittington replaced his prop and flew back to Florida. The Datsun sits in an unused hangar at Stead AFB.

Now that the bleachers, the fences, the ticket booths and concession stands are all gone it's practically the only thing left from the 14th National Championship Air Races — that and Darryll Greenamyer's new speed record.

STARFIGHTER

Story and Photography by Scott Rayburn

Reprinted by Permission AIR PROGRESS (January 77)

UPDATE: The speed run described in this report remained an unofficial record. Technical problems with the filming of the start and stop of the run caused the critical points to be non-discernible on the film.

A second attempt at the record was made on October 24, 1977. A record-setting speed of 996.99 mph (1604.1 kph)has been claimed. The National Aeronautic Association is awaiting documentation of the record run for review. — JSAY

Silence, utter silence. This is a lonely, desolate place, the mountains and foothills surrounding it as scarred and barren as the sage-covered basin floor.

There is a dry lake bed at the eastern end of this valley known as Mud Lake. It's accessible only by a twelve-mile stretch of dusty desert track. The lake itself has worn the cracks and ruts of an arid nakedness for ages, the mud so sun-baked and hard-packed that the normally lethal bullets of now ancient weapons literally bounced off its surface when it was used as a gunnery range for a wartime air corps.

It is the size and nature of this broad, level expanse of dried earth that led to its selection for what was about to take place this cold weekend of October 2nd.

But Nature had intervened. A plague of bitter storms struck this valley repeatedly in the weeks prior. The once dry lake bed now rested beneath several inches of slimy clay and murky water; a continuing satanic procession of heavy, vengeful black thunderheads providing a constant reminder that Nature is still master here, still to be challenged.

Four passes, two in each direction, are required in the FAI certified attempts. At speeds in excess of 1100 mph, Greenamyer pounded the craft along the 3 km straight line course, remaining well within the 300 meter altitude envelope.

The silence is overwhelming. There is wind, blowing across the rippling sheets of water, punctuating the stillness of the early afternoon. There are the distant voices of the handful of men and women calling out to one another in preparation for what is to happen very soon. And, occasionally one can hear the glopping sounds of a boot being pulled free of the sticky, oozing mud that forces the activity to take on an agonizingly slow pace.

The wind carries a new, subtle sound now, one so very alien to this primal arena. It is the faint, yet unmistakable whine of a powerful jet engine, somewhere off among the clouds.

Eyes begin to search the horizon, scanning the rich brown earth of the surrounding mountains to pick up the source. One small group of men, positioned on a nearby hill-top, a panorama of the vast brown lake spread out beneath them, pay a special attention to this new sound. They have a special responsibility in this place, for they must send out the warning to the tiny figures on the lake's surface that the time is upon them.

These spotters are equipped with radios and, as the noise becomes clearer and steadily increases, they prepare to transmit.

On the lake bed, all eyes are peering towards the south, waiting for the first glimpse of the red and white bullet that is due any moment.

The radios come alive with the crackle of electronic voices. The spotters have done their job, and the responsibility now falls on the crews shivering in the biting cold of the lake as they prepare to switch on the maze of electrical equipment and high-speed cameras.

Continuing assaults by freak thunderstorms plagued timers and officials who now had to wade ankle deep in mud over the several miles of the normally dry lake bed. Delays and technical complications because of the bogged environment were held as partial causes for Greenamyer's record remaining unofficial.

"They've got him! Here he comes!" yells a radio operator who has picked up the transmission. Immediately, everyone focuses on the spot on the horizon, the spot where a tiny speck against the backdrop of the mountains will appear at any second. The entire episode will last but a few moments. There will be no sound to forewarn them — the speck is well beyond the limits of sound.

"There he is!" Sure enough, dropping down low, barely distinguishable against the darkened hills behind him, the shadow of a passing cloud obliterating most of the light and contrast, comes the airplane — no longer a speck, but now a growing presence, its graceful lines and stubby wings taking on definition and form.

It appears in the field of vision like a thin pencil gliding along a wire strung along the surface of the lake. There is no sound; only a presence.

The airplane streaks by the first camera station, flying only a few feet above the water's surface. Its trajectory is as smooth and effortless as a dream as it continues its path along the submerged course line. Now it is past the last station, nearly two miles away, and begins to arc gently off the course, the glowing fireball of its afterburner pinpointing its diminishing size. The silence remains.

A few put their hands over their ears. Others wince and brace themselves in anticipation. They all know what comes next; yet they will never quite get used to it. It comes: a shotgun, a stick of dynamite, a terrible clap of sound explodes in the ears of each person watching, waiting. The retort is deafening, frightening — momentary and sudden.

Now all sound returns. The screaming whoosh of the engine struggles to catch up with its source; a low, steady grumbling as the secondary shock waves bounce off the surrounding terrain.

A momentary look of shock, and then a smile. Everyone turns to his neighbor, and affectionate laughter breaks out. Darryl has done it — he has broken the record. It is over.

In a span of less than seven seconds. Darryl Greenamyer brought to a climax over ten years of work and worry. He has made a successful assault on the world jet speed record.

One of the observers on the lake bed, operating a portable ham radio inside of his motor home, is talking to a neighbor of his in Baja California during the run. As Greenamyer approaches, Ed Everett tells his neighbor to stand by; he is going to key the microphone to transmit the event live over the airwaves. The blast of the sound barrier being shattered is carried over the frequency. Everett speaks into the mike to his friend: "You've just heard history in the making."

"That was real nice, real nice," answered his neighbor.

Indeed. Across this flooded dry lake bed twenty miles outside of the small Nevada desert town of Tonopah, Greenamyer flew an arrow-straight course at less than 100 feet above the rippled surface at a speed of well over 1,000 mph — more than enough to make him the new holder of this tough and significant speed mark.

The greatest significance lay with the aircraft he used. No mere flying machine, this. It was the specially built, specially modified Lockheed F-104 Starfighter that Greenamyer spent the past decade of his life constructing with a painstaking thoroughness — a labor of love aimed at this one moment in time. (It was also a labor that nearly cost him his eyesight when he inadvertantly arced two high-voltage cables only inches from his face during final assembly. Even with dark glasses, his eyes were excruciatingly sore for days.)

The final result was an exquisitely beautiful machine; the brilliant red

Greenamyer kissing the earth at the normally quiet Tonopah Airport strip, his fuel nearly exhausted. In an effort to reduce excess weight aboard the craft, no drogue chute had been installed, necessitating some very "hot" landings.

with white and gold trim paint scheme masking the brute power of the massive J-79-10 engine encased in her smooth shell.

Airplane and man were after what is officially categorized as the Class C-1, Group III low altitude jet speed record; a category established by the international organization which has officiated, sanctioned, and governed all aviation and space records since the inception of manned, heavier-than-air flight: the Federation Aeronautique Internationale (FAI).

Requirements for this class call for an "unaugmented" (non-rocket) jet-powered aircraft to make four consecutive passes, two in each direction, along a 3 km (1.86 miles) straight-line course at less than 100 meters (328 feet) above the surface.

In addition, at the end of each pass, when the pilot must wrestle his supersonic bird in a tight 180 for the next attempt, the aircraft must be kept within a 500-meter (1,640-foot) altitude envelope. This is to prevent the obvious advantage of beginning the run at extremely high altitudes, thus allowing an exaggerated speed to be reached at the end of a descent.

(Watchdogging the Starfighter at the two key altitudes are, in a flash of contrast, a Cessna 150 and 172 with official observers aboard.)

The course itself — a mere straight line — is relatively simple to prepare. At least it would have been without the annoying interference from the days of rain. Surveyors dispatched to set up the course arrived at the peak of the storm activity and were forced to wade through ankle-deep mud as they lined up their scopes and plumb lines.

At each end of the course are the camera stations; 16mm high-speed cameras focused exactly perpendicular to the course-line at the 3 km marks. Tied into the cameras is a digital timer. The whole setup is akin to the system used in the Olympics: millisecond time readouts from the timers are superimposed on the film, thus marking the precise beginning and end of each pass. This film is later analyzed and the time converted into a corresponding speed figure. Only then can the attempt be certified an official success, and Greenamyer's name added to the roster of record holders.

(This will mark Greenamyer's second entry in the FAI record listings. He successfully flew his six-time Reno Air Race champion F8F Bearcat to a new absolute speed mark for propeller driven aircraft in 1969, with an average of 482.462 mph. Ironically, that record will soon be assaulted by another of Greenamyer's sponsor's fleet of exotic aircraft; the sleek, contra-rotating, modified Mustang RB-51.)

The current record for this class stood at just under 903 mph, set in 1961 by a Navy F4H Phantom. To be awarded the new record, Greenamyer would have to better that speed by 3 percent, or a 931 mph average. No sweat, thought the former Lockheed SR-71 test pilot — he was consistantly breaking the thousand-mile mark during his test runs earlier in the week.

"Only 1,000 miles per hour?" asks one incredulous soul upon hearing of the attempt. "Gee, jets fly that fast, and faster, every day."

To be sure, our military friends go supersonic daily in their sleek, sophisticated machines. But, pause and consider that Greenamyer's Starfighter would have to make its attempt on the deck. Even at the 5400-foot elevation of Tonopah, the air is very dense; considerably more so than the 30- and 40,000-foot environments such aircraft usually frequent. Furthermore, just flying at over a grand only 100 feet or so above Mother Earth is an achievement in itself, let alone a record.

"And you say this guy Greenamyer built the thing himself?" continues the skeptical questioner.

Yep. The 38-year-old test pilot began assembling parts a decade ago, scavenging from all over the country the exotic hardware necessary to make his dream a reality (including sections of fuselage salvaged from four wrecked Air Force Starfighters).

He set up his operation at Van Nuys Airport, north of Los Angeles, and began to methodically piece together his own jet interceptor.

It was in the last few months of Spring, 1976, that the bare aluminum hulk began to take a final shape. Now it was affectionately known as "the project," and crewmembers Roger Davies and Phil Greenburg, with the help and technical expertise of men like Bob Allen, Bob Hariston, Bruce Boland, Pete Law, Jim Black, Randy Scoleville, and others — all swept up in the mystique of this unique affair — put in double overtime to get the final nuts and bolts in place.

A flawless paint scheme was added in late May, a magnificent job of application performed by a local auto shop owner known as Fudge.

The premiere unveiling of the Starfighter took place during the California National Air Races at Mojave in early June.

All was now ready — with one small exception: a gaping cavity remained in the aft section; a cavity that should have housed the jet's mighty engine.

Starfighters still make up the air defense fleet for many nations, and a variant of its General Electric-produced powerplant is used to power the Phantom fighter, among others. So, engines are plentiful. But how many civilians try to buy one?

Two dates for the record attempt — July 4 and August 15 — were cancelled; an engine still had not been procured. Greenamyer toured the country, talking with the brass in the Pentagon, and with the corporate giants who either manufactured, serviced, or used the J-79 for test purposes. No-go all the way down the line. Speculation had it that no one in the top echelons wanted to assume any sort of liability in allowing this one man, this civilian, to have one of their engines.

Federal laws prohibited Greenamyer from attempting a purchase from the military inventory of a foreign goverment. So the ten years of work and sweat, not to mention the money, were about to go down the tubes.

Meanwhile, Greenamyer had signed with Ed Browning, the owner of the Idaho Falls, Idaho-based Red Baron Flying Service (the folks who brought you the Red Baron Air Racing Team) in a sponsorship agreement in early May. ("Darryl was in a pinch, and I happened along at the right time," said Browning in a recent interview.) With Browning's financial and logistical support, Greenamyer made one last, desperate effort.

The engine was finally installed in late summer at the Starfighter's new home in Idaho Falls. The source of the powerplant, which was only loaned for the record attempt, remains a closely guarded secret; a secret that is reported to be an integral part of the loan agreement.

No matter. The bird had her feathers now. She lifted off for the first time in mid-September.

"It's been 15 years since the Navy set the record. Why such a long interval?"queries our skeptic.

Well, until Greenamyer, no one had an aircraft capable of such speeds and performance except the military. And why should they want to break their own record?

Indeed.

The speed mark is only the first of Greenamyer's targets. His sights are now set on shattering the world altitude record, this one held by the Soviet Union in their sophisticated MiG-23 "Foxbat," which flew to 118,000 feet in 1973. (One source has reported that the service ceiling for a stock F-104 is a mere 60,000 feet.)

Greenamyer's biggest glitch in his record attempt (aside from a

reluctant APU starting unit which had people running for cover) was a fuel flow problem. During one test run, his deceleration upon leaving the course was so great that the aircraft flamed out (jet talk for an engine abruptly quitting), the remaining fuel having been thrown forward in the tanks, uncovering the boost pump intakes. He successfully re-lit and made for home.

Later, he would be plagued with an oscillation in the afterburner ("it chugged along like a Model T," he quipped afterwards), again, the probable victim of a suspect boost pump system.

Fuel was a key factor throughout. Although topped off with over 1100 gallons (6,800 lbs.), the Starfighter would be flying on fumes upon its return to the usually quiet Tonopah Airport strip only 15 minutes later — the high power settings and low altitude gobbling up every drop in the tanks (tip tank nacelles were absent in the effort to streamline and lighten the 19,500 lb. gross aircraft).

Furthermore, the borrowed dash-10 engine was nearly 500 lbs. heavier than the standard dash-7 powerplant that Greenamyer had been counting on. To compensate, an equal amount of lead was added in the nose cone — 1,000 lbs. of dead-weight for the stubby-winged craft.

As this one, lone man taxied out and applied power for a thundering takeoff, one question remained: why?

Few people will ever get a direct answer from Greenamyer himself. He is a reserved, private individual — with only a very small circle of close friends and advisors privileged to hear his inner thoughts.

To answer this final, and perhaps most important, question, one must speculate, relying upon this quiet man's record. He is a doer, a challenger, familiar and comfortable with the world of high performance and the unknown. He is not a "playboy," or an adolescent-minded *nouveau riche* by any means. Greenamyer has worked hard for everything he has ever accomplished — and will always continue to do so.

For the chance to pit his skills in a calculated attempt against the odds, strapped in the driver's seat of a glistening red missile, he has weathered the constant storm of tight money, military aloofness, government indifference, and cooporate power and influence — all of which were obstacles. He ultimately overcame and beat those odds, without sacrificing his own pervasive and strong convictions of the individual.

A generation ago, a famous columnist eulogized the passing of another conqueror of the unknown. Although these are not his exact words, these are the thoughts which can apply equally well here:

Man is not mere dust on this Earth; no mere cog in the daily machinery of life. If his spirit is allowed to be free to search the unknown, for those higher levels of achievement, he then becomes a glowing ember, occasionally fanned into brilliant flame by the winds of the endless sky.

OSHKOSH '77

Reprinted by Permission
AVIATION NEWS (September 77)

Edited for Space

OSHKOSH, WI — When you get to the Oshkosh area, ATC tells you: "Report over Omro, then follow the two lane highway to the Howard Johnson's, then turn left to the blue water tank." Then, "Will the yellow Breezy follow the green Cessna that's following the white Rockwell 690A." Then, "North Central 727, you're cleared to land but watch for the 1911 Curtiss Pusher that's out over the lake."

Honest to goodness, that's the way the ATC dialog goes around Oshkosh at the time of the Experimental Aircraft Association's convention each July.

Once you've surmounted the hurdles of operating in an environment in which there may be literally hundreds of aircraft in a very small piece of airspace over this small city, you're ready to land at the convention.

You turn final, and there before you lies the greatest panorama of general aviation anywhere in the world. Thousands of experimentals, antiques, classics, warbirds and visiting aircraft of every description cover Oshkosh's Wittman Field. Plus, more people, cars, trucks and campers than you've ever seen in any one place before.

This year, like last, the weather gods were not kind to Paul Poberezny, EAA's founder and only president.

The persistent rain failed to dampen the spirits of the EAA'ers and their guests, however, for on the first day, 600 display aircraft arrived. The traffic was so heavy that the tower couldn't log all of the aircraft movements.

They came from all over. A Ryan PT-22 from Sarasota, FL, a Howard DGA from Milliken, Ontario, Canada, an Ercoupe from San Marcos, CA, a P-51 Mustang from Dallas, TX, and countless others.

By the end of the convention's second day, the display aircraft total had risen to 874. Movements totalled 9343 and visitors were pouring in.

By the end of the fifth day, EAA officials estimated that attendance had topped the 320,000 mark, a convention record. Original projections called for *only* 300,000 through the whole eight days of the convention. Over in the transient parking area, more than 5000 airplanes were tied down. Ten thousand people were installed in dormitories at the University of Wisconsin's Oshkosh campus. Twenty to thirty thousand people were camping, and every motel within 20-30 mi was full.

Fifty Australians had made the long voyage from Down Under, plus others from Britain, France, Germany, Spain and Latin America. At the convention's end, foreign visitors totaled 584 people from 41 countries.

To handle this enormous influx of people and airplanes, Wittman Field became a substantial, self-sufficient city in its own right. There were hospitals, first aid stations, fire services, movie theaters, ham radio network, CB network, grocery stores and everything else needed to sustain a modern, if temporary, city. This city was manned almost exclusively by more than 500 EAA volunteers.

Plus, of course, the incredible FAA tower crew which operated 24 hours a day. To handle the load, an additional 27 highly qualified controllers were brought in, and the standard FAA procedures bent a bit to fit the special circumstances.

FAA tower chief Russell Lincoln and his people handled more than 75,000 aircraft movements in eight days. In FAA ranks, to serve as a temporary at Oshkosh during convention time, is an honor, indeed.

This was the 25th anniversary of the EAA which has grown to be the largest and most influential sport aviation organization in the world today. EAA was formed in January 1953, and since its inception, has issued more than 115,000 membership cards. Active members are reported to total more than 50,000 today, with members scattered in 550 chapters around the world.

EAA activities include political lobbying in Washington, extensive educational activities for its members and operation of the world's largest non-government funded aviation museum at Franklin, WI.

There is a certain attitude among the sport aviation enthusiasts who make up EAA which is hard to describe. Perhaps it can best be called *pride*. Pride, not only in the aircraft that they build or restore, but in things like keeping the convention grounds immaculate. In spite of nearly 350,000 people trooping over Wittman Field in poor weather, there was scarcely a scrap of paper to be seen anywhere.

1977 Warbird Grand Champion John Harrison's North American AT-6A —Ted Koston photo

To better serve the specialized needs of its members, EAA operates three subsidiary divisions: Antique/Classic, International Aerobatic Club and Warbirds of America.

Each of these divisions has its own program of activities at Oshkosh. The Warbirds gather at the west end of the field, the antiques and classics at the east, while in the middle is the huge gathering of experimentals.

One of the great joys of visiting Oshkosh is wandering through the display areas. There, you'll see almost every type of aircraft of which you've ever dreamed — probably in immaculate condition. The spectator areas hold their own riches, for many with interesting airplanes don't join in the formal activities. Aircraft of every age and type are parked there, ranging from the old familiar flivvers to the expensive modern transportation types

1977 Custom Built Grand Champion Roger Rourke's Starduster Too —Dick Stouffer photo

like Beech King Airs and 690 Commanders.

The EAA has an extensive competition program to determine who has the best of type whether warbird, antique, classic or experimental. These prizes are eagerly sought and the competition produces standards of construction and finish which put many factory-built products to shame. Many experimentals feature technical innovations to challenge the imagination of any pilot. Antiques, classics and warbirds are restored to better than new condition.

The Oshkosh umbrella covers many interesting side events, such as the EAA-AC Flight Rally. On July 30, pilots converged on Oshkosh from four points: Flint, MI, Kansas City, MO, Minneapolis, MN and Dayton, OH. Sponsored by the AC Spark Plug Division of General Motors, the rally serves as a kickoff for the convention, with participants competing for various trophies awarded for piloting skill and aircraft capabilities.

Another interesting event was the flight of the Ryson Cloudster from San Diego, CA to Oshkosh. The Cloudster flew 1647 mi in 32.16 hrs of flying time — 13.82 hrs of which were under power and 18.34 hrs with the engine off and the prop feathered. The average ground speed for the entire flight was 51.21 mph, with altitudes ranging from sea level to 16,800 ft.

The Cloudster is a motorglider developed by T. Claude Ryan, the aviation pioneer whose firm built Charles Lindbergh's Spirit of St. Louis.

While the Cloudster flight was at the forefront of aviation technology, demonstrating incredible efficiency, EAA also reached into its supply of nostalgia to commemorate the 50th anniversary of Lindbergh's flight. The EAA Museum built a replica of the historic aircraft which, accompanied by a Stinson SM-8A, is making its way around the 48 continental states. Naturally, a stop was made at Oshkosh so that the visitors could see the aircraft.

EAA also commemorated the early days of air racing with the "Golden Age National Air Races Day." Honored guests included many of the great pilots and designers of the 1920s and 1930s.

Among the notables was Cliff Henderson, manager of the National Air Races during their long run. Other figures from air racing history included Roy Russell, Jimmy Haizlip, Gordon Israel, Al Menasco, Matty Laird, Bob Granville and Harold Neumann.

Of course airplanes of the period were there, too, including Bill Turner's fine Miss Los Angeles replica and the original Steve Bonzo.

EAA and the Confederate Air Force collaborated to produce a pair of warbird spectaculars, including

"Twelve O'Clock High" and "Baa, Baa Black Sheep." Pappy Boyington was there, along with the P-51s, B-17, P-38, Hellcats, Wildcats and Corsairs that were the real stars of the show.

The sight and sound of these airplanes brought back many memories to the veterans in the crowd, especially the sight of the B-17 making a low approach with one main wheel down, one up and smoke pouring from a feathered engine.

At one point, the CAF's B-17 produced some unintended excitement for its final landing was made with a real emergency. No play acting. One of the four mighty engines had gone to sleep of its own accord. Such is the power of the CAF, however, that the big bird flew away the next morning with all four running.

Almost all the airshow performers in the Western world were on hand. Art Scholl, Bob Hoover, Bob and Pat Wagner, Debbie Gary, Bob Lyjack, Duane Cole and all the others. Show aircraft ranged from the ubiquituous Pitts to the Bellanca Viking and Bob Lyjack's venerable 48 year old Waco Taperwing. The Canadian Snowbirds' Avro Tudor jet trainers provided counterpoint to the big Waco's rumbling radial.

One of the most entertaining events of the convention was the "History of Flight" in which aircraft of all periods paraded in stately fashion before the huge assembled crowd. Participants in the flybys ranged from a 1911 Curtiss to the latest exerimentals and factory-built ships.

There was a serious side to Oshkosh, too. The National Association of Sport Aircraft Designers (NASAD) held its annual meeting at the convention. This group, made up of distinguished sport aircraft designers, reviews and grades plans for sport aircraft. Their aim is to provide standards for judging both the technical quality of an experimental aircraft and the ease with which it may be built. To date, 14 aircraft have received the NASAD seal of approval.

EAA's Air Museum Foundation also sponsored a series of workshops, including beginning and advanced welding, wood working, metal working, synthetic coverings and fabric work.

Eight days of compressed aeronautical experience. Eight days of feasting for the sport aviation buff, the antique/classic lover, the warbird worshiper.

What was most exciting this year? Well, there were so many, appealing to so many different tastes, that it's difficult to say.

The ultralights were popular. The Raz Mut, a product of Centre de Recherche, Drummondville, P.Q., Canada impressed with its relatively high performance. The ultralights offer a low-cost entry into aircraft construction and flying that is not available with conventional aircraft.

Frank Christensen introduced his spectacular new Eagle II aerobatic ship. Intended for amateur construction, the Eagle II is available in kit form with a choice of 200 and 260 hp powerplants.

It is said to be a competent competition machine, yet docile enough for relatively low-time pilots.

Since Pitts is also about to introduce a 260 hp machine, the competition for aerobatic aircraft sales should heat up considerably over the next few months.

For those would-be or former fighter pilots, Windstar, Boise, ID offered an approximately one-half scale F-80 to be powered by the Davis "Cold Jet" powerplant. The Cold Jet is a fanjet-like propulsion system which, rather than having a jet engine as its core, uses a 350 cid Chevrolet automotive engine to drive the fan. Bob Nikkel, Windstar president, told us that the engine is producing about 300 lbs of thrust which will translate into 300 mph speeds in the aircraft. Its power-to-weight ratio is said to be 4.3:1, or better than the Bede jet's.

Another highlight was Joe Alvarez' 200 kt turbocharged Revmaster powered Polliwagen. This tiny two-seater employs a sailplane-like Wortmann airfoil for efficiency, plus flaps which can be reflexed from -10 to +80-deg in flight. If 200 kt speeds on 4.5 gph were not enough, Alvarez is also planning space-age instrumentation which will provide digital display of all important temperatures and pressures, plus fuel, time and electrical system data.

And while the newcomers attracted much of the attention, the hardy perennials also received their share of admiration. The opportunity of talking to veteran experimenters like Percival H. Spencer, designer of the Republic Sea Bee amphibian, and more recently, of the Spencer Air Car, is priceless.

That's a major part of the attraction of Oshkosh: interesting, accomplished people. Not only famous designers, airshow performers and air racing greats, but also men and women who've worked for years to build the airplane of their dreams, or to restore a fine vintage machine to its former splendor. It's interesting, too, to talk to the visitors who've come by air or highway, to see what brought them to Oshkosh, to see what they got out of the show and to see if they're going to become involved in the sport aviation movement.

That's what Oshkosh is: involvement, in one of the fastest-growing segments of aviation.

Flying for fun is a religion whose time has come and whose disciples are everywhere. Almost 350,000 were at Oshkosh. They came from 42 countries around the world to experience the greatest feast that sport flying has to offer. And, they went away satisfied.

TODAY, THE WORLD. . . TOMORROW??

by Jack Cox
from tapes by Don Taylor

Reprinted by Permission
SPORT AVIATION (Feb—May 1977)

Edited for Space

On August 1, 1976 Don Taylor lifted his T-18 off Runway 18 at Oshkosh, Wisconsin's Wittman Field . . . and early in the afternoon of September 30 he touched down again on Runway 22, completing the first round-the-world flight in a homebuilt aircraft.

Don and his Thorp, *Victoria '76,* had flown almost 26,200 miles over a period of 61 days. This is the story of what transpired between that take-off and subsequent landing at Wittman Field, taken largely from tapes made by Don after his return to his home in Hemmeet, California.

—Jack Cox

THE DECISION

When did Don decide to make a second attempt at a round-the-world flight? Those who know him say probably about two seconds after he wisely decided not to try bucking seven layers of clouds loaded with ice and snow between Japan and the Aleutians that late summer day in 1973. Don is a lot of things, but one thing he is not is a quitter.

Oshkosh '76 became the target date. . .August 1, to be precise. N455DT also got a new name. "Victoria" became "Victoria '76", for obvious patriotic reasons.

OSHKOSH

At Wittman Field, Don was being interviewed, photographed and generally fussed over by the working press . . . then, inevitably, it was Sunday, August 1.

Victoria '76 was pushed out to the taxiway and a fuel truck was backed into position. Very carefully the various tanks were filled, accompanied by a lot of rocking of the wingtips to be sure gasoline was filling every nook and cranny inside the sealed wing panels.

At 9:30 A.M. the Lycoming barked to life and Don wasted little time taxiing out on Wittman Field's wide north/south runway and giving the little bird the needle. The heavily laden Thorp had a take-off roll of 25 seconds . . . not bad at all . . . and lifted off just after 9:35 A.M. CDT.

August 1st — Oshkosh, Wisconsin to Burlington, Vermont
730 Miles — 4 Hours 45 Minutes

Overnight . . . last night on U.S. soil for 51 days.

August 2nd — Burlington to Moncton, New Brunswick
400 Miles — 2 Hours 40 Minutes

Cleared Canadian Customs and had the aircraft checked for the North Atlantic crossing.

"I got approval to leave Canada, fly the Atlantic and to re-enter Canada at the end of the flight."

August 2nd — Moncton to Goose Bay, Labrador
470 Miles — 3 Hours 40 Minutes

Held up by weather for two days. Met two couples flying to Europe in a light twin — would leave together for Greenland on the fourth.

August 4th — Goose Bay to Sondre Stonde (Sonderstrom), Greenland
1150 Miles — 8 Hours 20 Minutes

Cleared to Narssarssuag, Greenland, Don climbed out to begin his first significant overwater leg of the flight. This day would see the second longest flight he would have to make . . . it would be a day he would recall more vividly than all others when at last he had circled the earth.

"Last time I just sailed up the fjord, circled the field and landed . . . that's all there was to it. This time I was IFR the last 100 miles in. I let down on the Sumatak beacon . . . turning back out to sea as I descended.

"I could see about 1/2 mile, so I turned back into the beacon and flew back about half as long as I had flown out.

"I never was able to see the cliff walls of the fjord . . . so I turned back and headed out to sea again. I called Sumatak and told them (the Danish controllers) I would have to divert to my alternate, Sonderstrom."

During this aborted approach to the mouth of the fjord leading to Narssarssuag, Don had been in radio contact with the pilot in the twin he had met earlier at Goose Bay. That aircraft had left shortly after his own departure and was holding at 9,000 feet to see if Don could grope his way in for a landing.

"The twin kept asking how I was doing. I told him I couldn't make it and that I thought we ought to head north. He said, 'Thank you,' and started calling for a change in his clearance. The Danes instructed him to maintain 9000 up the coast to the Kook Islands and from there to head inland to Sonderstrom. 9000 feet is

the minimum for overflying the Greenland icecap, so I crawled up from sea level to that altitude also and headed north. Pretty soon I began to notice some ice building up on the airplane . . . then a little more . . . and then it began to snow.

"Things were beginning to get uncomfortable. It was 400 miles to Sonderstrom . . . and I couldn't tell the Danes about the ice because this would panic them. At that point one person in panic was enough! Besides, I probably couldn't have gotten in a word anyway. The pilot of the twin had his wife and another couple on board and was really hollerin' about the ice he was picking up.

"So, I just turned 90° to the ice cap and went out and down for about 10 minutes. At about 4,000 feet the ice melted, just like that, and I finally broke out at 3,000. Off to the right were the Greenland cliffs . . . a pretty sight, I can tell you.

"I could see for four or five miles, so I turned and began to follow the coastline north. I was quite happy about how things were going . . . but my friend in the twin was still hollerin'! I called him (and the Danish controllers) saying I had found a hole and had descended to 3,000 where I was in the clear.

"At the Kook Islands a Danish helicopter pilot came up on the radio saying he had just come over from Sonderstrom and that at 5,000 feet you could see 5 or 6 miles. He also told me the heading from the Kook Islands to Sonderstrom was 54°. I looked at my chart . . . and found that it stopped at the Kook Islands! Because of the crowded conditions of the Thorp's cockpit, I stored the charts I wasn't using under the seat cushion and sat on them — and at the weight I was flying on this trip, I couldn't turn loose of the stick long enough to retrieve them. So, I had little choice but to turn to 54° and head up and over the glacier . . . and hope the helicopter pilot wasn't feeding me a line.

"At this point I realized I hadn't asked how far it was to Sonderstrom . . . but it really didn't matter because I was committed. I had now been in the air for over eight hours and with the fuel I had left there was no going back to Canada. After about 30 minutes I tuned my ADF . . . and hoped. Then 'zip' the needle locked on and about 10 minutes later there the runway was — right over the next hill. What a beautiful sight!"

August 6 — Sonderstrom to Keflavik, Iceland 850 Miles — 5 Hours 40 Minutes

Don's route to Iceland was straight up and over the great Greenland ice cap. He climbed to 10,000 and due to blowing snow off the glaciers below kept creeping upward, eventually to 15,000 feet. Here he went on oxygen.

"I was cruising along and everything was O.K. But after a while a big cloud showed up right in my path. I didn't want to fly through it because I knew I would get icing again — but I couldn't get over or around it and when I looked back,

A chunk of the Greenland icecap afloat in the Atlantic. An iceberg as seen from 12,000 feet off Greenland.

clouds were building up there, too. Finally, I noticed my OAT and I read -27. I just sailed right into that cloud . . . it was so cold it just couldn't ice! All that happened there was that my toes were real cold. I had on lots of clothes and 3 pairs of socks, but after 4 hours I thought I was going to lose my toes."

August 7 — Keflavik to Glasgow, Scotland 830 Miles — 5 Hours 10 Minutes

Stayed with the U.S. Navy at Keflavik. This was to be a speed record attempt and as luck would have it Don had a good tailwind. He has claimed a record over the course of approximately 160 knots.

"Yeah, but it was downhill all the way!" Don says today.

August 8 — Glasgow to Leeds-Bradford, England 150 Miles — 1 Hour 20 Minutes

This was a short leg to Leeds-Bradford. His departure was carried on British television . . . a very important key to the ultimate success of the world flight, Don was later to learn.

August 9 — Leeds-Bradford to Munich, West Germany 700 Miles — 5 Hours 49 Minutes

Don filed for Venice, Italy and started out across Europe. Near Munich he ran into rain, which quickly turned to sleet, then ice. Deciding that the philosopher who said "discretion is the better part of valor" was surely a single engine instrument pilot, Don landed at Munich.

August 12 — Munich to Kerkira, Greece 700 Miles — 5 Hours 5 Minutes

Leaving from Munich, there was little point in making the short hop to Venice, so Don filed for Brindisi, a city on the Adriatic almost at the end of the heel of Italy.

Once past the Alps the weather became beautiful. Don leisurely cruised down the west coast of Italy, but as he approached Brindisi, a near frantic controller started calling him. The message in a heavy

Italian accent came out something like, ''nogassabrindisi, nogassabrindisi.'' Don finally realized he was being told there was no fuel at Brindisi . . . so with a good reserve on board, he simply refiled for Kerkira, Greece, an island just across the Strait of Otranto.

August 13 — Kerkira to Larnaca, Cyprus 700 Miles — 5 Hours 24 Minutes

A beautiful flight across Greece and the storied Agean Sea and on to the troubled island of Cyprus, landing at Larnaca.

August 14 — Larnaca, Cyprus to Diyarbakir, Turkey 485 miles — 3 Hours 30 Minutes

Before leaving California to fly around the world, Don had talked to a pilot who had attempted to fly a Piper Super Cruiser along a similar route. One of the things they discussed was the problem of Cyprus. The Greeks control the western side of the island and the Turks the east. Neither country will clear aircraft into the other. The Super Cruiser pilot had found a way, however, . . . file for an intersection 60 miles off the coast of Cyprus, cancel the Greek flight plan by radio, turn left and fly back into Turkish air space!

This Don did and the flight went smoothly . . . until he approached Diyarbakir. The airport there is a military field and at first the tower refused permission to land. However, a lot of chatter on the radio eventually convinced the Turkish controller that Don did indeed have prior clearance.

August 15 — Diyarbakir to Tehran, Iran 700 Miles — 5 Hours

The flight to Tehran was long but uneventful . . . until Don entered the pattern for landing.

"Tehran was bad 3 years ago, but today it is even worse. Before, I had trouble getting in but this time I darn near got myself killed! I was cleared for a final approach — and then was asked to do an ADF approach. Heck, I didn't have the plate out and I didn't know the approach, so I couldn't comply. So he says, 'O.K., go . . . go! go!'. All at once, swish, a Hercules went right under me! I turned and looked over my shoulder and could see two F-4s coming around the corner . . . and a 747 behind them!

"I decided I better get the heck on the ground and out of the way. I got down and turned off in about 1000 feet.

"Once on the ground I had a little trouble with my clearance, understanding the locals and in paying my bills, etc. . . . and then there was NOTAM 151100. Before I left the states I had been cleared to land at Tehran but I was told I would have to comply with NOTAM 151100. The FAA tried for two weeks to find out what the NOTAM said, but couldn't. I never did learn what it said . . . until I landed. Then they tell me I can't park the airplane there overnight. Great! Here I am in the middle of nowhere

Arid desert landscape stretches away below the wing of Victoria 76.

and they tell me I can't stay overnight!

"Well, I just hid the T-18 under some large Iranian airplane and, fortunately, no one noticed me."

August 17 — Tehran to Zahedan, Iran
710 Miles — 5 Hours 8 Minutes

"Last time I had cleared direct to Zahedan — right out over the desert. Well, this time I had met an Irishman in Tehran who had been flying in the area for about 30 years. He told me, 'No way! You don't fly across that desert.' Then he started telling me a lot of horror stories about the worst desert in the world!

"Well, gee, I thought — last time, in ignorance, I had just cruised across without giving it a thought. But now he had me worried! So, I took the Irishman's advice and detoured 100 additional miles around the desert, overflying three cities on the way.

"At 15,000 I was in such turbulence that I once almost lost the airplane. And haze! I've never seen anything like it. I was solidly on the gages for about 500 miles. Most deserts are blessed with clear air, but this one . . . well, the Irishman was right, it is the worst desert in the world!"

August 18 — Zahedan to Karachi, Pakistan
490 Miles — 3 Hours 12 Minutes

"Troubles start now. I'm back in the Middle East, see. Well, it's the same old thing, except that now they have improved on their tactics! It took me 3 hours to get into Karachi and 5 hours to get out.

"It got so bad that, finally, I went to one of the guys there and said, 'I just don't know how to get out of your airport.' He then took me to an 'agent' — who said, 'Yes, I can get

you out of here . . . for $30(!).' He literally took me by the hand and led me through the bureaucratic obstacle course to its conclusion. It took about an hour.

"The agent's fee, of course, included a little grease for the palms of those officials who wouldn't give me the time of day when I was muddling through on my own.

"When in Rome . . . etc., etc."

August 19 — Karachi to Ahmadabad, India 370 Miles — 2 Hours 25 Minutes

"In Ahmadabad it was the same place, same guy, same taxi fare, same old problems."

August 20 — Ahmadabad to Nagpur, India 430 Miles — 2 Hours 50 Minutes

Hot! And more of the same red tape on the ground.

August 21 — Nagpur to Calcutta 605 Miles — 4 Hours 15 Minutes

"By this time I was getting to be an old hand in Middle Eastern customs. Hired myself an 'agent' (who appeared before I could get out of the cockpit!) and breezed right through the red tape jungle!"

August 22 — Calcutta to Bangkok, Thailand 1,000 Miles — 7 Hours 13 Minutes

"Met 3 Australian gals in a restaurant in Bangkok — tourists on a package tour. Right away one of them asked me if I were married.

"'Yes,' I said.

"'Are you sure,' she said.

"'Yes!'

"'To which one of the others said 'at'ts me mother and ain't she awful!'"

August 23 — Bangkok to Kula Lumpur, Malaysia 770 Miles — 5 Hours 25 Minutes

"Last time I went through some of the worst weather in the world on my way down the west side of the Maylay Peninsula — 4 hours on instruments during which I thought I was going to lose the bird. So this time I cut over and flew down the east side of the peninsula. I cut the bad weather about in half . . . only about 2 hours of bad stuff.

"On the first trip an English airline captain flying for Malasian Air Service, Jim Bartlow, had left his card saying that I should look him up if ever in Malaysia again. Well, upon arrival at Kula Lumpur I did just that.

"I spent a wonderful evening with the Bartlows — saw the retractable gear version of the BD-4 the captain was building and met the only other EAA member in Malaysia."

August 24 — Kula Lumpur to Kuching, Malaysia 615 Miles — 4 Hours 5 Minutes

"At the airport at Kula Lumpur, Jim asked how I was flying to Kuching, to which I replied that I planned to fly down over Singapore

and then across open water to Kuching.

"'No way!,' he said. 'Singapore is a madhouse from ground level to 10,000 feet. Take a 100 mile detour out to sea and avoid it at all costs.

"Well, I wasn't too keen on an additional 100 miles over the ocean, but Jim seemed to know what he was talking about . . . so I took the detour.

"On the way I tuned in on Singapore and, boy, was he right. It was a madhouse! I would take the detour anytime."

August 25 — Kuching to Kota Kinabalu, Malaysia 505 Miles — 3 Hours 48 Minutes

"Making good time now — same airport, same hotel, same nice people."

August 26 — Kota Kinabalu to Zamboanga, Philippine Islands 500 Miles — 3 Hours 50 Minutes

"Somehow the people at Zamboanga knew I was coming. They had me in and out in 30 minutes — including two press interviews!"

Victoria 76 in front of the main terminal building at Yap.

August 26 — Zamboanga to Davao, Philippine Islands 252 Miles — 1 Hour 51 Minutes

"At Davao I had a scheduled program with the Rotary Club, but being a week late, they didn't wait!"

August 27 — Davao to Yap Island 900 Miles — 6 Hours 38 Minutes

"Starting with this leg I began the long overwater flying. I couldn't get a forecast out of Davao, so I set out by dead reckoning. I flew to Yap by way of Palau — with a 30° ADF correction to make when I finally raised Yap.

"Yap was my first landing on a coral runway. Only one airplane was based there, a big twin called the Evangel. The pilot let me put the Thorp in his hangar that night."

August 28 — Yap to Guam 515 Miles — 3 Hours 56 Minutes

"I landed at Agana, Guam and had four wonderful days there. The base commander, Capt. Williams, put all his facilities at my disposal for the maintenance I needed to do on the airplane. Members of the Guam-Agana Naval Aero Club couldn't do enough for me. Their mechanic actually made a set of brake pucks for the T-18!

"Finally, on the 31st I headed for Truk. Now, I'm a fiddler. No matter how far out I am, I start fiddling with the ADF . . . I always want it to come in sooner! About 100 miles out of Guam I tried Truk and got nothing — which was expected.

Then, just for something to do I tried Guam. Again nothing. I hear Guam's signal but the needle won't track. I make the only decision I can make out here — return to Guam. I hate to do it but there was no choice. The ADF was my lifeline out here.

"Back at Guam I discovered the problem was my fault. When I had installed the loop on the bottom of the airplane and had installed the plug into it, I had not pushed it in all the way. The result was about an 1/8 of an inch of the pins exposed. These had corroded. The fine radio man there helped me clean everything up and it worked fine.

"I was extremely fortunate this happened where it did. Any other place, and I probably would still be out there somewhere . . . ."

September 1 — Guam to Truk
625 Miles — 4 Hours 44 Minutes

September 2 — Truk to Ponape
435 Miles — 3 Hours 2 Minutes

"Just out of Ponape, I called in and was told the airport had light rain. I landed and the rain was light at the numbers. At mid field it was moderate. At the turnoff it was pouring so hard I couldn't see the nose of the airplane! I managed to taxi to the ramp — in front of a Continental air liner — but the tower would not tell me where to park. Finally, I just stopped right in front of the big jet. Then they told me what I could do with it!

"Here the educated people won't work with their hands, so no one would help me park the Thorp. The pilot and co-pilot of the jet finally helped me."

September 3 — Ponape to Wake Island
1030 Miles — 7 Hours 31 Minutes

"The long legs were coming up now. My course was laid out to fly to Wake via Eniwetok where the Air Force has a beacon. This way I would have a definite fix before heading out over a really long stretch of overwater flying.

"As I was approaching Eniwetok I was fiddling with the ADF again, listening to the identifer when right in the middle of it — zap! right off the air. Nothing!

"Boy, what a mess! Here I am 300 miles from nowhere. My first thought was that the ADF had gone out on me . . .but I wasn't ready to admit that yet. I had been in contact with Guam, so I called him and asked that he confirm that Eniwetok was off the air.

"He soon came back with the news that Eniwetok was, in fact, off the air. I said to him, 'Please ask Eniwetok to put the station back on the air.' Shortly, came the news, 'They are pedalling!' — meaning they had rigged up a temporary generator and were on again — which I was able to confirm as my ADF needle jerked into action.

"Over Eniwetok I talked to the Air Force and they assured me they would stay on the air until I was on the ground at Wake.

"I picked up Wake from 400 miles out and just sat there and sang songs, praised the Air Force and all that until the island came into view.

Victoria 76 in flight over the Pacific Ocean. Photo by Jack Gilbert

"Wake is a nice little island manned by 20 Air Force personnel and about 200 Filipinos who do a terrific job keeping the island going. The facilities were great . . . but here my real *problems were just beginning!"*

September 5/September 4 — Across the International Date Line from Wake Island to Midway 1140 Miles — 8 Hours 49 Minutes

When Don left the United States on his round the world attempt, he did not have permission to land at the Navy installation at Midway . . . or to depart there for Adak in the Aleutians. This was the critical part of the entire flight. It was the only route across the Pacific within the range capability of the T-18. Don had left on faith that his supporters at home could get the Midway/Adak clearance for him by the time he arrived there. Try as his friends might, however, the Navy could not be made to relent, so 34 days later Don found himself on a tiny island in mid-Pacific — Wake — with some very big decisions to make. He could fly back home reversing the route he had been flying . . . he could fly direct to Hawaii and ship the T-18 home (not enough range to buck the Easterlies to the mainland) . . . or he could fly to Midway without a clearance and take his chances with the powers that be that he would be allowed to leave. For Don only one of the choices was acceptable.

In order not to get the Air Force base commander at Wake in hot water, Don filed for Hawaii, overflying Midway on the way. He had made one last call to CINCPAC in Honolulu in hopes of obtaining permission to land at Midway, but was turned down again. Apparently, however, they knew his intentions for when he reported over Midway, Don was given landing instructions!

On the ground Don was accosted by the authorities even before he could get out of the cockpit . . . "You've violated our airspace" . . . "You have no right to be here" . . . etc. Arrest, fines, confiscation of the airplane and other miscellaneous fates worse than death or even taxes were mentioned over the next few days.

Exactly what transpired over the next 14 days Don has decided should not be told in gory detail . . . because in the end he *was* allowed to go on his way with only an official slap on the wrist. Suffice it to say that his plight went all the way back to the Joint Chiefs of Staff in Washington, to the FAA and the halls of Congress — with pressure being applied by EAA Headquarters, AOPA, Don's home district Congressman, his sponsor and one heck of a lot of friends who wanted very much to see him have a chance to realize his dream. Finally, the Navy went to the NAA in Washington and worked out an arrangement by which persons on official, NAA sanctioned record flights would be able to land at Navy installations . . . and Don would be the first to obtain such permission!

Having already learned in an early morning weather briefing that headwinds were too strong, Don said, "I'm not going."

The following day the winds were still too high and again Don announced his intention to remain an unwelcome guest of the Navy. Next morning, however, the forecast was for no more than 5 knot headwinds along his course . . . and the meteorologists concluded that this was about the best Don could ever hope for.

Shortly after landing at Midway, Don had met some British pilots and their support crew who were operating two Vulcans from the Navy's facilities on a highly classified mission.

"Say, aren't you the bloke we saw on TV in England just before we left to fly out here — the one trying to fly around the world in some sort of backyard aeroplane ?"

From that point Don and the Limeys were as thick as thieves. Shortly after receiving the favorable weather report, one of the British pilots asked Don when he was leaving.

"In the morning," he replied.

An hour or so later a Vulcan took off . . . and ghosted back in some 7 hours later.

"Where ya been?"

"Oh, we just flew a little weather recon up to Adak and back."

The British turned their weather info over to the Navy, who ran it through their computers to prepare a complete winds aloft analysis for the entire 1650 mile leg from Midway to Adak, for 3, 6 and 9 thousand feet!

Don had lost an hour of daylight during his 14 day delay on Midway — down from 13 to 12 hours — so he was out before sunup to prepare for a crack of dawn take-off. Nearby was a Lockheed P-3 also being prepared for flight.

"Where you fellows going?"

"We've got a tactical mission later this morning . . . to Adak."

The P-3 would take-off several hours after the departure of the T-18, overtake him on a parallel course and be in Adak when he arrived . . . a "coincidental mission" for which Don was truly grateful.

The Midway/Adak leg was the emotional high point of the world flight. From a navigational standpoint, it was a near-perfect 11 hour run, but considering the flowage rate of adrenalin, it was something else. At 1650 statute miles it was the longest leg by a margin of 500 miles; it was entirely over water; and, perhaps, as significant as any other factor, it was the great psychological hurdle of the flight. This was the leg that was being substituted for the Japan/Aleutians segment of the 1973 attempt . . . where that flight had ended in failure. This was the barrier that had to be broken. This was Don Taylor's greatest personal challenge.

"Jumping off from Midway without a fix . . . looking over my shoulder and watching that last speck of land vanish over the horizon . . . I did wonder — wonder if I would make it."

That moment of wonder and wondering soon passed, however, as the demands of accurate navigation began to take almost all of Don's time. His enemy was a crosswind component over the entire 1650 miles that varied from 10 to an incredible 50 knots . . . but his weapon to combat it was the Navy winds aloft forecast. Every 5° of latitude Don cranked in the wind correction factor and altered his course accordingly.

The potential loneliness of an 11 hour flight was also precluded on this flight by a friendly voice on HF that started checking in right after his take-off and thereafter every 30 minutes until he reached Adak. Actually it was two voices, sometimes a man and sometimes a woman.

"Thorp Delta Tango, Anchorage."

"Roger, Anchorage, this is Thorp Delta Tango."

That was the extent of the communication . . . and it was all that was necessary.

Morning became afternoon . . . afternoon became evening and the moment of truth was at hand. A flick of a switch would reveal the fate of one very small airplane and its solitary occupant. Would the ADF needle lock right on Adak, indicating that the RAF/Navy weather analysis — and Don's use of it — was on the money? Or would it remain lifeless, dooming Don to a frantic race with darkness, the fabled Aleutian fog and an ever diminishing fuel supply to find some chunk of rock somewhere in the North Pacific upon which to land?

"When the ADF started homing on Adak, I was within 1° of the heading to the airport! That weather analysis I had received was the most accurate one I've ever seen. 1650 miles and right on the money — it was fantastic!"

Don landed with about a half hour of daylight left. Even before darkness the fog rolled in and all but closed everything down. But . . . he had done it. He had beaten the Pacific.

Before leaving Adak Don was able to get some insight into the problems he had been having with the Navy. It seems he wasn't the only person flying the Pacific in a tiny airplane this summer season. Only a couple of months before the same P-3 that had shadowed him on his way to Adak had located another would-be-round-the-worlder in the drink near the Aleutians. An Australian pilot had landed without clearance . . . in a Grumman American Trainer! And Peter Garrison had only recently made a precautionary (and wise) landing at Adak on his way home from Japan in Melmoth. As much as anything Don had simply been the straw that broke the camel's back to the undermanned Navy personnel charged with the search and rescue responsibity for the North Pacific.

September 22 — Adak to Cold Bay, Alaska
650 Miles — 4 Hours 20 Minutes

"I stayed with the Air Force at Cold Bay. The FAA drove 40 miles to get me at the airport and the Air Force the 40 miles to take me back."

September 23 — Cold Bay to Anchorage, Alaska
690 Miles — 5 Hours 30 Minutes

"I pulled a boo-boo here. I tried to go up over the mountains between King Salmon and Homer and got into icing conditions again. I was a pretty miserable rascal for about 20 minutes.

"The head of the FAA at Anchorage met me, and introduced me to an attractive young lady and a very pleasant fellow. We talked for a minute or two and suddenly it came to me.

"I know these people!

"The FAA guy laughed and said, 'These are the two who followed you by radio all the way from Midway to Adak.

"You can believe me when I say those two are my buddies for life!"

September 25 — Anchorage to White Horse, Yukon
520 Miles — 4 Hours

"Well, I had to get down into the flat country of Canada and the Dakotas before I would feel comfortable, so for a change, I did just exactly what the local bush pilots told me to do. I got right down over the Alcan Highway and followed it faithfully. "

September 26 — White Horse to Fort St. Johns, B.C.
640 Miles — 4 Hours 23 Minutes

More of the same.

September 27 — Ft. St. Johns to Edmonton, Alberta
370 Miles — 2 Hours 9 Minutes — And — September 28 — Edmonton to Edmonton(!)
120 Miles — 1 Hour

"After an overnight in Edmonton, I departed from the downtown Industrial airport and after about half an hour the engine got very rough. I turned back for Edmonton and with the situation getting worse by the minute, I decided to declare an emergency and take advantage of

the 10,000 ft. runways at Edmonton International.

"What a reception I got once on the ground — the airport manager, the Royal Canadian Mounted police . . . and seven *fire trucks!*

"In the shop of an FBO on the field, a sharp young mechanic — with a broken arm — diagnosed my problem in about two minutes flat. Fouled plugs. I had flown for the last 70 hours or so on 115/145 av gas — all I could get at the military bases — and the little Lycoming just couldn't handle the high lead content. My broken-wing mechanic (he was better on one arm than most with two good arms) solved the problem by throwing away my plugs and selling me a good set out of his own airplane!"

September 29 — Edmonton to Minot, North Dakota 660 Miles — 4 Hours 20 Minutes

"Back in the good ol' USA! I got a tremendous welcome and a big party."

September 30 — Minot to Oshkosh, Wisconsin 690 Miles — 4 Hours 30 Minutes

Don had many trials and tribulations on his two trips around the world — but Thursday, September 30, 1976 made up for all of them. The morning dawned sparkling clear from the Rockies practically to the east coast. Even winds aloft were in his favor. By 1:00 P.M. TV cameras were sprouting all over the place and newspaper reporters were checking their tape recorders — just a few yards from the parking spot occupied by Don's T-18 two months earlier. Everyone was huddled around a walkie-talkie tuned to the Oshkosh tower frequency, hoping to hear the initial radio contact.

About 1:30 P.M. a familiar voice broke the silence.

"Oshkosh tower, Thorp Delta Tango. Ten Northwest. Request permission for a low pass across the field and a time check for record purposes."

All eyes strained for the first glimpse of the T-18 . . . and, finally, there it was — lower than expected — aimed right for the tower. A cheer went up as Don cruised by. Thorp Delta Tango squeaked on at 1:48 P.M., about 12 minutes ahead of his ETA . . . and the deed was done. Don Taylor had become the first person to fly a homebuilt airplane around the world . . . which also made him the first to make the flight solo in a homebuilt and the first to do it in an airplane constructed by the pilot, each a separate entry in the record books.

Records are made to be broken, as they say. Someday, someone will fly around the world in a homebuilt a little faster than did Don. Someone may even do it non-stop.

But no one else can ever be first.

THE FIRST BFA NATIONAL CONVENTION

by Dick Brown

Reprinted by Permission
BALLOONING (July-August 77)

Edited for Space

On the eastern slopes of the Sierra Nevada range, where the Truckee River trickles through downtown Reno, more than 350 aeronauts gathered for the First BFA National Convention, held June 1-5, 1977. With the Holiday Inn (complete with Monte Carlo Casino) as convention headquarters and snow-capped mountains as a back-drop, balloonists from 20 states unpacked their bags and settled in to a week of carousing, balloon flying, and federation business.

"High plains drifting" became a big part of the convention activities at mile high Stead Air Base, site of the National Air Races traditionally held each September. Calm early morning conditions allowed about 100 balloons to drift aimlessly over abandoned runways and interconnecting taxiways at the northwest corner of Stead Field. Landings were easily accomplished amid western Nevada sagebrush and rattlesnake hide-outs. Although the theme of Convention '77 was fun-flying and balloon seminars, there were two BFA-sanctioned tasks, with BFA President Don Kersten and Vice President Sid Cutter "suspiciously" receiving first place honors.

Social activities included two buffet-style luncheons, an informative gambling clinic (although a bit late for some), a western rendezvous at the Ponderosa Ranch, an eye-catching balloon trade show, and a Chuck Wagon farewell brunch. At one luncheon, attended by Nevada Lieutenant Governor Bob Rose, Malcolm Forbes related some of his more memorable ballooning experiences while at a second luncheon, Dr. Dale Topping discussed ballooning and airship history. An evening bus caravan across the Ponderosa (location of TV's legendary Bonanza series) on the north shore of Lake Tahoe ended at the Silver Dollar Saloon (perhaps temporarily known as the Balloon Saloon) where thirsty aeronauts gulped down a few tin cups of western hospitality.

At the convention banquet, keynote speaker Ward Tunte Van Orman of Akron, Ohio described some of his precious ballooning experiences, including his long-held secret of balloon navigation — calling down to the first person you see and asking "where are we," then marking the response on a map, thus developing a track as the balloon drifts from town to town. Van received both his free balloon and airship pilot licenses in 1918, crossed the South Atlantic from Germany to Brazil on the 'Graf Zeppelin' in 1932, competed in 11 National Balloon Races (and won five), and competed in 10 James Gordon Bennett Races (and won four).

An engineer and inventor, Van Orman retired from Goodyear in 1962. He has recently completed a soon-to-be-published book titled "Wizard of the Winds." In his talk, Van reminisced old times, including the time he was forced to make an emergency landing in the dark on the deck of a German freighter during the 1925 Bennett race. He mixed ballooning romance, humor and tragedy and climaxed his tales with a death-defying crashlanding in the Canadian wilds.

On the business side of the convention, the BFA conducted committee meetings and six informative workshops.

Joe Wood, a Denver balloonist and CPA, conducted a workshop on Balloon Accounting which included methods of setting up corporate books, tax advantages, balloon business organization, annual cost analysis, and reporting and filing for balloon businesses, partnerships, and joint ventures.

Jim Winker, Raven Vice President and balloon engineer, chaired a Maintenance and Repair Workshop involving all the need-to-know inspection techniques, maintenance procedures, logbook entries, repair responsibilities, legalities, and the FARs.

Dale Topping, past president of the LTA Society, discussed gas balloons, sport airship projects, and some LTA history at his Airship Workshop. Dale was accompanied by Dave Smith; George Wright of the Association of Balloon and Airship Constructors (ABAC); Roger Wolcott, a noted gas balloonist who recently participated in filming a gas balloon sequence for TV's Code R series; and Zenon Hansen, author of a new book "The Goodyear Airships," who presented a slide show of the historical development of non-military airships.

The earning potential of ballooning as a profession, the hazards of flying commercially, the effectiveness of various types of promotions, and the techniques for calculating expenses were the primary topics of Tom Oerman's workshop on Commercial Ballooning.

Dave Mesnard's workshop on Homebuilts explored the pros and cons of building your own balloon, the pertinent FAA requirements, and some technical information on design, materials, and construction.

Fred Krieg, the chairman of the BFA-FAA Liaison Committee, directed an AOPA/BFA Instructor Seminar. Two sessions were required to explain the proposed BFA Instructor Rating, to present the highlights of the AOPA flight instructor refresher course, to present excerpts from "Fundamentals of Instruction — Part I," and to review the pertinent FARs.

The general consensus was that the BFA's first convention has been a great success, thus paving the way for future conventions elsewhere in the country. A large part of the success can be attributed to the organizers and volunteer workers who poured so much time and energy into the concept, then followed through with detailed planning and just plain hard work. These individuals were honored at the Saturday night banquet.

THE BALLOONATIC ASYLUM

by J. Scott Hamilton

Reprinted by Permission
PARACHUTIST (April 77)

Free fall photography by
Greg Pike

Photographs by
Stan McGrew
Vickie Hendricks
and the author

Looking back on any great adventure, it is usually impossible to put your finger on any single event, decision, or person responsible. Rather, it usually seems that what at first appears to be a series of unconnected events and decisions in the lives of people who may not know each other and who may be scattered all over the continent suddenly and unexpectedly begin to coalesce, gathering momentum, and finally leading inexorably toward the adventure.

My first balloon jump was an adventure like that.

Except for being there and making the jump, I didn't really have much to do with it. I can't really say "It all began when suddenly. . ." First, there was a character named Link (whom I have not yet even met) who would play an important role. A Denver attorney who owned a patch of ground in south central Colorado, he had become a genuine aficionado of ballooning, over the past ten years. Link had a friend, David Levin. During this time, David was working on a Master of Laws degree in tax law in Boston and jumping out of airplanes for grins at the PI Center at Orange. David was in the habit of visiting Link in Colorado each winter for a skiing vacation. Meanwhile, in South Florida, Frank Rider, a commercial balloon pilot, was operating a commercial sport "balloon port" and had, on occasion, flown jumpers in his balloons to help pay the rent. At the same time, back in Colorado, Vicki Hendricks was going to school at the University of Denver and jumping as much as her educational obligations and the usual financial pinch faced by the students everywhere would allow, and bugging every balloon pilot she met for a chance to jump out of his balloon. (The replies were uniformly negative, from the pilot put-off to rude put-downs.)

David was about to graduate, and mentioned to Link that he really didn't have it in his heart to settle down and dig in the practice of law, just yet. Link shared with him an idea he had regarding that patch of ground in southern Colorado, and soon David had bought adjoining acreage at the mouth of the little cottonwood-filled canyon which opened out onto the broad, vast mountain-rimmed expanse of the San Luis Valley near the sleepy farm town of Del Norte. It was a busy summer, as a luxurious lodge and basic supporting structures for balloon storage, classrooms and other necessary facilities were built from scratch. Meanwhile, David was on the spot supervising things and learning how to fly a balloon, himself.

That same summer, Frank brought his balloon to Colorado to fly around the state with friends. He was taken

with the state's natural beauty and the people he met, and began to consider moving up from Florida. He had found a North Carolina balloon manufacturer who was considering opening a blimp manufacturing firm in Denver, and cultivated the relationship.

Any new enterprise seems always to take at least twice as long to get cranked up as the most pessimistic predictions. The Balloon Ranch was no exception, and it was fall before the lodge and support facilities were ready, a full commercial balloon pilot had been hired, and a massive advertising campaign launched. The original idea had been that the spot would be a resort for balloonists, that is people who were already "into" ballooning, to gather. Somewhere along the line, the emphasis changed to attracting people who had never been in a balloon to come out for their first ride. Advertising touted "The Balloon Ranch Adventure" which could combine not only your first exciting balloon ride, but also cross-country skiing, snow-mobiling and horseback riding on the Ranch, and downhill skiing at the nearby Wolf Creek ski area.

About the same time the advertising hit the papers and other media, it turned cold, and the commercial balloon pilot quit and the Ranch was getting more snow than the ski area. The handwriting was on the wall that if the balloonatics expected to survive the winter, they were going to have to get out and furiously beat the bushes for paying customers. Between the demands of the new business and the simple fact that it's a *long* way from the Ranch to any established drop zones, David was getting pretty horny for a jump. Somehow Frank learned that the Ranch was looking for a replacement commercial balloon pilot, and got himself hired.

The University of Denver, formerly the Methodist Colorado Seminary, was founded in 1864, and its hilltop campus is graced by many massive, elaborate stone buildings dating back to its early days. The Balloon Ranch is 200 miles from there, halfway to Albuquerque. About the time that things were getting started at the ranch, Vickie was on her way across campus, to class, when she rounded the corner of a building only to be confronted by a 5-story high, two-tone green balloon. At that moment, David, who was there on a promotion for the Ranch, turned to notice that one of his spectators was a pretty girl with an SCR patch on her jeans.

"*You're a jumper!*" he exclaimed with considerably more delight than Vickie was accustomed to seeing from balloonists. Thus encouraged, they fell into conversation about the possibilities of jumping out of the Ranch's balloons. A relatively modest price (I say "*relatively*" because balloon envelopes deteriorate so rapidly from the heat of the burner and the ultra-violet bombardment of the sun, particularly at this altitude, that their operation is extraordinarly expensive, compared to say, your Cessna 180, which means that hauling anything in them gets pretty expensive) was agreed upon and a December date set. David solemnly guaranteed perfect weather for that weekend, but, of course, no one took him seriously.

In any event, that is how I come to find myself arising at 3:30 A.M. and motoring south in the icy darkness, wrapped as best I can in my down sleeping bag (my VW's heater having chosen this occasion to fail). The sun's first light finds me speeding down the final pass into the silent flatness of the San Luis Valley, covered with about 6 inches of fluffy powder snow. The signs lead me a few miles over bumpy roads to the lodge at the mouth of the canyon. Pulling into the parking lot, I am greeted by four of the hugest, furriest, blackest, and (thank goodness) friendliest Newfoundland retrievers I have ever seen, who are almost immediately joined by a thin fellow with a mischievous smile who says "*You must be one of the jumpers*," in precisely the same wistful manner an insane asylum inmate might greet a new arrival ("*I know you're crazy too, so we can share our little secret in safety*"). He is co-owner David Levin, and he has to be a jumper too, although I do not confirm it until later.

In the lodge thawing frozen feet before the fireplace and warming the innards with a cup of coffee, other jumpers who arrived last night begin stumbling in, along with a mixture of hired hands and local folks coming out for a look.

The overall effect is mellow. Everyone is pleasant, and seems to be quietly enjoying themselves, yet looking around I notice that without a single voice having been raised or order shouted, we are organized, the balloons have been loaded on the trucks, a NOTAM filed, and we are ready to roll.

Normally, a consistent light downslope morning breeze allows the balloons to be launched from the back yard of the lodge. Just to be sure we are jumping over open territory, and not in the rugged foothills surrounding the Ranch on three sides, however, we truck about

Balloon jump coordinator "Miss Vickie" demonstrates the highly infectious balloonatic grin.

a mile out into the valley, to set up for a launch.

The weather this morning is precisely as guaranteed. (Do these guys have a connection, upstairs?) The sky is absolutely clear, and the breeze is 0 to 2. Methodically the balloon "envelopes," as the colorful gas-bags are called, are hand-deployed from their huge storage packs, and attached to the wood and wicker gondolas. The balloons are partially inflated with a small portable fan, then the burners ignite with a roar, and the inrushing hot air completes the job of inflation. We will jump the gigantic pink, red, and blue AX-8, a 105,000 cubic foot displacement balloon which will carry Frank, the pilot, plus four jumpers. David will fly chase and take pictures from the smaller two-tone green AX-7. On a first-come, first-served basis, I will be on the second load, so I will help the ground crew pick up the jumpers from the first load.

The first load is quickly chuted-up and compulsory group photographs in front of the balloon taken. Frank has given all an exit briefing, and with a great roaring of burners and shouting and waving and picture-taking, the balloons rise. They seem to be going no place in a hurry, gradually rising directly overhead. Anticipating a mad cross-country dash to keep them in sight, we are disappointed. Finally, they begin to gradually drift southeastward but we estimate their altitude at only around a thousand feet, and since they have so much farther to climb, there's no need to leave yet. Almost on a whim, one of the ground crew calls them on the CB. They report at

Balloonatics take off for flight across San Luis Valley.

3,500 feet, almost ready to drop. Oops! Suddenly everyone is scrambling into the trucks and we roar off in a cloud of snow, in hot pursuit. The pilot's spotting has been good and everyone lands close to a road. While exuberant, they seem also awed and somewhat overwhelmed by the experience. I am further intrigued.

Soon, those of us in the second load are chuting up and crowding into the balloon's wicker gondola. It is a tight fit, and turning around requires the cooperation of the other occupants. Inches from our heads, the burner roars, making conversation virtually impossible. We begin to rise: one inch, two, a corner of the gondola gently bumps the snow, and there is no more contact. The waving, smiling, yelling, picture-taking ground crew slowly falls

Aeronaut soon to become parachutist examines the land below.

away. The sensation is eerie. It is spectator-like, as though this were not the real world, but its projection on the globe's largest panoramic color TV screen. There is no sensation of motion. Since we drift precisely with the gentle breeze, there is absolutely no feel of wind. A sense of detachment from reality is strong and disturbing. I notice I am tightly gripping one of the gondola's posts. I force myself to release it, can't stand it, and grab it again. The drop zone drifts below, first one direction, then another, and it appears that we will be able to jump right back onto the launching point, until, somewhere around 2,500 feet, we begin that same steady drift to the southeast.

Some prominences below poke through the snow-cover, giving a photographic negative appearance to the landscape. Long afternoon shadows trail from fences, trees, and telephone poles, exaggerating the three-dimensional appearance of the landscape. We are approaching 4,000 feet, and Frank shouts against the roar of the burner that we will be dropping soon and points out a road near which he suggests we try to land. Greg Pike and I will be going out together on this "pass," with him going first and shooting pictures of my exit with his wrist mounted 35 mm. We are shouting at each other at the top of our lungs to communicate over the burner's roar that we will try to land near a farm house I point out. Abruptly, in mid-shout, Frank turns off the burner, and the sudden silence is overwhelming. In the contrast, conversation immediately goes to quiet whispers. It seems somehow profane to make a sound, at all. Frank will establish the balloon in a descent of about 700 feet per minute before we are allowed to leave. As we step off, the balloon, suddenly lightened by some 400 pounds will rebound into a 700 foot per minute ascent, power off. Failure to first establish the descent could overstress the balloon, with disastrous consequences.

Utter silence. The descent has begun, but still there is no breeze, no sound. Frank quietly calls out the increasing rate of descent: "*100 feet per minute. . .200 feet per minute. . . .*" There is no sensation of descending. Seconds pass and Frank says quietly: "*Okay, you can get ready now.*"

Greg climbs up, standing on the rim of the gondola, holding its posts, facing me. I climb up to face him and Frank murmurs: "*Okay, you can go anytime now.*" Adrenalin is racing wildly, Greg says in a tremulous whisper: "*Are you ready*?" The inflection is "*Are you sure you really want to do this*?"

I'm not ready, but, hell, we can't take another pass, so I whisper, equally tremulously: "*Ready.*" And Greg replies with an even lower-whispered: "*Ready; go*" and falls off backwards into space. In the weirdness of the situation, I am an instant slow in following, and the balloon's rebound almost throws me back into the gondola. It is like one of those dreams where you are trying so hard to run, but everything moves in slow motion; I strain against the posts as the delay, which is probably no more than a second, seems an eternity. Finally, I overcome the forces and tip over the side.

Greg is vertical below, and I see him shoot.

The sensations are indescribable. Absolute silence, no wind feel or noise, yet severe instant ground rush (we are at about 3,600 feet). Helplessness. The exit was perfect, and I am stable, but only if I don't move anything. There is no air to fly with, and none to take a pilot chute off my back even if I could not resist the overwhelming urge to pull. I am directly over Greg's back, and he is rolling back to horizontal position, in preparation for the pull. I have plans to count to ten and dump, and my count has reached six before there is the first slight sound and feel of moving air. It seemed as though it would never come, but when it finally does, I am able to track off Greg's back, for the pull.

Parachutes open, we whoop and holler all the way to the ground. Now, I understand the strange reticence of the others. Gathering up the gear and walking over to the road, the first attempt at conversation is: "Man, can you imagine what Owen Quinn felt when he jumped off the top of the World Trade Center?"

The jump was at least as sensational as my very first. This could be addicting. We crave more. I want to make another alone, to simply savor the sensations. I want to make others, higher, to try relative work on pure inertia in that airless void. (This could get to be a very expensive habit. . .)

A ground crew truck meets us with the traditional balloonists' champagne. Delicious: this drink, this day, this place, this experience.

That night, back at the lodge, after baking in the sauna, relaxing in the whirlpool, and feasting on a marvelous dinner, served family-style before the great stone fireplace, we agree verbally on what each of us knew when he landed: we must return.

The Author

J. Scott Hamilton has authored numerous articles for PARACHUTIST. He is a National Director, the Vice-President and Chairman of the extremely important USPA Safety and Training Committee. Hamilton is also an experienced pilot and was an instructor in the Army Special Forces HALO Group.

COMING IN ON A WING AND A SQUARE

Story and Photos by Kevin McCabe

Reprinted by Permission AIR PROGRESS (July 77)

Edited for Space

Three thousand feet above the field, and time to lose altitude. A hard turn — once, twice, three times around with the wing nearly vertical — and the altimeter unwinds. A quick check for traffic during the rollout onto downwind looks good, with only a 182 on short final far below. Just a touch of the controls and the wing slows, sliding easily onto base as the rate of descent picks up. No, too much — better stretch the glide to prevent landing short. Easy now, let it grab some air and pick up the speed again in the turn onto final. It's looking good at 400 feet. . .a bit high at 200, so slow up just a touch. . .75, 50, 35. . .NOW, full stall. . .and touchdown, only inches from the target. Not bad, especially without power.

And certainly not bad without even an airplane!

The "wing" used in that descent isn't a piece of Wichita aluminum, nor is it attached to an aircraft fuselage. It's the newest idea in parachute canopies, created entirely of fabric and inflated by ram air pressure to an airfoil shape. These nearly-square canopies fly much like an airplane without power, and outperform every other parachute in the sky. Combined with other equipment innovations, new training concepts, and an overriding concern for safety, the square canopies have revolutionized the sport of skydiving.

Round canopies, whether the military type modified for sport use or those designed specifically for skydiving, are basically drag devices. Only a small amount of lift can be created from the air passing over the canopy, along with an equally limited forward speed produced by air flowing through the rear drive slots and modifications. Ram air squares, though, produce lift and forward speed just like a standard wing. The fat airfoil shape, maintained by air entering from openings on five or more individual cells, gives a lift-to-drag ratio ranging from around 2.5 in full glide to 4.5 or more at stall. The result is an unprecedented ability to control the descent and landing, including the ability to fly a standard pattern and touchdown feather-light within fractions of an inch of a designated target.

The old attitude of, "If I hit the drop zone that's close enough," has been relegated to the "Dark Ages" of the 1960's. Using modified round military canopies, the drop zone was about all that could be hoped for. Even with the high performance round canopies introduced in the mid-60's for sport use, such as the Para-Commander, "dead center" landings weren't commonplace, though they were certainly more consistently possible. The square canopies have changed that — now the ten centimeter (roughly four inches) target disc is almost ridiculously easy for an experienced jumper to hit. Chuck Collingwood, a member of the Army's Golden Knights Parachute Demonstration Team, used a square to hit 136 dead centers in 141 jumps — and 70 of those dead centers were in a row! Accuracy competition has changed too; at the 1976 Nationals, only *one* round canopy was in the event, completing the changeover to squares begun only six years ago. Officials of the sponsoring organization, the United States Parachute Association (USPA), reportedly are looking at ways to make the event tougher for ram air canopy fliers in the future; ideas have ranged from an even smaller target, to tying the disc on the tail of a fast-moving cat (hopefully that last suggestion was made in jest. . .or was it?).

Of course, hitting the disc isn't possible without being somewhere near the drop zone when exiting the plane. The term "somewhere near" has also undergone redefinition because of the square canopies. With their high forward speeds in full glide, squares have a range of at least twice that of a round canopy making the inexact science of "spotting" the exit point less critical. Pity the poor jumper with the only round canopy on the load. While everyone else is relaxing back at the drop zone, there's a fair chance he'll be in the middle of a cornfield swearing, and wondering if he can hitch-hike back to the airport!

Even though accuracy competition was the first to benefit from the square canopies, other areas of the sport have accepted them equally as well. Style competitors, performing fixed routines of turns, rolls, and loops in freefall, have found the light weight and small size of a packed square to be of great benefit. In all freefall maneuvers, the volume, mass, and location of equipment is a crucial factor. Smaller canopy containers create less drag and turbulence, while lighter weight lowers the inertia to permit faster movements. Elapsed time for the fixed style routine is now as low as 6.6 seconds, roughly twice as fast as the best times only ten years ago.

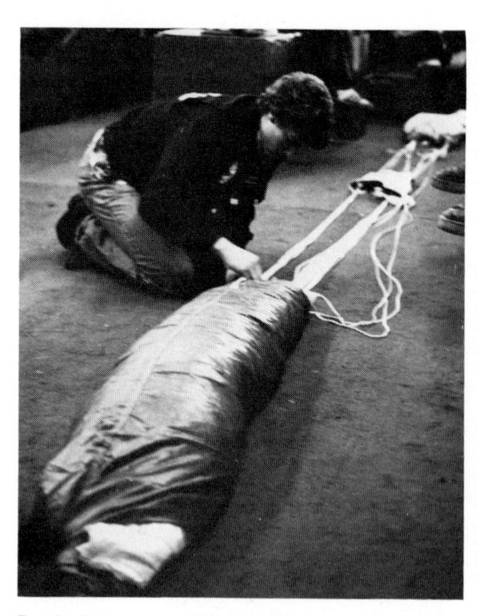

Paul Briggs packing his ram air main canopy. Unlike conventional round canopies, the square mains are often packed in a rolled form, instead of being folded or "flaked".

Even relative work — the art of formation flying in freefall — has benefited from the coming of the squares. The same size and weight considerations important to style competition apply equally to relative work. The square canopies use low-porosity fabric and light-

Toni Couturier checks the lines of a military-type round canopy, often used by sport jumpers without modification as a reserve.

weight suspension lines, reducing their size to the point where both main and reserve canopies can be worn on the back. With such a "piggyback" rig, there's no front-mounted reserve to cause turbulence or burble at terminal velocity. Equally important, such designs reduce the hazards of front-mounted reserves — inadvertent deployment and fouling on exposed hardware — to nearly zero. Other equipment changes, such as highdrag jumpsuits and lightweight harnesses, helmets, and boots, allow an amazing ability to control position and speed in freefall. Times for forming a four-man "star" are now consistently below 4 seconds in competition; even ten-man formations can be completed in less than 12 seconds from exit. Consider that as little as five years ago, even a 20 second four-man star was better than average! However, square canopies aren't entirely without drawbacks. When properly packed, a square is just as reliable as more conventional gear, but the totally different flight characteristics can trap an inexperienced or poorly trained jumper. Downwind landings under a round canopy can easily cause a few bruises; the same landing with a square in full glide can be like stepping from a train at 40 miles per hour, with rather unpleasant results. A number of jumpers have also learned the hard way not to stall the wing too early in the landing approach; unlike a 150, there's no spring steel gear to soak up the energy of bounce-it-on landing. Each of the manufacturers of square canopies recommends thorough training, not only in the characteristics of the particular canopy, but in the basics of skydiving as well. A well-trained jumper, like a well-trained pilot, is far safer and better equipped to handle any unusual situations. Squares *are* reliable — but if something does screw up, they have a reputation for doing a great imitation of a spinning brick. There's simply less margin for error, and less time to react before the situation becomes critical.

Fortunately, basic parachute training has kept pace with the equipment revolution. Back in the Dark Ages, training could be anything from weeks of overly-technical sessions down to "OK, here's your chute, there's the plane, let's go!" The lack of a standard training system, combined with the inability or unwillingness of many clubs to provide a full-time, competent instructor, led to far too many un-

Dave Sauve lands one of the newest *round* canopy designs. The long panels beneath the skirt line on each side act as stabilizers.

necessary accidents. Through the urgings of the USPA, a vast majority of parachute clubs have cleaned up their acts; most now offer a fairly standard course of ten to twenty hours, covering emergency procedures, equipment usage, and often packing as well. Increasingly, though, students are turning to a relatively new development: the commercial parachuting center.

Before the start of the sport's phenomenal growth in the 60's, clubs offered the only practical way to learn to jump. Most clubs were (and still are) fairly informal, formed simply to obtain a jump plane and a place to drop. Anything more — training buildings, rigging facilities, or large aircraft such as a DC-3 for large star loads — requires an investment that many clubs can't afford or don't need. Yet the demand for such facilities grew, as did the number of persons interested in learning.

Commercial parachuting centers have met this demand. The possibility of profits, as always, brought forward persons willing to risk capital. These commercial centers have discovered that good facilities, available aircraft, and safe operations attract both students and experienced jumpers, even at a price somewhat higher than offered by most clubs.

Air Progress traveled to two such centers to observe student operations: Ripcord West near St. Louis, and Parachuting Service outside of Detroit. These two organizations are typical of a growing number of commercial drop zones located throughout the country. Most operate year-round (weather permitting), and have a good mixture of both students and experienced jumpers. Safety isn't just something important at such centers — it's the MOST important factor. "Safety is what we sell," said Jim Cowan, owner of Ripcord West. "Everything we do here — training, packing, and jumping, whether it's students or experienced jumpers — is safety-oriented. We're not just paid to teach, or to provide equipment and aircraft; we're paid for our judgment, our knowledge, and above all our concern for each jumper's safety." That concern has paid off; there has never been a student fatality or serious injury at Ripcord in nearly 2,500 first jumps.

Both Jim and his wife, Myrna hold Expert ratings from USPA. Each of the eight other regular instructors is nearly as experienced; many in fact, received their initial training at Ripcord and joined the saff after their experience grew. Even though their occupations and lifestyles vary greatly, each instructor has the same goal of safety, and the same willingness to teach proper techniques. "Everybody who jumps here," said Jim, "is still a student in one sense, no matter how many hundreds of jumps they might have. Each of us is still learning how to have even more fun, and how to do it safer than it's ever been done before. If any jumper — or even a spectator — shows a willingness to learn by asking questions, then it's our job to do whatever's necessary to teach them the right way, the *safe* way." Jim is most certainly a nut on safety — and there isn't a single jumper at Ripcord who'd have it any other way.

The learning process at Ripcord, like that used at Harold Lange's Parachuting Service, leaves nothing to chance. Each instructor follows a standardized course syllabus, similar in scope to Jeppesen's First Jump Course developed in cooperation with USPA. Each course includes training in landings, canopy control, aircraft exits, equipment familiarization, and some basics of spotting and packing. Ripcord, like most other commercial centers, provides additional training as the jumper progresses from static line to first freefall, and also prior to each increase in freefall duration. This later training includes individualized instruction in freefall control methods, advanced spotting, and recurrent instruction in the basics previously taught.

Roughly half of the seven hour first jump course at Ripcord is devoted to perhaps the most important aspect of training: Emergency procedures. Ripcord does everything possible to prevent hazards, no matter how minor, through set procedures for training, equipment maintenance, and jump operations. All student mains, for example, are packed under the direct supervision of a Senior Parachute Rigger. The canopies themselves are of the simplest and most reliable type, mainly modified 32'-diameter military rounds; square canopies, while great for experienced jumpers, would be a bit too much for a student to safely handle. Each reserve parachute is equipped with an automatic opener, as well as a radio used to assist students with normal canopy control. An additional device consists of a lanyard attached to the main canopy and to the reserve chute ripcord; if the main is released ("cut away") after a partial malfunction, the lanyard pulls the reserve ripcord to deploy

Paul Briggs makes a feather-light landing under his square canopy. Note the open leading edge, which forces ram air into the canopy in flight to maintain the airfoil shape.

the second canopy. But Jim doesn't rely on only mechanical devices. Each jumper is trained to rely on his *own* abilities first. "The backups are there if he needs them, if he's in over his head — but there's not much our students can't handle on their own, after going through our training course."

Each student learns how to deal with total and partial main canopy malfunctions in "hanging harness' simulators, along with techniques to handle off-airport landings — though with the radios on each reserve, there's little chance of any student unexpectedly encountering tree, wire, or water hazards. While in this simulation of "under canopy" conditions, the student experiences the feel of the harness, control responses of a normal canopy, and the physical motions required to correct or cut away a partial malfunction, and to deploy the reserve after a total malfunction or cut-away. Each procedure is repeated until both the instructor and the student are completely satisfied with the results. Should a student then get one of the extremely rare malfunctions on a jump, there's no panic; he's already learned that it can be corrected quickly and safely, and moreover that *he* has the ability to take the proper actions just as he did in training. Without such knowledge and confidence, a student might hastily deploy a reserve on a minor, non-crital, partial malfunction, or fail to take any action at all on a total malfunction through reliance on the backups. But with the training, he's learned to rely on his own abilities. Not only does this increase safety, but it also helps the student to relax and enjoy the jump, secure in the knowledge of his own abilities, and the backups, which are there if he needs them.

Who takes these first jump courses? Just as in flying, there's no such thing as a 'typical' jumper. Ripcord has trained lawyers and linemen, salesmen and students. Nor is there a typical age. "A lot of our students are young, but we've had ones that were middle-aged desk jockeys and even one grandmother," said Jim. The mixture of people is similar at Parachuting Service, though the nearby campuses of half a dozen colleges and universities tend to bring in a higher percentage of teens and students.

One such student is Debbie Spychalski, a psychology student at the University of Toledo. After completing eight hours of training at Parachuting Service, we asked her about her upcoming first jump.

Are you scared by the prospect of jumping from a plane?

Only about the plane ride, really. I've been in a light plane once, and I'm afraid of heights.

If you're afraid of heights, why would you want to leave an airplane at 2800 feet?

My friends told me I couldn't do it, so at first I just wanted to show them I could. Now, though, it's more like something I have to prove to myself. And of course, I'm also doing it for fun!

After the training you've received, do you feel that you can handle any problems?

If I didn't think so, I wouldn't do it. At first, I was scared by all the stories you hear from people that don't jump, but after training I've got it all together.

Did the training cover all that you wanted to know, or is there something you're not sure of?

About the only thing that bothers me is getting out of the plane and onto the step. I've heard that the wind blast is more than expected, and of course there's no way to really simulate that on the ground.

Other than the exit, then, you feel ready, right?

I can't wait!

Low, scud, though, made the wait longer than expected; it was the following weekend before Debbie made her first parachute descent under a modified military canopy. Her exit, according to the jump-master, was letter-perfect, as was her canopy control and landing. With only a few radioed suggestions from the ground, Debbie came down within twenty feet of the target disc. Later, she talked about her experience.

From the grin on your face, is it safe to assume you enjoyed your first jump?

It was fantastic! I don't know how to begin to describe it, it was all so great.

Was there any specific thing about the jump that made the greatest impression?

I guess the quiet, after the static line opened my main canopy. Except for the wind rustling the canopy, there's no sound — no cars, not even the jump plane. It's really beautiful and peaceful.

Now that you've proved to yourself that you can do it, are you going to jump again?

Are you kidding? I'm already on the manifest for another one!

WHAT TO LOOK FOR, AND HOW MUCH TO PAY

Skydiving, compared to flying, is the best bargain in the air. The new Streakeagle 280 that you've been drooling over could set you back seventy grand; learning to jump costs about fifty. Dollars, not thousands.

How can a prospective jump student find a safe drop zone? Commercial centers and most clubs are listed in the phone directory, along with equipment dealers who will gladly offer advice. The United States Parachute Association (806 15th St., N.W., Suite 444, Washington, D.C. 20005) offers a drop zone directory for $1.00, along with several other helpful publications. The local FAA GADO can also help locate a nearby drop zone.

Before deciding to take a first jump course, by all means visit the drop zone. Spend a few hours observing and talking with both students and experienced jumpers. There's no absolute method to tell whether the particular drop zone is safety-minded, but there are a few things to check:

Established drop zone — A "jump-by-night" operation is not the place to learn safe procedures. Stick to drop zones with established reputations.

Attitude — Are the jumpers safety conscious? Do they answer all of your questions frankly and willingly? Do the experienced jumpers look upon students as future equals, or merely present nuisances?

Course length — Safe procedures take *a minimum* of four or five hours to learn. Conversely, courses lasting more than 20 hours may well be too technically oriented for the average student.

Standardization — Is there at least one person designated as a full-time instructor? Is a written syllabus used? A syllabus is a strong indication of care and concern for safety.

Maintenance — Is there a regular program of maintenance for student and rental equipment? Is a licensed parachute rigger available? Check the jump planes too — would you want to fly in them *without* a parachute?

Continuing training — Do students receive further training as they progress to freefall and relative work? Are packing classes available to students who own their own gear? Is recurrent training in basic emergency procedure required for students who haven't jumped in over a month?

Student canopies — Most drop zones stick with the reliable modified military gear, though a few now permit students to use higher performance round canopies such as the Para-Commander. *NO* student should be allowed to jump a ram air square canopy without considerable experience under a round, additional training, and the approval of both his instructor and jumpmaster.

Safety backups — All reserves should be equipped with automatic openers. If the course includes training in main canopy cutaways (not all do), each main should have a lanyard from the canopy to the reserve ripcord. If there are hazards nearby — trees, wires, hangars, or open water — there should be some means of ground-to-air communications, whether by radio, bullhorn, or even a large arrow to show proper headings.

This list is not meant to be all-inclusive or absolute. In jumping, as in flying, there's no substitute for common sense; if yours tells you that the drop zone doesn't meet your personal safety standards, remember that there are hundreds more.

Good luck, and good jumping!

GETTING TOGETHER UP HIGH

by Michael Leeds

Photographs of Nationals by Jerry Irwin and Tom Dunn

Reprinted by Permission PARACHUTIST (October 77) Edited for Space

"There it is! At 10 o'clock," Hawthorne, the pilot, cried. He banked the DC-3 over to the left and we had our first view of Tahlequah, Oklahoma.

Here it was Thursday, two days before the official opening of the 1977 National Parachuting Championships and others had already been practicing the whole week. Without wasting any more time we stormed out of the plane grabbed our rigs and ran over to the equipment inspection tables under the big tent.

Within the hour we were on our way up for our first and last fun jump before practice began.

After our "quickie" we had a chance to talk to the people who had arrived earlier and to head to town to arrange sleeping arrangements.

We stopped for a little partying at one of the motels where jumpers were staying and then headed over to the dorms. There was plenty of room at the Northeastern Oklahoma State University and most of the building was air-conditioned; we needed the air considering the 105-110 degree temperatures we were stuck with the three weeks of the Nationals. From the dorms it was a quick walk into town for dinner and then we made the rounds of parties in the dorm. And that pretty much is how the evenings went in Tahlequah.

Saturday was the official practice day. It was time for the briefing. Judges explained the 10-man rules and competitors seemed to understand them. But there were a lot of doubts as to whether the event could be judged. After talking through every 4- and 8-man jump, judges and competitors seemed to have come to an understanding.

Sunday morning brought the beginning of competition. The sky was cloudless, but hazy. The haze would interfere with competition every day.

After a week of competition, broken only occasionally by low clouds or a thunderstorm, the competition was over. Three teams had scooped up the gold, one team had broken a world record and national records had been established in 4- and 8-man sequential.

Air Freight. This team from Elsinore, Calif., won the 10-man event with a fine average of 11.1 seconds. Although their first jump resulted in a blown-up star and a slow time, Lady Luck was on their side as the judges missed it and awarded them a rejump. Air Freight scorched this one in 9.8 seconds, the fastest time ever recorded in nationals' competition.

"Dirt Dives are Free — And Dirt Divers are Easy." 4-Man Champion Dirt Divers left to right: Wilson Rudd, Dave Wallace, Jim Reilly, Bungee Wallace.

The team, captained by Don Hansen (surfers will recognize his name from the world-famous surfboards his company manufacturers), made about 150 practice jumps preparing for the speed star event. Air Freight finished fifth in 1976, and worked hard for their championship medals. It was reported that a few of their practice times were in the high eights.

Air Freight had no real competition, in spite of the presence of several other experienced 10-man teams. The Seagull Squad, from Logan, Utah, and captained by Jerry Bird, cruised to second place with a 13.9 second average. The team didn't exactly train very hard, making only 20 practice jumps before the nationals. Bird's team showed what can be done with ten experienced divers and a little last-minute effort.

The third place team was Mirror Image, from Pope Valley, Calif. Mirror Image wasn't really serious about 10-man — they had put their training efforts toward the 8-man event. Their average was 14.6.

The Sky Pirates, also from Elsinore, finished fourth. The nucleus of this team was four members of the three-time U.S. champion Captain Hook and the Sky Pirates. Mitch Poteet captained the team, which like Air Freight, had been training exclusively for the 10-man event. The Sky Pirates were doing well this year until a blown jump (and a 29 second time) dropped them out of contention for a medal.

It's About Time from Pelicanland, Maryland finished fifth. The team was captained by Bob Holler, who did his 4000th jump at the Nationals.

The Slots Are For Tots, from DeLand, Fla., had times comparable to Air Freight until they were zapped for being judged as putting a 4-man out the door (the rules allow up to a 3-way hook-up). The Slots also averaged 11.1 for their five scored jumps, although their best time of 10.6 seconds could not equal Air Freight's fine 9.8.

Exitus, from Marshall, Mich., a speedy team which was zapped in 1976 for exit rule violations, was zapped again this year, this time in the second round.

An all-woman team entered 10-man this year. "Snoots Are For Toots," from DeLand, billed themselves as a "semi-serious" team and finished 11th out of 20 teams.

Competitors and members of the Competition Committee, meeting after the Nationals, agreed that the current 10-man rules are not "judgeable" and should be changed again. Most of the competitors feel that the exit portion of the jumps cannot be clearly judged by telemeter-equipped judges and that a "no show, anything goes" exit is needed. (This means that although the entire team must be completely inside the aircraft after the stand-by command is given by the judges, it can do anything else — hold on, pile on top of each other, etc.). Participants also felt that a videotape recording system would be quite handy to give the judges a second or third opportunity to evaluate a jump.

Although the 10-man event still rightfully carries a lot of prestige, the sequential events attracted most of the attention this year. The winners of either new event took home all the marbles: a slot on the U.S. Team and a trip to the World Meet.

And unlike 10-man, which Air Freight led from their first observed jump, the outcome of both 8- and 4-man was not determined until after the last jump in each event. The excitement was electric during the last few rounds, especially in 4-man where the lead changed several times. The pressure on the top teams was equally tremendous.

Mirror Image out-skydived the 20 other 8-man teams to win by two points. (A point is given for each maneuver successfully completed during the 50 seconds of working time. Mirror Image averaged about 4.7 maneuvers per jump for the six rounds.)

The winning team was built by B. J. Worth last May. He recruited jumpers from his old '75 U.S. Freefall Exhibition Team, from the ranks of Pope Valley's locals, and from the old Casa Grande, Ariz., gang. The team essentially practiced full-time during the week, using movie film records of their practice jumps as well as a lot of careful dirt diving. After about 160 practice jumps, Mirror Image demonstrated unquestionable excellence in the event.

Two other pre-meet favorites, Clear Eye Express from Issaquah,

National 10-Man Champions and World Record Holders: Air Freight from Elsinore, California.

Wash., and Slots-8 from Deland, finished the scheduled six rounds tied for second place with 26 points. In accordance with the rules, a jump-off round was conducted and Clear Eye won it and the silver medal.

Terminal Chaos, from Elsinore, was fourth with 20 points.

Considering that it was the event's first time around, 8-man went pretty well. Confusion regarding the rules allowed at least one jump to be "rescored," without a re-jump, but it was a minor glitch.

Competition in the 4-man event was even greater than the 8-man; out of 35 teams that entered, nearly 10 seemed to be in contention for the gold after several early rounds. Perhaps because of the hot competition, 4-man went very poorly, with a plethora of protests, rejumps, and frustrated skydivers. In fact, the final results of the event were not known until several hours after the last jumper had landed, due to a pending protest.

The Dirt Divers, from Franklin County Parachute Center in Franklinton, N.C., emerged the victors with 44 points, an average of 7.3 per round. The team is captained by Wilson Rudd, an engineer with Steinthal Parachute Company.

Kinetic Sunshine, from Palatka, Fla., did some excellent skydiving in the final few rounds to finish second with 43 points. The same team was seventh in the hogflop in 1976.

Reflection, whose members jump on the Mirror Image 8-man team, won the bronze medal with 41 points.

After leading for several rounds, Sierra, from Yolo, Calif., had a poor last jump to tie with Michigan Freefall.

As expected, there were several protests. The judges decided most of them and a few went to the jury. In 10-man, the jury denied a rejump and left the results final. Eight-man competition had its share of protests, but no one could question the dominance by Mirror Image. And in 4-man sequential a rejump awarded by the judges gave the Dirt Divers a last chance and a victory.

Along with the close of competition and the beginning of the Boogie, Friday was the night of the Awards Banquet. Bill Ottley chaired the ceremonies and everyone had a lot of fun (and some decent food!). After the competition medals had been awarded the winners of the overall relative work awards were named. Members of the Mirror Image team picked up both the individual and the team awards.

The Boogie was a smashing success. World records were made, new movies were shown, and equipment manufacturers and dealers showed off their latest inventions.

The jumping, of course, was the focus. For two days after the competition ended, large stars continually filled the skies.

The biggest attention-getters at the boogie, were three "megaformation" loads. The 30-man triwedges that went so well at Zephyrhills, Fla., last Easter made it obvious that much bigger formations were within reach of today's skydivers.

The goal of a 50-man formation was set early last May, and a series of three jumps was organized to work up to "half-century" formation: a 40-man box, a 45-man penta-diamond, and a 50-man pentagon. The dives were static formations only; no sequential was attempted as was done at Z-hills.

The 40-man box was considered to be a warm-up load, and several divers turned down invitations to get on the manifest because they were skeptical of its chances for success. They soon regretted that decision, as the box was built cleanly from 15,000 feet on the first attempt. It is *not* an understatement to say it was easy — the ability of the 40 jumpers and the careful organizing made it that way.

The box was a national record and an unofficial world record. Although the 40-man was the largest complete formation ever built, the FAI has yet to finalize a "large formation" category.

The 45-man penta-diamond was attempted the next day. Again, there were endless walk-throughs in the hot sun. The organizers ran themselves ragged coordinating aircraft, rounding up jumpers and supervising dirt dives.

Again, a strung-out exit from 15,000 feet and a long dive to a building formation that was literally swarming with divers closing cautiously in their assigned slots. The penta-diamond wasn't quite as stable as the box, but it was never in trouble, either. Whoops and yells could be heard in free-fall as well as under canopy — a 45-man!

Bad news on the ground, however: the formation wasn't built correctly

Snoots are for Toots 10-Woman Team.

as one diver had closed in the wrong slot, resulting in an asymmetrical formation (and therefore not eligible for a record).

Still, enthusiasm was up for the 50-way load, and the organizers (Mike Barber, Mike Gennis, Tim Heneghan and Ed Mosher, mostly) went to work again.

By this time, the jumpers were tired and the enthusiasm, which was electric after the 45-man, began to fade under the hot sun. The 50-pentagon funnelled quite nicely at 7,000 feet or so. This last jump quenched the megaformation madness.

Air Freight set a world record 10-man time, clicking one off in 9.4 seconds, beating their best competition time and eclipsing the 10.8 set by the Soviets last year. The Snoots tried several times to capture the women's record (held by the U.S.S.R. at 17.9 seconds) but didn't succeed. However, 22 women did get together in a near-miss for a 24-woman cluster-flake.

A GATHERING OF THE CLAN

by Bill Ottley

Reprinted by Permission PARACHUTIST (October 77)

It was cool, the first less-than-100° day in two weeks. It was Thursday, July 28th. It was practice day for the 18th Annual USPA Style and Accuracy Nationals, and suddenly flared bells and batwing jumpsuits were gone. Out came the black Pioneers, tucked tight for style, ready for accuracy. A look around the briefing tent generated a real sense of *déjà vu* — we were all old friends, 63 men, 15 women, 78 in all, ready to go. The group was subdued, hair was shorter, the official briefing took only twenty minutes — we'd all heard it before.

As in every past year there was a mystery competitor, and for 1977 the "Who's He?" was Dennis Wise, from Lubbock, Texas, who — according to Rumor Control — had been turning 5.5s in practice. Could he do it in competition? There were even a few new faces — like Mike Mingold and Kevin Roll from the Air Force Academy, where they still teach style and accuracy to their cadet jumpers — but mostly it was good buddies together again.

Jack Bergman wore the Meet Director's hat, while also doubling as night bartender. George Applegate honchoed accuracy, Dick Barber style; in charge overall was Chief Judge John Mirus. Chief of Judges Training Larry Casada shepherded a crew of fledglings.

The forest of billowing tents which had housed the RW teams just a few days earlier had given way to but two, now sited along the fence in lonely splendor: the Army Team had theirs (gold and black, naturally) and there was a big red one with huge block letters on it spelling out "RAEFORD." Another thing was missing too: of round canopies there were none. Truly the square revolution was complete at last.

Friday morning Meet Director Bergman called for a 9 o'clock start, and Judge Don Strickland used the first wind dummy jump to log his three-grander, toeing the disc as he landed. It was the first of many 0.00 scores.

Then Mirus called for style, round one, against a sky of cotton-candy clouds. On his first jump Jim Hayhurst turned a left series in six seconds flat, with a minor — 6.2' total. And this kind of jumping was delivered by the top competitors all through the Meet. By early afternoon the puffies got a little closer together, the pit crew switched to accuracy, and the all-seeing, no-foolin' Parapad made its debut in U.S. national competition. At 5:25 p.m. the winds were gusting; thunderstorms filled the horizon, beautiful but treacherous; "Apples" threw the smoke and we quit for the day.

Saturday dawned clear and hot, and again Mirus and Bergman ordered up an early start. Perhaps it was the small size of the group, perhaps it was the level of national competitive experience ("we're

Probably one of the most versatile skydivers at this year's Nationals, Jack Brake won accuracy and was a member of the second place 10-Man team.

having fun, we're the last of the renaissance men and women, we *know* what we're doing," said Paul Sitter at one point, which sort of summed things up) — but, whatever the reason, S&A '77 moved with incredible smoothness.

At the end of two rounds of style Hayhurst led with 6.2 and 6.8, Cheryl Stearns was No. 1 among the ladies with a 7.6 and a 7.8; and it took only two accuracy rounds for the fearsome Parapad to drop all but ten men and two gals (Linda Miller and Julie Allbritton) from the perfect score column.

Sometime early Saturday afternoon, Jack Brake — probably the finest all-around jumper in the United States today, if not the world* — wandered ruefully up to the scoring area, acknowledging with a half smile, "I guess I got a real burger on that one." He'd been tagged with a major gig, which bumped his time from 7.8" up to 8.3"; and in the end that single slip likely cost him the national overall gold medal.

But at sunset Saturday Perry Hicks' 11 centimeters and Linda Miller's 15 led women's accuracy, while Jack Brake held a one cent edge in the men's division. Former accuracy champion John Wolfe and Golden Knight Matt McManus were hot on his heels, tied for second place, both with a total of 2 centimeters. The mystery man, Dennis Wise, was living up to his advance billing: fourth in style.

Sunday morning it was all over — one of the slickest and surely one of the shortest Nationals ever. The Parapad had done its job nicely: there were only two ties in men's accuracy (for fifth and eighth places) instead of the interminable jump-offs which had been needed in 1976 to finally find a winner. Brake protected his one centimeter total through eight electronically measured jumps, to earn first place. McManus was second with two cents, Dana Englestad third with three, Jim Hayhurst fourth with four, Tony Fugit and John Wolfe tied for fifth, each with five. Perry Hicks led women's accuracy with a 12 centimeter total, followed by Linda Miller (21) and Cheryl Stearns (88).

**He had already earned a silver medal in RW competition the week previous. — Ed.*

National Overall Champion Jim Hayhurst

At the close of the men's style event Roy Hatch and Jim Hayhurst shared the same blistering average for four jumps: 6.65 seconds. Doug Christian followed with 6.9, and then, just a tiny 5/100ths of a second slower there was a three-way tie between Dennis Wise, Mark Limond, and Dana Englestad.

On the women's side Perry Hicks held a tenth-of-a-second lead, with 7.8 as her four-jump average. Cathy Collingwood, Cheryl Stearns, and Debby Schmidt tied with 7.9, and Suzi Joerns, who's been a top performer ever since the middle '60s, showed she had not lost her touch: her average was 8 seconds flat.

It was tie-breaking jump-off time. The crowd gathered around the judges' circle at the peas, and heard Dennis Wise score a lightning 6.1, followed by Mark Limond (7.2), and Dana Englestad (7.7) to settle their three-way battle. Roy Hatch was clocked a half-second faster than Jim Hayhurst (6.2 versus 6.7) to anchor the men's gold medal firmly around his neck; and Cheryl Stearns beat out Debby Schmidt, by 7.5 over 8.6, to break the tie in women's style.

At the Awards Banquet on Sunday night Jim Hayhurst became our new overall men's champion, beating out Roy Hatch by five centimeters in accuracy (their style scores were identical). Perry Hicks was the new women's champion, having won both style and accuracy, a first in women's competition.

The 1978 U.S. Team, selected from the overall standings, was announced: Men — Doug Christian, Dana Englestad, Roy Hatch, Jim Hayhurst, Mark Limond, Dwight Reynolds, Bob Von Dereau, and Dennis Wise; Women — Cathy Collingwood, Perry Hicks, Suzi Joerns, Linda Miller, Judy Patrick, Roslyn Royster, Debby Schmidt, and Cheryl Stearns.

Another Nationals was history.

(Photo by Jackie Haynes-Smart; published in the NEW ZEALAND & SOUTH PACIFIC AVIATION DIGEST.)

Andrew Belche, 1976 New Zealand competition winner, prepares to jump during competition over Mount Maunganui, New Zealand.

SOUTHERN ILLINOIS UNIVERSITY WINS NIFA MEET

by Yearbook Staff

Most of the awards in the 1977 National Intercollegiate Flying Association Air Meet were won by the team from Southern Illinois University. The 1977 meet was held at Stillwater, Oklahoma, and hosted by the Oklahoma State University Flying Aggies.

Southern Illinois University was awarded both the Grover Loening Silver Cup for being the outstanding collegiate flying club in the nation, and the Champion Spark Plug Team Trophy for their performance during the meet.

Awards and standings for the 1977 meet are as follows:

NATIONAL CHAMPIONSHIP

Sponsored by Champion Spark Plug Company, Toledo, Ohio

1. Southern Illinois University, Carbondale, Illinois
2. Parks College-St. Louis University, Cahokia, Illinois
3. San Jose State University, San Jose, California
4. The United States Air Force Academy, Colorado Springs, Colorado
5. University of North Dakota, Grand Forks, North Dakota

THE LOENING TROPHY (Best overall record for the year)

Southern Illinois University, Carbondale, Illinois

THE JUDGES TROPHY (For best overall team performance)

Sponsored by Trans World Airlines, Inc.

Parks College-St. Louis University, Cahokia, Illinois

TOP MALE PILOT TROPHY

Sponsored by Alpha Eta Rho and the AOPA Foundation

Bob Allison, University of North Dakota

TOP WOMAN PILOT TROPHY

Sponsored by All-Ohio Chapter, The Ninety-Nines, Inc.

Jean Wu, University of Michigan

TOP TWENTY-FIVE PILOTS AWARDS

Sponsored by Jeppesen Sanderson, Denver, Colorado

Bob Allison, University of North Dakota
Alan Fenter, Metropolitan State College
Mark Cassen, San Jose State University
Mark Beck, San Jose State University
Jean Wu, University of Michigan
Greg Clasen, United States Air Force Academy
Kevin Airey, Parks College-St. Louis University

Bev Dorskoff, San Jose State University
Emily Seiler, Western Michigan University
Gerald Molidor, Parks College-St. Louis University
Scott Davis, Southern Illinois University
Ronald Johnson, United States Air Force Academy
Todd Almand, United States Air Force Academy
David Hill, Central Texas College
Jerry Zimmerle, Orange Coast College
Paul Librizzei, Florida Institute of Technology
Stace Byon, University of North Dakota
Valerie Thal, Cochise College
Susan Staples, Southern Illinois University
Penny Van Riper, Dowling College
James Roth, Western Michigan University
Robert Duboise, Orange Coast College
Dale Dietz, Southern Illinois University
Lewis Thorp, Southern Illinois University
James Reaney, Southern Illinois University

AVIATION YEARBOOK INDEX

a

b

c

d

e

n

o

p

r

S

t

u

V

W

y